STATISTICS:

AN INTUITIVE APPROACH

Lincoln L. Chao

California State University
Long Beach

SCIENCE RESEARCH ASSOCIATES, INC.
Chicago, Palo Alto, Toronto
Henley-on-Thames, Sydney, Paris

A Subsidiary of IBM

In Memory of
My Mother

Printed in the United States of America

Library of Congress Catalog Number: 73-89380

ISBN   0-574-19010-4

preface

At the time this book was being contemplated, the following announcement was made by a large state university to the statistics instructors of the colleges that provide most of its transfer students: their statistics courses would be accepted by the university as long as they put emphasis on the theory of probability and statistical inference rather than on traditional descriptive methods. This announcement illustrates two recent trends in the teaching of elementary statistics. One of these is a change in emphasis from descriptive statistics to probability theory and statistical inference. The other relates to the level at which beginning statistics courses are being taught. Elementary statistics, which has traditionally been an upper-division course for most students, is now being offered as a lower-division course in an increasing number of institutions.

This book was written with these two trends in mind. Its primary objective is to introduce the basic concepts of probability theory and the methods of statistical inference to students who have no mathematical background beyond one or two years of high school algebra.

The approach is essentially nonmathematical. The student is expected to know only how to apply statistical techniques in a decision-making process; the mathematical derivation of those techniques is not introduced. In most instances, the student is urged to follow the theoretical developments intuitively; in others, he is asked to take the derivations on faith. I hope that this approach will make statistics, which is often considered a great hurdle in a student's college career, an easier and more enjoyable course.

Each chapter starts with an introductory section explaining its scope, basic concepts, and objectives. Each theory or method is amply illustrated with examples taken from a variety of fields, including business, the social and natural sciences, and educational psychology. Similarly, the exercise problems provided after each major section of the text are interdisciplinary in nature. Short answers to most of the problems are listed at the end of the book. A solutions manual is provided for the instructor.

I wish to express my deep appreciation to my colleagues in quantitative methods at California State University, Long Beach, for their generous help and suggestions. I am especially grateful to Dr. Harry G. Romig, professor emeritus, who made many constructive criticisms, and to Dr. Perri Stinson for her invaluable comments. Heartfelt thanks are due also to Professor James Curl, of Modesto College, whose critical review and illuminating suggestions are in great part responsible for the present form of the text. I am also indebted to Professor Jerald T. Ball, of Chabot College; Kenneth Goldstein, of Miami-Dade College; Sidney Katoni, of New York City Community College; Barron G. Knechtel, of Orange Coast College; and Joseph Deken, of Stanford University, for their comments and assistance. Last but not least, I wish to express my gratitude to my wife and sons for their encouragement and sacrifices; without them this project could not have been undertaken. I wish to extend my special thanks to my son, Martin, for preparing appendix D, the table of square roots.

Lincoln L. Chao
Long Beach

contents

1-1
what is
statistics?

Most words have multiple meanings; the word "statistics" is no exception. For example, "statistics" may be used to denote a collection of figures or numbers ("Here are the statistics for the first half." "Have you seen the latest statistics on unemployment?"), or the term may be used to designate a discipline, or branch of learning. These two meanings of statistics are actually closely related because statistics as numerical data is an essential part of statistics as a discipline.

The origin of statistics as a discipline can be traced to two areas of interest: *government records* and *games of chance.* When national states began to emerge at the end of the Middle Ages, it became necessary to collect information about the territories under their jurisdiction. This need for numerical information about citizens and resources led to the development of techniques for collecting and organizing numerical data. By the seventeenth century, surveys similar to our modern census were already in existence. At the same time, insurance companies were beginning to compile mortality tables for determining life insurance rates.

In the early stages of its development, statistics involved little more than the collection, classification, and summarization of numerical data. Even today, these activities are still an important part of statistics. (Think of all the tables, charts, and graphs you have studied.) Because the objective of this kind of data handling is to describe the important features and characteristics of collected information, we generally refer to it as *descriptive statistics.* The following is an example of descriptive statistics.

A school psychologist wants to know the average IQ of the five foreign students in his district. An intelligence test is administered to the students, and the IQ's are found to be 101, 103, 105, 107, and 109. The statistical method the psychologist uses to find the average IQ is the calculation of the arithmetic average. (The computation of the arithmetic average, simple as it is, is an important part of descriptive statistics.) The process is limited to the data collected in this particular case, and does not involve any inference or generalization about the IQ's of other foreign students. The method is descriptive in nature, because the average, which is 105, summarizes and describes the collected information.

The other source of statistics is found in gambling. In the seventeenth century, Europeans were intensely interested in games of chance, an interest that led to the development of the *theory of probability*. The study of probability, in turn, led to the development of a new field of statistics called *inferential statistics,* or *statistical inference.* Statistical inference is a technique by which conclusions or generalizations are drawn about the characteristics of a totality on the basis of partial or incomplete information. The following brief examples illustrate the meaning of statistical inference and its importance.

A drug manufacturer claims that a new cold vaccine developed by his company is 90 percent effective; that is, 90 out of every 100 persons who use the vaccine will survive the winter without catching cold. Let us assume that 30 persons have received the vaccine; of the 30, 25 survived the winter without catching a cold. Should we accept the manufacturer's claim?

Consider the problem of deciding whether to accept or reject a shipment of purchased goods in a manufacturing plant. A portion of the shipment is inspected to determine whether the shipment should be accepted. If 20 units are inspected, and two units are found to be defective, should the plant reject the shipment and return it to the supplier?

Let us assume that two different teaching methods are used to instruct two groups of students in a given subject. The two groups are believed to be comparable in ability. At the end of the instructional period, a standard test is administered to both groups. On the basis of the average test score for each group, can we evaluate the relative effectiveness of the two methods?

These are typical of the questions handled by inferential statistics. They involve a decision or choice between alternative courses of action in the light of empirical evidence. This is why statistics has been often referred to as a body of techniques dealing with decision making on the basis of observed data.

In summary, statistics as a discipline, or branch of learning, refers both to the treatment of numerical data and to the methods and theories that are used to handle numerical data for inferential purposes. More specifically,

> Modern statistics refers to a body of methods and principles that have been developed to handle the collection, description, summarization, and analysis of numerical data. Its primary objective is to assist the researcher in making decisions or generalizations about the nature and characteristics of all the potential observations under consideration of which the collected data form only a small part.

Thus, descriptive statistics and statistical inference are two areas that make up the discipline of modern statistics. Of the two, the latter has become increasingly more important.

1-1 exercises

1. Most newspapers contain a "vital statistics" section. What does the word "statistics" mean in this particular context?

2. In what context is the word "statistics" generally used on radio and television?

3. Why was statistics sometimes referred to as "political arithmetic" in the past?

4. In an economics examination, three freshmen received grades of 90, 85, and 80; and three sophomores received grades of 89, 86, and 92. From the following statements made on the basis of these figures, identify those that are derived from descriptive methods and those that are derived from statistical inference.
   a) The average grade of the three freshmen is 85, and the average grade of the three sophomores is 89.
   b) The three freshmen's grades fluctuate more than the three sophomores' grades.
   c) In economics, the average grade received by sophomores is generally higher than that received by freshmen.
   d) In the next economics test, freshmen will probably receive lower grades than sophomores.
   e) Freshmen's grades in economics usually fluctuate more than sophomores' grades.

5. Four brand A automobile tires and three brand B tires are tested to determine their service lives. The service lives for brand A are 29,000, 33,000, 37,000, and 41,000 miles; for brand B, they are 30,000, 32,000, and 34,000 miles. From the following statements made on the basis of these figures, identify those that are derived from descriptive methods and those that are derived from statistical inference.
   a) The average service life of brand A tires is 35,000 miles, whereas that of brand B tires is 32,000 miles.
   b) The service life of brand A tires varies more than that of brand B.
   c) If the price of brand A tires is the same as that of brand B tires, you should purchase brand A tires.
   d) The four tires of brand A vary more in service life than the three tires of brand B.
   e) The average service life of the four brand A tires is longer than that of the three brand B tires.

6. Use your own words and examples to illustrate the difference between descriptive statistics and statistical inference.

1-2
basic
concepts:
population
and sample

population

The totality of "all potential observations" mentioned in the previous section is generally referred to as the "population." (Some statisticians refer to this concept as the "universe.") The term "population," like the word "statistics," has various meanings. In common usage, it refers to the number of people in a certain region or locality. (The population of China is 800 million; the population of the United States is 220 million; the population of New York City is more than 10 million.) In statistics,

> "Population" is defined as the totality of all potential measurements or observations under consideration in a given problem situation.

As shall be seen shortly, the term does not necessarily refer to the number of human beings in a given locality.

Each different problem situation involves a different population. If the problem is to ascertain a particular characteristic of all elementary school age children in California, then the measurements of that characteristic in that group of children constitute the population. If the purpose of investigation is to determine what proportion of all machine parts produced by a certain manufacturing process are defective, then the population consists of the measurements of the quality of all the machine parts produced by that process. If the problem is to determine the probability that heads will occur when a coin is tossed, the population consists of the outcomes of a presumably infinite number of tosses; similarly, in determining whether each of the six sides of a die has an equal chance to turn up when a die is rolled, the population consists of the outcomes of a presumably infinite number of rolls.

Populations are generally classified into two categories: finite and infinite. Finite means "capable of being reached or surpassed by counting"; infinite means "lacking limits or bounds" or "extending beyond measure or comprehension." A *finite population is one that includes a limited number of measurements or observations*. For instance, the total number of students presently attending colleges and universities in the United States, all calves born alive in the state of Wisconsin in 1973, and all the institutions of higher education in North America are finite populations. Some finite populations consist of only a few units, whereas others consist of millions. But as long as the total number of all potential measurements is capable of being reached, the population is considered finite.

A population is said to be infinite if it includes a large number of measurements or observations that cannot be reached by counting. For instance, the population of all live births of the human race is infinite because there is no limit to its number. Similarly, the population of digit numbers taken *with replacement* from an urn containing ten balls, each of which is marked with a different one of the ten digits, is infinite, because the number of balls that can be drawn is beyond limit or measure.

Similarly, the populations in the three examples mentioned earlier (the manufacturing process, the coin, and the die) are also infinite. They are infinite populations because, at least hypothetically, there is no limit to the number of observations each of them can include.

Population characteristics are generally called *parameters,* and population parameters are usually considered *true values.* For instance, the average IQ of all first graders in the country is a characteristic, and hence a parameter, of the population of the IQ's of first graders. It is the true average of the IQ's. Similarly, the proportion of all TV viewers who watch a particular program at a given time is a parameter of the population of the TV viewers; it is the *true proportion,* or *population proportion.* It is absolutely impossible to compute the true value of any parameter of an infinite population. In most cases, it is also impracticable to compute the true value of any parameter of a finite population. As a result, it is necessary to make inferences about population parameters from information contained in a small portion—or sample—of that population.

sample

A sample is a collection of measurements or observations taken from a given population. The number of observations in a sample is, of course, smaller than the number of possible observations in the population; otherwise the sample would be the population itself. Samples are taken because it is not economically feasible, though possible in some cases, to make all the potential observations in the population.

If, for example, we wanted to estimate the average annual expenditure of the American college student, we would probably draw a sample of, say, 1000 students, find out the annual expenditure for each of them, add the figures, and then find the average. We do this because we simply do not have the time and resources to contact all college students, even though it is possible to do so. On the basis of the sample average thus obtained, we make an inference about the average expenditure of all college students. Similarly, we may draw a sample of, say, 100 calves born in Wisconsin in a given year as a basis for estimating the average weight of all calves born in that state during the same year.

To estimate the value of a parameter of an infinite population, it is absolutely necessary to use sample information. Thus, to determine the proportion of defective parts produced by a production process, quality-control technicians examine a batch of parts to find out the number of defective parts contained in it. Such a batch, which constitutes a sample, is usually taken at regular periodical intervals. Tossing a coin 300 times to determine whether it is fair, and rolling a die 600 times to determine whether it is balanced, are other examples of sample taking from infinite populations.

A sample characteristic is generally called its *statistic*. For example, the average IQ of 1000 first graders selected at random from among all first graders in the country is a statistic; it is a characteristic of the sample of the IQ's of 1000 first graders. The proportion of a sample of 300 TV viewers who watch a particular program is also a statistic. Such a proportion is called a *sample proportion*. Statisticians use sample statistics to make inferences about population parameters.

1-2
exercises

1. Suppose that the freshman class of your college consists of 5000 students, all of whom have taken a standard aptitude test that was administered to all entering freshmen in the country. Explain the circumstances under which the 5000 scores received by these students can be considered (a) a sample and (b) a population.

2. During a certain week 2000 customers were served by a restaurant. Explain the circumstances under which these 2000 customers can be considered (a) a sample and (b) a population.

3. Suppose that a balanced coin is tossed 100 times, and 60 heads are obtained. Answer the following.
   a) What is the sample proportion?
   b) What is the population proportion?
   c) What is the size of the sample?
   d) What is the size of the population?

4. Suppose that 60 percent of all registered voters of a given country are members of a certain party and 40 percent are not. From a sample of 50 voters, it is found that 25 belong to the party. Answer the following.
   a) What is the sample proportion of voters who belong to the party?
   b) What is the population proportion of voters who belong to the party?
   c) What is the population? Is it finite or infinite?

5. A poll is conducted to determine voting preferences in a presidential election. Answer the following.
   a) What constitutes the sample?
   b) What constitutes the population?
   c) Is the population finite or infinite? Explain.

6. A survey is conducted to determine whether American housewives prefer one brand of detergent to another. Answer the following.
   a) What constitutes the sample?
   b) What constitutes the population?
   c) Is the population finite or infinite? Explain.

7. For each of the following statements, define the population that is being sampled, and tell whether it is finite or infinite.
   a) One thousand college students are interviewed about their opinions on the legalization of marijuana.
   b) Five hundred nonunion workers are interviewed to determine their attitudes toward unionization.
   c) We select a sample of screws produced by an automatic machine and test them to determine the proportion that are defective.
   d) We select a sample of hogs and obtain their birth weights.

8. Give an example to illustrate each of the following: (a) a finite population, and (b) an infinite population.

1-3 areas of application

Statistical methods and principles have found applications in many fields: business, the social sciences, engineering, and the natural and physical sciences.

The growing complexity of the economy has created a tremendous degree of uncertainty about the future operations of any business enterprise. As a result, more and more companies are using statistical analysis as a decision-making tool, especially in such areas as market research, forecasting, and analysis of economic trends.

Statistics also plays an important role in education and psychology. For instance, an educator may want to find out if, for a certain group of students, there is a correlation between achievement-test scores and grade-point averages. If there is a correlation between the two, he can make predictions about grade-point averages on the basis of achievement-test scores.

The need to analyze and interpret numerical data has made it necessary for educators and psychologists to have at least some basic understanding of statistical methods. Almost without exception, courses in statistics are required for education and psychology majors. In fact, the psychologist's need for special statistical tools has led to the development of new statistical techniques in recent decades. (Chapter 13 deals with some of these techniques.)

Because statistical techniques have been applied to a wide variety of research projects involving the study of individuals and groups, statistics has become, in most schools, a required course in sociology, anthropology, and related behavioral sciences.

Statistical methods have also been used extensively in the areas related to the biological sciences. In agriculture, they are used to determine the effects of seed strains, insecticides, and fertilizers on yields. In medicine, they are used to determine the possible side effects or effectiveness of drugs and to provide better methods to control the spread of contagious diseases. Statistics is a highly recommended, if not yet required, course for students majoring in these areas.

In recent years, statistics has found increasing applications in the physical sciences, where it has been used for handling the collection of data and the testing of hypotheses. The research needs of physical scientists have expanded the arena of the statistical technique of experimental design. It is hardly necessary to mention that in engineering the application of statistical principles to quality control has been an accepted practice for several decades.

One reason for the rapid growth in the application of statistics in recent decades is the increasing ease with which large quantities of numerical data can be handled. Electronic computers have made it possible to analyze, in a relatively short time, large quantities of data. The contribution of the desk calculator should not be ignored; it, too, has made the work of the statistician easier.

1-3 exercises

1. In what way is the electronic computer responsible for the increasingly widespread application of statistical techniques?

2. Explain why statistical techniques have found increasing application in business and economics.

3. Illustrate the use of descriptive statistics with an example from any field in the social sciences.

4. Illustrate the use of descriptive statistics with an example from the natural or physical sciences.

5. Illustrate the use of inferential statistics with an example from the social sciences.

6. Illustrate the use of inferential statistics with any example from the natural or physical sciences.

1-4 scope of the text

As mentioned in the preface, this is a textbook for beginning students who have never had a course in college mathematics. The approach is intuitive rather than mathematical; theoretical discussions are reduced to a minimum.

Most students have been exposed to the basic mathematical notations in high school mathematics courses; nevertheless, a brief review of the summation sign and various useful notations will help them understand better the material they will encounter in the text. These notations are reviewed in chapter 2. Chapter 3 deals with the concepts of events, the counting of events, and probability; chapter 4 discusses the axioms and basic rules of probability, the idea of a probability function, and discrete probability functions.

Descriptive statistics is treated in chapters 5 and 6, with emphasis on the arithmetic mean and the dispersion of observations for the sample as well as the population.

The rest of the text is devoted to a study of inferential statistics. Chapter 7 presents the normal distribution and the central limit theorem. These concepts are of paramount importance because they are invariably involved or implied in the techniques and principles covered in later chapters. Chapters 8 through 13 cover the most basic and important topics of statistical inference, including the methods of inference about a single mean, the difference between two means, and the equality among three or more means. They also deal with tests about goodness of fit, statistical independence, and simple regression and correlation coefficients. The text ends with chapter 14, which deals with the rudiments of modern decision theory.

2-1 introduction

In statistics, the same expression is often referred to several times in a given discussion. To avoid writing out these expressions over and over again, statisticians use a sort of mathematical shorthand. Many beginning statistics students are handicapped by their unfamiliarity with the mathematical symbols and notations that make up this shorthand. They often fail to grasp the precise meaning of the notations, and thus are unable to use them in a correct manner. Because these notations are an essential part of statistical expressions, and because a knowledge of them will help the student better understand the discussions and presentations he will encounter in this text, we will review them in this chapter.

Three kinds of notations will be discussed: (1) summation notation, (2) set notations, and (3) functional notations. The first and the third are used most frequently in this text; the second is worthy of inclusion because of its usefulness in the discussion of probability.

2-2 summation notation

In the preceding chapter, we discussed the concept of a sample. We may consider a sample a series of observations, with n designating the number of observations (size of the sample). Although the order of the observations in a sample is immaterial, we usually arrange them in some order. Thus, there is the first observation, the second, . . . , and the nth. The notion of a series is introduced here because it provides us with a basis for handling collected data by the use of summation notation.

Each individual observation in a series is usually identified by a serial number, such as 1 for the first observation, 2 for the second, . . . , and n for the nth. Each serial number is written as a subscript.

Consider, for example, the ages of all the students in your class. The age of each student is an observation. We may start with any student and consider his age the first observation, the age of the student sitting next to him the second, and so on. Let the first observation be 20, the second

21, the third 19, the fourth 20, the fifth 25, . . . , and the nth 23. We will have a series as follows:

$$20 \quad 21 \quad 19 \quad 20 \quad 25 \ldots 23$$

Let X designate the age of an individual student, and let a subscript serial number indicate the order of each observation. Then the above series of numbers (ages) can be presented as follows:

$$20 \quad 21 \quad 19 \quad 20 \quad 25 \ldots 23$$
$$X_1 \quad X_2 \quad X_3 \quad X_4 \quad X_5 \ldots X_n$$

Observe that the numbers, and therefore the X's, are ordered according to their position in the series and not necessarily according to their magnitude. Observe also that the same number may appear more than once in the series; that is, some of the X's may designate the same value.

The lower-case letter i (or some other letter such as j or k) is frequently used to designate the serial number. Thus X_i (which is read "X sub i," or simply "Xi") denotes any one of the X's shown above, since i can be 1, 2, 3, . . . , or n. In other words, X_i is the ith observation of the series.

Because series are used so frequently in statistical expressions, a shortcut notation has been devised to express them. The summation sign Σ (the Greek capital letter sigma, which corresponds to S in the English alphabet) forms part of the notation. Thus, if there are n observations in the series for X, the sum of the observations $X_1 + X_2 + \ldots + X_n$ can be expressed by the notation

$$\sum_{i=1}^{n} X_i$$

This notation is read as "the summation of n times the value of X sub i beginning with $i = 1$ and ending with $i = n$." That is,

$$\sum_{i=1}^{n} X_i = X_1 + X_2 + \ldots + X_n \tag{2-1}$$

EXAMPLE 2-1   In the above series, the summation of the first three observations is

$$\sum_{i=1}^{3} X_i = X_1 + X_2 + X_3 = 20 + 21 + 19$$

and the summation of the first five observations is

$$\sum_{i=1}^{5} X_i = X_1 + X_2 + X_3 + X_4 + X_5 = 20 + 21 + 19 + 20 + 25$$

EXAMPLE 2-2   Let X represent any one of the following 10 digits, taken in order: 0, 1, 2, 3, 4, 5, 6, 7, 8, 9. Find the value of

$$\sum_{i=1}^{10} (X_i + 2)$$

Since $X_1 = 0$, $X_2 = 1$, $X_3 = 2, \ldots, X_{10} = 9$, we have

$$
\begin{aligned}
\sum_{i=1}^{10} (X_i + 2) &= (X_1 + 2) + (X_2 + 2) + (X_3 + 2) + (X_4 + 2) + (X_5 + 2) \\
&+ (X_6 + 2) + (X_7 + 2) + (X_8 + 2) + (X_9 + 2) + (X_{10} + 2) \\
&= (0 + 2) + (1 + 2) + (2 + 2) + (3 + 2) + (4 + 2) \\
&+ (5 + 2) + (6 + 2) + (7 + 2) + (8 + 2) + (9 + 2) \\
&= 2 + 3 + 4 + 5 + 6 + 7 + 8 + 9 + 10 + 11 = 65
\end{aligned}
$$

For the sake of simplicity, we shall, in most future discussions, use $\Sigma\, X$ instead of the elaborate expression $\sum_{i=1}^{n} X_i$ to denote the sum of all the observations of X in the series. The subscript i and the serial numbers 1 and n are omitted whenever the series of values to be included in the sum is perfectly clear.

rules of summation

Below are some simple rules to follow when performing the computations involving summation notation. They are concerned with either constants or variables or both. A *constant* is a quantity that does not vary throughout a discussion; it is opposed to a *variable*, which is defined as follows:

> A variable is a symbol used to stand
> for any one of the possible values or
> objects in a given discussion.

For instance, in example 2-2, the variable X stands for any one of the 10 digits. It is not a fixed value, but may assume a number of different values.

RULE 2-1   If c is a constant over n observations, the sum of the c's is equal to the product of the constant multiplied by the number of observations.

$$\sum^{n} c = c + c + c + \ldots + c = nc \tag{2-2}$$

EXAMPLE 2-3   Suppose that six observations are considered, and each observation involves the constant 9, the sum of the six observations is then

$$\sum_{}^{6} 9 = 9 + 9 + 9 + 9 + 9 + 9 = 6(9) = 54$$

EXAMPLE 2-4   If $c = 5$, the sum of four observations each multiplied by 10 is

$$\sum_{}^{4} 10c = 10(5) + 10(5) + 10(5) + 10(5) = 4(50) = 200$$

RULE 2-2   If c is a constant and X is a variable, the sum of the constant multiplied by the variable is equal to the product of the constant multiplied by the sum of the variable. That is,

$$\sum_{}^{n} cX = cX_1 + cX_2 + \ldots + cX_n$$
$$= c(X_1 + X_2 + \ldots + X_n) = c \sum_{}^{n} X \qquad (2\text{-}3)$$

EXAMPLE 2-5   Suppose that $c = 6$ and the series for X is 1, 3, 5, 7, and 9. Then the sum of cX is

$$\sum cX = 6(1) + 6(3) + 6(5) + 6(7) + 6(9)$$
$$= 6(1 + 3 + 5 + 7 + 9) = 6(25) = 150$$

RULE 2-3   If a and b are constants and X is a variable, then

$$\sum (a + bX) = (a + bX_1) + (a + bX_2) + \ldots + (a + bX_n)$$
$$= (a + a + \ldots + a) + b(X_1 + X_2 + \ldots + X_n) \qquad (2\text{-}4)$$
$$= na + b \sum X$$

EXAMPLE 2-6   If $a = 2$ and $b = 3$, and the series for X is 1, 3, 5, 7, and 9, then

$$\sum (a + bX) = [2 + 3(1)] + [2 + 3(3)] + [2 + 3(5)] + [2 + 3(7)]$$
$$+ [2 + 3(9)]$$
$$= 5 + 11 + 17 + 23 + 29 = 85$$

or

$$\sum (a + bX) = na + b \sum X = 5(2) + 3(1 + 3 + 5 + 7 + 9)$$
$$= 10 + 3(25) = 10 + 75 = 85$$

RULE 2-4   If X and Y are two variables, then the sum of all the values of $X + Y$ is equal to the sum of X plus the sum of Y. Symbolically,

$$\sum (X + Y) = (X_1 + Y_1) + (X_2 + Y_2) + \ldots + (X_n + Y_n)$$
$$= (X_1 + X_2 + \ldots + X_n) + (Y_1 + Y_2 + \ldots + Y_n) \quad (2\text{-}5)$$
$$= \sum X + \sum Y$$

EXAMPLE 2-7   If X takes on the values 1, 3, 5, 7, and 9, which are paired, respectively, with the values 0, 2, 4, 6, and 8, then

$$\sum (X + Y) = (1 + 0) + (3 + 2) + (5 + 4) + (7 + 6) + (9 + 8)$$
$$= 1 + 5 + 9 + 13 + 17 = 45$$

or

$$\sum (X + Y) = (1 + 3 + 5 + 7 + 9) + (0 + 2 + 4 + 6 + 8)$$
$$= 25 + 20 = 45$$

RULE 2-5   In many situations, some operation has to be performed on each of the values of X before the summation is carried out. In the case of the summation of X^2, for instance, we have

$$\sum X^2 = X_1^2 + X_2^2 + \ldots + X_n^2 \qquad (2\text{-}6)$$

EXAMPLE 2-8   Given the series of X: 1, 3, 5, 7, and 9, the summation of X^2 is

$$\sum X^2 = 1^2 + 3^2 + 5^2 + 7^2 + 9^2 = 1 + 9 + 25 + 49 + 81 = 165$$

Observe that

$$\sum X^2 \neq \left(\sum X\right)^2$$

for

$$\left(\sum X\right)^2 = (1 + 3 + 5 + 7 + 9)^2 = 25^2 = 625$$

RULE 2-6   If X and Y are paired observations, then

$$\sum XY = X_1 Y_1 + X_2 Y_2 + \ldots + X_n Y_n \qquad (2\text{-}7)$$

EXAMPLE 2-9   Given an X series of 1, 3, 5, 7, and 9, and a Y series of 0, 2, 4, 6, and 8, the sum of the products of X and Y is

$$\sum XY = 1(0) + 3(2) + 5(4) + 7(6) + 9(8)$$
$$= 0 + 6 + 20 + 42 + 72 = 140$$

Observe that

$$\sum XY \neq \left(\sum X\right)\left(\sum Y\right)$$

for

$$\left(\sum X\right)\left(\sum Y\right) = (1 + 3 + 5 + 7 + 9)(0 + 2 + 4 + 6 + 8)$$
$$= 25(20) = 500$$

2-2 exercises

1. Write out $\sum\limits^{n} c$ and find the sum for

a) $n = 3$ and $c = 5$
b) $n = 4$ and $c = -2$
c) $n = 5$ and $c = -\frac{2}{3}$

2. Write out the following summations.

a) $\sum\limits_{i=1}^{5} X_i$ c) $\sum\limits_{i=1}^{3} (X_i + Y_i)$

b) $\sum\limits_{i=1}^{4} (X_i + a)$ d) $\sum\limits_{i=1}^{3} X_i Y_i$

3. Let $X_1 = 2$, $X_2 = 4$, $X_3 = 6$, $X_4 = 8$, $X_5 = 10$, and $X_6 = 12$. Find

a) $\sum\limits_{i=1}^{3} X_i$ c) $\sum\limits_{i=2}^{4} X_i$

b) $\sum\limits_{i=1}^{5} X_i$ d) $\sum\limits_{i=3}^{6} X_i$

4. Write out $\sum aX$ and find the sum for

a) $a = 5$ and $X = 1, 2,$ and 3
b) $a = -2$ and $X = 1, -2, 3, -4,$ and 5
c) $a = 6$ and $X = 1, 2, 3, 4, 5, 6, 7, 8, 9,$ and 10

5. Consider two series for X and Y that are paired as follows:

i:	1	2	3	4	5	6	7	8	9
X_i:	9	12	15	8	16	10	5	13	12
Y_i:	5	13	12	9	12	16	8	16	9

a) Find $\sum (X + Y)$.
b) Find $\sum (X - Y)$.
c) If $a = 3$ and $b = 4$, find $\sum (aX + bY)$ and $\sum (a^2 X - b^2 Y)$.

6. Consider two series that are paired as follows:

$$X: 1 \quad 3 \quad 5 \quad 7 \quad 9$$
$$Y: 2 \quad 4 \quad 6 \quad 8 \quad 10$$

Find the following:

a) $\sum XY$

d) $\sum XY^2$

b) $\left(\sum X\right)\left(\sum Y\right)$

e) $\left(\sum X^2\right)\left(\sum Y\right)^2$

c) $\sum X^2Y$

7. Find the value of the following summations.

a) $\sum_{i=0}^{9} i$ 　　b) $\sum_{i=1}^{6} (i^2 + 3)$ 　　c) $\sum_{i=1}^{5} (i + 3)^2$

8. Use summation notation to abbreviate the following expressions.

a) $Y_1 + Y_2 + Y_3 + Y_4 + Y_5$
b) $cY_1^2 + cY_2^2 + cY_3^2 + cY_4^2 + cY_5^2$
c) $a(Y_1 + 1) + a(Y_2 + 2) + a(Y_3 + 3) + a(Y_4 + 4) + a(Y_5 + 5)$
d) $(Y_1 - 1^2)1 + (Y_2 - 2^2)2 + \ldots + (Y_n - n^2)n$

9. Use summation notation to express each of the following sums.

a) $X_5 + X_6 + X_7 + X_8 + X_9 + X_{10}$
b) $5X_5^2 + 6X_6^2 + 7X_7^2 + 8X_8^2 + 9X_9^2$
c) $3(X_3 - 1) + 4(X_4 - 1) + 5(X_5 - 1) + \ldots + 20(X_{20} - 1)$

10. Let the series of X, Y, and Z be as follows:

$$X: 0 \quad 2 \quad 4 \quad 6 \quad 8$$
$$Y: 1 \quad 3 \quad 5 \quad 7 \quad 9$$
$$Z: 10 \quad 20 \quad 30 \quad 40 \quad 50$$

Show that

a) $\sum (X + Y + Z) = \sum X + \sum Y + \sum Z$

b) $\left(\sum X^2\right)\left(\sum Y^2\right) \neq \left(\sum X\right)^2\left(\sum Y\right)^2$

c) $\sum XYZ \neq \left(\sum X\right)\left(\sum Y\right)\left(\sum Z\right)$

11. Evaluate the following summations.

a) $\sum_{X=0}^{5} (X + 2)$ 　b) $\sum_{X=1}^{4} (X^2 - X + 3)$ 　c) $\sum_{X=2}^{3} (X^3 + X^2 - 4X - 2)$

12. Let $\sum_{i=1}^{9} X_i = 45$ and $\sum_{i=1}^{9} X_i^2 = 285$, and find

a) $\sum_{i=1}^{9} (X_i - 3)$ 　　b) $\sum_{i=1}^{9} (X_i - 5)^2$ 　　c) $\sum_{i=1}^{9} (2X_i - 4)^2$

17

2-3 some set notations

A set is a collection of objects, events, or numbers. The individual elements of a set are usually listed or described within braces { }. A few set operations and their commonly used notations will be reviewed in this section. Included here are the notations of *union, intersection,* and *complement.* Just as operations on numbers lead to new numbers, such as $2 + 3 = 5$ and $4 \times 6 = 24$, operations on sets result in new sets. Union, intersection, and complement are set operations, and they lead to new sets.

complement

In order to understand the idea of a complement, it is necessary to introduce the concepts of a *universal set* and a *subset.* A universal set (U) is a set of all the elements or events under consideration in a given problem situation; a universal set may be made up of one or more subsets (S). A set S is said to be a subset of U if *all* its elements are also elements of U. The complement of S, designated by S' (read "non-S" or "S prime"), is the set of all the elements that are in U and *not* in S. Generally,

> The complement of a subset consists of all the elements of the universal set that are not included in the subset.

The sets S and S' are *mutually exclusive* because they do not have any element in common. They are also known as *collectively exhaustive* because together they include all the elements of the universal set U. (When the subsets of U are mutually exclusive and exhaustive, they are said to form a *partition* of U. For instance, let U be $\{1, 2, 3, 4, 5, 6\}$. Then U may be partitioned into the subsets $\{1, 2\}$, $\{3, 4\}$, and $\{5, 6\}$, which are mutually exclusive and exhaustive.)

EXAMPLE 2-10   A family consists of seven children: David, Elsie, Frank, George, Mary, Paul, and Ruth. These seven children form the universal set. The set S consists of the three girls, that is,

$$S = \{\text{Elsie, Mary, Ruth}\}$$

and the set S' consists of the four boys, or

$$S' = \{\text{David, Frank, George, Paul}\}.$$

Figure 2-1 shows the set S, its complement S', and the universal set U. This diagram and those in the next five figures are known as Venn diagrams. In Venn diagrams, a rectangle usually represents the universal set; and circles, parts of circles, or other designated areas within that rectangle represent subsets, or simply sets.

U

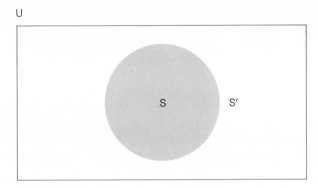

FIGURE 2-1
The complement of a set

intersection

The symbol ∩ is generally used to signify *intersection*. Let us consider two sets *A* and *B*. The notation *A* ∩ *B* (which is read "*A* cap *B*" or "*A* intersect *B*") is a new set, the intersection of *A* and *B*. The concept is defined below.

> The intersection of two sets *A* and *B*, or *A* ∩ *B*, is the set of all the elements that belong to both *A* and *B*.

In other words, *A* ∩ *B* is the set of all the elements that the two sets have in common.

EXAMPLE 2-11   Let *A* be the set of the six digits 0, 1, 2, 3, 4, and 5; and let *B* be the set of the six digits 4, 5, 6, 7, 8, and 9. That is, $A = \{0, 1, 2, 3, 4, 5\}$ and $B = \{4, 5, 6, 7, 8, 9\}$. Then,

$$A \cap B = \{4, 5\}$$

or

$A \cap B$ is the set of the two digits 4 and 5.

The set *A* ∩ *B* is shown as the shaded area in figure 2-2. Observe that these two digits belong to both *A* and *B*; they are the elements which *A* and *B* have in common.

EXAMPLE 2-12   Let $A = \{0, 1, 2, 3, 4\}$; $B = \{2, 3, 5, 6, 7\}$; and $C = \{3, 4, 5, 8, 9\}$. Then,

$$A \cap B \cap C = \{3\}$$

The intersection of the three sets *A*, *B*, and *C* is shown as the shaded

19

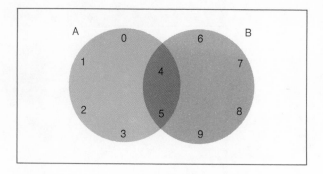

FIGURE 2-2
Intersection of two sets $A \cap B$

area in figure 2-3. The new set $A \cap B \cap C$ has only one element, the digit 3, in common. Hence, the intersection of the three sets is the set $\{3\}$.

If sets A and B have no common element, then the intersection of A and B is a *null,* or *empty,* set, usually denoted by $\emptyset$, a symbol that indicates a set without any element. As shown in figure 2-4, the intersection of A and B is a set without any element. We still call it a set just as we still call zero a number. When sets have no element in common, we refer to them as *disjoint,* or *mutually exclusive,* sets. The two sets shown in figure 2-4 are disjoint, or mutually exclusive.

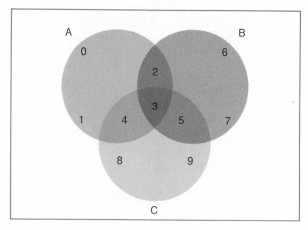

FIGURE 2-3
Intersection of three sets
$A \cap B \cap C$

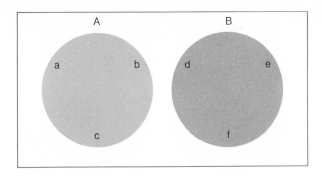

FIGURE 2-4
Two disjoint sets

union

Another operation on sets is the *union* of two or more sets. Let us consider two sets A and B. The notation $A \cup B$ (which is read *"A cup B"* or *"A union B"*) designates the union of A and B. The concept is defined below.

> The union of two sets A and B, or $A \cup B$,
> is the set of all the elements that belong
> to either A or B or both.

EXAMPLE 2-13   Let $A = \{a, b, c, d\}$ and $B = \{c, d, e, f\}$. Then,

$$A \cup B = \{a, b, c, d, e, f\}$$

Note that although the two elements c and d belong to both A and B, they are listed only once in the new set $A \cup B$. Figure 2-5 shows the union of A and B.

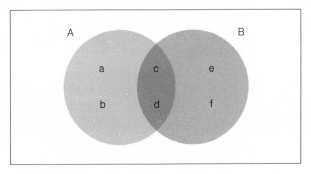

FIGURE 2-5
Union of two sets $A \cup B$

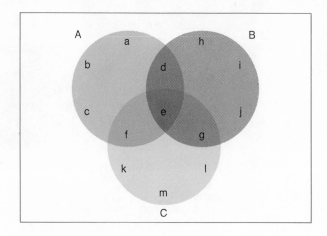

FIGURE 2-6
Union of three sets $A \cup B \cup C$

EXAMPLE 2-14 Let $A = \{a, b, c, d, e, f\}$; $B = \{d, e, g, h, i, j\}$; and $C = \{e, f, g, k, l, m\}$. Then,

$$A \cup B \cup C = \{a, b, c, d, e, f, g, h, i, j, k, l, m\}$$

Figure 2-6 shows the union of A, B, and C.

The union of a set S and its complement S' is equal to the universal set U. That is,

$$S \cup S' = U \qquad (2\text{-}8)$$

In fact, if a universal set is partitioned into n subsets, then the union of the n subsets must be equal to the universal set.

2-3 exercises

1. Let $U = \{0, 1, 2, 3, 4, 5, 6, 7, 8, 9\}$; $A = \{0, 2, 4, 6, 8\}$; $B = \{1, 3, 5, 7, 9\}$; and $C = \{0, 1, 2, 3, 4, 5\}$. Verify each of the following.

   a) $A \cup (B \cap C) = (A \cup B) \cap (A \cup C)$ c) $(A \cup B)' \neq A' \cup B'$
   b) $A \cap (B \cup C) = (A \cap B) \cup (A \cap C)$ d) $(A \cap B)' \neq A' \cap B'$

2. Let U be partitioned into four subsets represented by four regions as shown in the Venn diagram below. The subset A is made up of the regions 1 and 2, and the subset B is made up of the regions 2 and 3. Use the symbols A and B and set notations to represent each of the four regions.

U

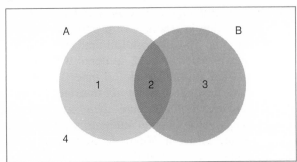

3. Let $U = \{a, b, c, d, e, f, g, h, i, j, k\}$; $A = \{a, b, c, d, e\}$; $B = \{d, e, f, g, h\}$; and $C = \{c, e, g, i, j\}$. Find

a) $A \cap B$ e) $A \cup B$ h) $A \cup B \cup C$

b) $B \cap C$ f) $A \cup C$ i) $A \cap B'$

c) $A \cap C$ g) $B \cup C$ j) $B' \cap C'$

d) $A \cap B \cap C$

4. Let $A = \{a, b, c, i, 2\}$; $B = \{c, d, e, 2, 3\}$; $C = \{c, f, g, 1, 3\}$; and $U = \{a, b, c, d, e, f, g, h, 1, 2, 3\}$. Are A, B, and C mutually exclusive? Are they exhaustive? Draw a Venn diagram to represent each of the following sets.

a) $A \cap B \cap C$ d) $B \cap C$

b) $A \cup B$ e) $(A \cup B \cup C)'$

c) $A \cap C$

5. In a survey of model preference for compact cars, prospective buyers were asked "In the past year, have you considered purchasing any or all of the following models: Model X, Model Y, or Model Z?" The responses of the 100 prospective buyers interviewed were as follows:

Model X	Model Y	Model Z	No. of respondents
No	No	No	30
No	No	Yes	13
No	Yes	No	17
Yes	No	No	11
Yes	No	Yes	9
Yes	Yes	No	8
No	Yes	Yes	7
Yes	Yes	Yes	5

Draw a Venn diagram, placing the number of respondents shown above in the appropriate region.

23

6. One hundred college seniors were interviewed about their plans after graduation. It was found that 50 of them will continue their education in graduate or professional schools (G); 50 will get married (M); and 50 will work full time (W). Further analysis of the collected data revealed that 12 students will continue their education *and* get married, but not work full time; 15 will continue their education *and* work full time but not get married; 20 will get married *and* work full time but not continue their education; and 10 will do all three things. How many students will do none of these three things? (Hint: Draw a Venn diagram, and place the correct number of students in each of the disjoint eight regions.)

7. An economics class consists of 40 students, of whom 25 are English majors (E) and 9 are on the Dean's list (D); among those on the Dean's list, four are also English majors. How many students are English majors or not on the Dean's list? (Hint: Use a Venn Diagram.)

8. A committee has 11 members: There are three Democrats, six Republicans, and two independent members. There are three women on the committee: one is a Democrat and the other two are Republicans. Find the number of people in each of the following sets. (D = Democrats, R = Republicans, W = Women, and I = Independents.)

   a) $(D \cap R)'$
   b) $(W \cap R)' \cup (D \cap R)'$
   c) $(D \cup W) \cap (R \cup W)$
   d) $(D \cup W)' \cap (R \cup W)' \cap (D \cup R)'$

9. Referring to problem 1 above, answer the following.

   a) Do A, B, and C partition U?
   b) Which sets are mutually exclusive and exhaustive?

10. Let U be $\{a, b\}$. List all the subsets of U. Let S be the set $\{1, 2, 3\}$. List all the subsets of S. (Hint: A null set is a subset of any set, and a set is a subset of itself.)

11. If a set S contains four elements, how many subsets of S are there?

12. If a set has n elements, how many subsets are there? How many subsets contain only zero elements? How many subsets contain exactly n elements? How many subsets contain exactly one element?

2-4 functional notations

One of the most important concepts involved in statistical discussions is the notion of a function. The primary purpose of this section is to review the symbols and notations that are commonly used in expressing functions. First of all, let us consider the meaning of function.

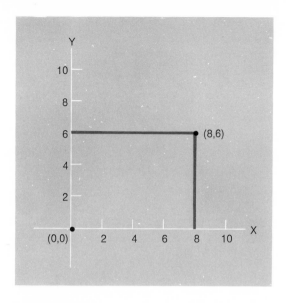

FIGURE 2-7
A two-dimensional plane

function defined

Consider two sets of elements. Each element of the first set is designated by X, and each element of the second set is designated by Y. Then there may exist a useful mathematical relationship known as the *functional relation* between X and Y.

> A function is a special association between an element (X) of one set and an element (Y) of another set in which each X element is paired with one and only one Y element.

The elements X and Y are often considered an *ordered pair* (X, Y). It is "ordered" because the order in which the two elements are listed is essential. In fact, a mathematical function is frequently referred to as a collection of ordered pairs of numbers. Such ordered pairs can be shown graphically in a two-dimensional plane. (See figure 2-7.) Two perpendicular lines in such a plane are taken as the axes. The *horizontal line* is the X axis, and the *vertical line* is the Y axis. The point where the two axes intersect is called the *origin* $(0, 0)$. Each point in the plane represents a pair of numbers called *coordinates*. The first coordinate, or X, measures the horizontal distance from that point to the Y axis; the second coordinate, or Y, measures the vertical distance from that point to the X axis. The set of numbers designated by X, the first element of the ordered pair, is referred to as the *domain*, and the set of numbers designated by Y,

25

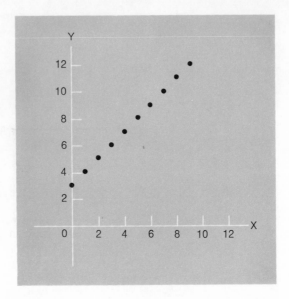

FIGURE 2-8
A function $Y = 3 + X$ where X
designates any one of the ten
digits

the second element of the ordered pair, is referred to as the *range* of
the function. Thus, graphically speaking, the domain of a mathematical
function is represented by the X axis, and its range represented by the
Y axis.

EXAMPLE 2-15   Let $Y = 3 + X$, where X is an element of the set of 10
digits: 0, 1, 2, 3, 4, 5, 6, 7, 8, and 9. Our task is (1) to find the range of this
function, and (2) to represent the function graphically.

1) The domain of the function is already given, namely, 0, 1, 2, 3, 4, 5,
   6, 7, 8, and 9. Since the value of Y is 3 plus X, that is, $3 + 0 = 3$,
   $3 + 1 = 4$, $3 + 2 = 5, \ldots$, and $3 + 9 = 12$, we have the range of the
   function: 3, 4, 5, 6, 7, 8, 9, 10, 11, and 12.

2) The graphic representation of this function is shown in figure 2-8.
   Note that each point in the graph represents an ordered pair; and the
   ten ordered pairs are (0, 3), (1, 4), (2, 5), . . . , (9, 12). The ten points
   form a function, a collection of ordered pairs.

EXAMPLE 2-16   Show graphically the range of the function $Y = X + 2$
for $0 \le X \le 5$; that is, for X greater than or equal to 0, and less than or
equal to 5.
   Since the domain of the function is $0 \le X \le 5$, the range must be

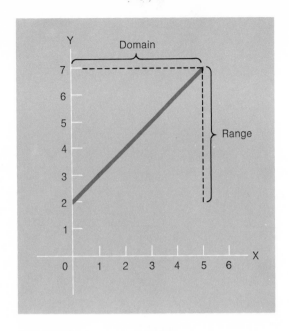

FIGURE 2-9
Domain and range of $Y = X + 2$
for $0 \leq X \leq 5$

$0 + 2 \leq Y \leq 5 + 2$, namely, $2 \leq Y \leq 7$. Figure 2-9 shows the domain and range of this function.

EXAMPLE 2-17   Consider two functions $Y = 15 + 0.5X$ and $Y = 30 - X$. What is the domain and range of each of the two functions?

The two functions are partially shown in figure 2-10. The domain of each function is the set of all numbers, and so is the range of each function. That is, the domain of each function is $-\infty \leq X \leq +\infty$ and the range is $-\infty \leq Y \leq +\infty$. Observe that in each of the functions in the above three examples, each value of X is associated with *one and only one* value of Y. Furthermore, a functional relationship exists if two or more values of X are all paired with the same value of Y. The following example provides an illustration.

EXAMPLE 2-18   Consider the equation: $Y = X^2$. Is this a function?

First, let us see the relationship between X and Y as shown in figure 2-11. As indicated, two values of X, one negative and one positive, but with the same absolute magnitude, are paired with one value of Y. This is still a function, because for each value of X there corresponds one and only one Y value, which satisfies the definition of a function. For instance, if $X = 2$, $Y = 4$; and if $X = -2$, $Y = 4$.

27

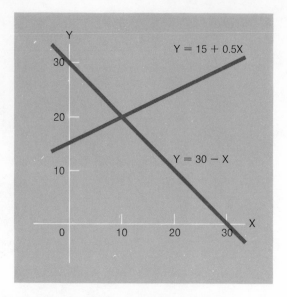

FIGURE 2-10
Two functions

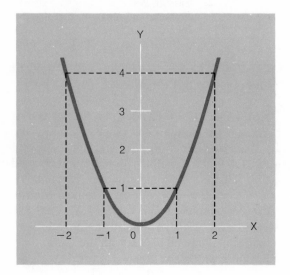

FIGURE 2-11
A function: $Y = X^2$

If you are still not sure whether a certain relationship between X and Y is a function, simply plot it on a two-dimensional plane. Then draw a vertical line in such a way that it will intersect the graph. If every vertical line that intersects the graph intersects it at only one point, the relationship

28

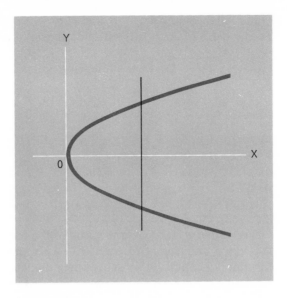

FIGURE 2-12
A graph showing a curve that is
not a function

represented by the graph is a function. It is not a functional relation if the line intersects the graph at more than one point. Thus the curve in figure 2-11 is a function and the curve in figure 2-12 is not.

notations

The letter f is frequently used to designate a function, and $f(X)$ (read "f of X" or "the value of f at X") is called a *functional notation*; $f(X)$ must not be confused with "f multiplied by X." The expression $Y = f(X)$ signifies that there exists a *function rule* which pairs each value of X with one and only one value of Y. When the function rule and the value of X are known, the value of Y is determined.

Consider the function rule: $Y = 3 - 2X + X^2$. It is understood that $f(X) = Y$; that is,

$$f(X) = 3 - 2X + X^2$$

Instead of writing $3 - 2X + X^2$ for Y, we simply write $f(X)$. Thus, if $X = 3$, we have

$$f(3) = 3 - 2(3) + 3^2 = 3 - 6 + 9 = 12 - 6 = 6$$

and if $X = 5$, we have

$$f(5) = 3 - 2(5) + 5^2 = 3 - 10 + 25 = 28 - 10 = 18$$

Other letters such as g, h, P, and F, or in many cases double letters such as FR and RF, are also used to designate functions.

EXAMPLE 2-19   Given the following function rules:

$$f(X) = 3 - 2X + X^2 \text{ and } g[f(X)] = f(X) + 2,$$

Find (a) $f(1)$,   (b) $g[f(X)]$,   and (c) $g[f(1)]$.

a) $f(1) = 3 - 2(1) + 1^2 = 3 - 2 + 1 = 4 - 2 = 2$
b) $g[f(X)] = f(X) + 2 = 3 - 2X + X^2 + 2 = 5 - 2X + X^2$
c) $g[f(1)] = f(1) + 2 = 3 - 2(1) + (1)^2 + 2 = 3 - 2 + 1 + 2 = 6 - 2 = 4$
or

$$g[f(1)] = f(1) + 2 = 5 - 2(1) + (1)^2 = 5 - 2 + 1 = 6 - 2 = 4$$

Observe that $g[f(X)]$ is a function of the function $f(X)$. In such a case, it is more convenient to treat $f(X)$ as a variable, and the value of the function g is obtained by substituting the value of that variable for $f(X)$. Referring to example 2-19 above, let $Y = f(X)$. Then,

$$g[f(X)] = g(Y) = Y + 2$$

When $X = 1$,

$$f(1) = 3 - 2X + X^2 = 3 - 2(1) + 1^2 = 3 - 2 + 1 = 4 - 2 = 2$$
and

$$g(2) = 2 + 2 = 4$$

The letters X and Y used in this section are *variables*. The value of the variable Y is obtained when a number is substituted for X in the function rule. Thus, variables can be classified as *dependent* and *independent*. The variable which stands for a value in the domain of a function, or X in the function $Y = f(X)$, is referred to as the independent variable. It is called independent because it is "free" to assume any value in the domain of the function. Once the value of X and the function rule is known, the value of Y is determined. That is why Y is referred to as the dependent variable.

2-4 exercises

1. Explain whether each of the following is a function.
   a) $Y = f(X)$, where X denotes any student and Y denotes a grandparent of the student
   b) $Y = f(X)$, where X designates any one of several children in a family and Y designates a parent of the children
   c) Let U be the set of all students in your college, and let $Y = f(X)$, where Y is the height of student X in U.
   d) $Y^2 = X$, where the possible values of X constitute the domain.

2. Let $Y = f(X) = 2 + 2X$. Show graphically the
   a) range of $f(X)$ if the domain is known to be $1 \leq X \leq 5$
   b) domain of $f(X)$ if the range is known to be $-10 \leq Y \leq -2$

3. Given $f(X) = X$, what is the domain of $f(X)$? What is the range of $f(X)$?

4. Suppose that the domain of the function $f(X) = X^2 - 2X + 4$ is $\{2, 4, 6, 8, 10\}$. What is the range of this function? If the range of the function $f(X) = X^2 - 4$ is $\{-4, -3, 0, 5\}$, what is the domain of this function? Show the two answers graphically.

5. Given $f(X) = X^2 - 2X + 4$, find
   a) $f(2)$ d) $f(a)$
   b) $f(-2)$ e) $f(-a)$
   c) $f(5)$ f) $f(c^2)$

6. Let $f(X) = (X^2 + 2)/X$, find
   a) $f(1)$ b) $f(-2)$ c) $f(2)$ d) $f(5)$ e) $f(c)$

7. If $f(X) = (1 + X)^X$ find
   a) $f(0)$ b) $f(a)$ c) $f(2)$

8. If $f(X) = (2 - a)^X$, find
   a) $f(0)$ b) $f(2)$ c) $f(a)$

9. Let $g(X) = 10$. Find
   a) $g(2)$ b) $g(-5)$ c) $g(10)$

10. If $g(Y) = Y^2 + 1$ and $f(Y) = Y + 2$, find $g[f(Y)]$, $g[f(1)]$, and $g[f(3)]$.

11. Let $f(X) = X + 1$ and $g(X) = X^2$. Find
    a) $f[g(0)]$ b) $g[f(-1)]$ c) $f[g(1)]$ d) $g[f(2)]$

12. Let $f(X) = aX + b$ and $g(X) = cX + d$.
    a) Is $f(-X) = -f(X)$?   c) Find $g[f(X)]$
    b) Find $f[g(X)]$ d) Does $f[g(X)] = g[f(X)]$?

events 3

and probabilities

3-1
introduction

One of the primary reasons for the existence of probability theory and statistical inference is the presence, in almost every aspect of life, of *random phenomena*. A phenomenon is random if chance factors determine its outcome; in other words, the outcome occurs haphazardly without aim, plan, or choice. All the possible outcomes may be known in advance, but the particular outcome of a single trial in any experimental operation cannot be predetermined. Nevertheless, some regularity is built into the process so that each of the possible outcomes can be assigned a *probability fraction*. (For example, the probability that a balanced coin will show up heads after being tossed is 0.5.)

The simplest example of a random phenomenon is the result of the toss of a coin. Nothing but chance factors determine whether tails or heads will occur in any single toss. There is no deterministic regularity here; that is, one cannot say for sure that heads, or tails, will come up on a particular toss. Similarly, in the roll of a cubic die, we cannot predetermine which side will turn up. Other examples of random phenomena are the number of pupils who will attend school on a certain day, the number of defective parts produced in a production run, the number of automobile accidents on the Los Angeles freeways during a certain month, and the number of deaths of a certain age group in a particular community. The existence of random phenomena in so many diversified fields has made it imperative for students in the social and natural sciences to study the theory of probability.

Outcomes such as those that result from the tossing of a coin are known as *random events*. In the section to follow, we shall discuss the concept of a random event and the totality of all the possible events in a given problem situation. The various rules of counting the total number of events will be discussed in section 3; and the concept of probability and its computation will be considered in section 4.

In brief, this chapter is about events, counting of events, and probability. The discussion will revolve around the idea of counting because the counting of events is all it will take to handle the probability problems to be considered in this chapter.

3-2 outcomes and random events

Before we attempt to choose the team that will win the NFC championship, we must know which teams are in the National Football Conference. We cannot predict or expect the number of heads to come up unless we know how many coins are used in the toss. Generally speaking, in order to make intelligent decisions we need to know at least all the possibilities, or possible outcomes, in any given problem situation. This explains why we shall, first of all, discuss the universal set of all possible events before we deal with any particular event or outcome.

sample space

Briefly, a *sample space is a universal set*. It is always related to an experiment. Thus,

> The sample space of an experiment is the set of all possible distinct outcomes of the experiment.

To understand this concept better, we need to clarify the meanings of some key terms. The words "outcome" and "experiment" are used by statisticians in a very broad sense. *An experiment is some process or operation that leads to well-defined outcomes.* The simplest example is the toss of a coin to see whether heads or tails will show up. The toss of a coin is an operation, or experiment, and "heads (or tails) showing up" is the outcome. It is "well defined" because all the possible outcomes are known.

An experiment may consist simply of checking to see whether potential customers like or dislike a new product. Other examples of experiments are counting the number of pupils absent from any class in a certain school on a given day, measuring the amount of milk produced by each of the cows on a farm in a given period, or observing the total number of spots that show up on the faces of two dice after they are rolled.

An outcome is the result of a single trial of an experiment. A trial is an act that leads to one of the possible distinct outcomes of the experiment. Any particular toss of a coin is a trial, which may result in tails or heads, an outcome. Corresponding to other experiments in the preceding paragraph, an outcome may be a simple Yes or No answer to the question whether a certain customer likes a new product, three students absent from an algebra class on the first Monday of October, fifteen gallons of milk produced by a specific cow, or a sum of seven spots showing on two dice after a particular roll. Each of these is called an *individual outcome* because it is the result of a single trial of an experiment.

A sample space is a universal set, because it is the set of all possible distinct outcomes that will be under consideration in a given problem situation. The following simple examples will make the concept clearer.

EXAMPLE 3-1   In the experiment of tossing three coins to see how many heads (H) or tails (T) occur, we have the sequence of three coins as an individual outcome. Thus THH forms one single outcome showing that the first coin comes up tails, and the second and third come up heads. The set of all possible distinct outcomes of this experiment is

$$U = \{TTT, TTH, THT, HTT, HHT, HTH, THH, HHH\}$$

Observe that there are no possibilities that might result from this experiment other than these eight outcomes.

EXAMPLE 3-2   In the experiment of rolling two dice to see the sum of spots on the sides that turn up after they are rolled, we have the sequence of two dice as an individual outcome. Let the first number in each cell of table 3-1 represent the spots on the first die, and the second number the spots on the second die. Then the two numbers in each of the 36 cells constitute an individual outcome. These 36 outcomes form the sample space, since each is distinct from the others and together they exhaust all the possibilities of the experiment.

Table 3-1   Sample space for the experiment of rolling two dice to see the sum of spots on the sides that turn up

First die	Second die					
	1	2	3	4	5	6
1	1 + 1	1 + 2	1 + 3	1 + 4	1 + 5	1 + 6
2	2 + 1	2 + 2	2 + 3	2 + 4	2 + 5	2 + 6
3	3 + 1	3 + 2	3 + 3	3 + 4	3 + 5	3 + 6
4	4 + 1	4 + 2	4 + 3	4 + 4	4 + 5	4 + 6
5	5 + 1	5 + 2	5 + 3	5 + 4	5 + 5	5 + 6
6	6 + 1	6 + 2	6 + 3	6 + 4	6 + 5	6 + 6

Frequently, an experiment may involve numerous or even an infinite number of distinct outcomes, in which case it will be necessary to resort to methods other than the one used above to describe a sample space.

random events

Observe that the outcomes of each of the above-mentioned experiments are all random phenomena. This is why an outcome of any experiment is frequently referred to as a *random event*. Random events may be classified as simple and compound.

> A simple random event is the outcome of a single trial in any particular experiment.

In example 3-1, *TTH* or *HHT* or any other three letter sequence in *U* is a simple random event. Thus we may say that a sample space is the set of all simple random events. Similarly, in example 3-2, the outcome of $2 + 3 = 5$ is a simple random event, and $1 + 4 = 5$ is another one. These two events are distinct, since the first one is formed with the first die showing two spots and the second die showing three spots, whereas the second is formed with the first die showing one spot and the second die showing four. Thus they are two distinct random events even though, in both cases, the sum of spots is the same—five.

A *compound* event, on the other hand, consists of at least two simple events. It is defined as follows:

> A compound event is a set of simple random events, or a subset of the sample space.

In example 3-1 above, the subset $\{THT, HTT, TTH\}$ is a compound event because it is made up of three distinct simple events. Thus the event of obtaining two tails in the toss of three coins is a compound event; so is the event of obtaining two heads in this experiment. Obviously, it is a subset of the sample space. In example 3-2, the event of obtaining a sum of five spots consists of four simple events, namely $\{(4 + 1), (3 + 2), (2 + 3), (1 + 4)\}$, and the event of obtaining a sum of seven spots consists of six simple events, namely $\{(6 + 1), (5 + 2), (4 + 3), (3 + 4), (2 + 5), (1 + 6)\}$. Thus both events, five spots and seven spots, are compound events. If any one of the four simple events that result in a sum of five spots shows up, we say that the compound event of five spots has occurred; if any one of the six simple events that result in a sum of seven spots shows up, we say that the compound event of seven spots has occurred. Furthermore, a compound event may be formed either by two or more compound events or by a combination of simple and compound events. Thus, in example 3-1, the subset $\{TTT, TTH, THT, HTT\}$ is a compound event of two or more tails; and the compound event of 10 or more spots includes the simple event $(6 + 6)$ and the compound events $\{(6 + 5), (5 + 6)\}$ and $\{(5 + 5), (4 + 6), (6 + 4)\}$.

Just as any element of a set may be considered a subset of the universal set, a simple random event may also be referred to as a subset of the sample space.

The above experiments are just two of innumerable situations that generate random events. Other examples of simple random events are the measurement of the height of a person, the measurement of the speed of an automobile, the response to a question raised in a marketing survey, and the number of defective parts found in any particular production run.

3-2 exercises

1. In the experiment of tossing two coins to see how many heads (H) or tails (T) occur,

   a) how many simple random events are there?
   b) what is the sample space of this experiment?
   c) is "two heads" a simple or compound event?

2. In the experiment of rolling one die to see which one of the six sides will show up,

   a) what is the sample space of this experiment?
   b) is "odd number of spots" a simple or compound event?

3. A family of four persons, two teen-age children and their parents, are interviewed by a market researcher to ascertain their like (L) or dislike (D) of a new product. Let the responses of the father, mother, first child, and second child form a sequence. How many possible sequences are there? What is the sample space of this experiment?

4. Leonard Carlson has a date with Elsie Johnson on Saturday evening. There are three, and only three possible entertainments available that evening. They may go to a football game (F), to see a movie (M), and/or attend a concert (C). Which of these three events will actually materialize will depend on Elsie's mood, which is highly unpredictable. Look at the Venn diagram at the top of the following page, and list the region or regions representing the following possibilities.

   a) The event that they will go to the football game only
   b) The event that they will do all three things
   c) The event that they will do none of the three things
   d) The event that they will neither see a movie nor attend a concert
   e) The event that they will not go to a football game

5. In problem 4 above, what is the sample space of all possible events? (Answer this question in terms of F, M, and C, and then identify each event by referring to the region or regions in the Venn diagram.

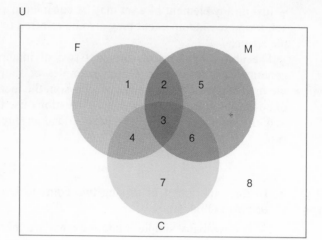

6. Referring to the Venn diagram above, use *F, M,* and *C,* and set notations, to represent each of the eight regions.

7. A sample space can often be represented by a geometrical figure. Suppose, for instance, that John and Mary are playing cards. John has three spades: king, queen, and jack; Mary has three hearts: also K, Q, and J. Each person's cards are shuffled and one of them is drawn at random. Let the numbers 1, 2, and 3 represent K, Q, and J, respectively. Then the sample space for each person is shown as three points on a number scale as follows:

King	Queen	Jack
1	2	3

where each point, which represents a card drawn, is a simple random event.

Since there are two persons involved in this experiment, we have a sequence of two cards as a simple random event. Let the X and Y axes of a two-dimensional plane be the two number scales for John and Mary. We then have nine points, or ordered pairs, representing the sample space as shown on the graph below. The point (1, 2) represents the event that John draws the spade king and Mary draws the heart queen, whereas the point (3, 1) represents the event that John draws the spade jack and Mary draws the heart king, and so on.

   a) Using the letters K, Q, and J, present the sample space of the nine events as a set of ordered pairs of cards.

   b) If we call any sequence of two cards with identical faces an event, is this event a simple or compound event? Why?

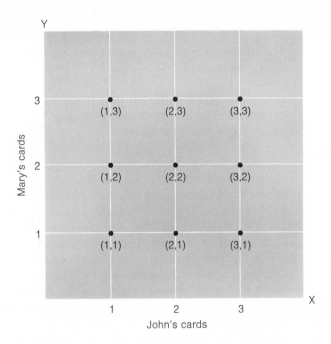

Y

Mary's cards

3   (1,3)   (2,3)   (3,3)

2   (1,2)   (2,2)   (3,2)

1   (1,1)   (2,1)   (3,1)

     1   2   3   X

John's cards

c) In problem 1 above, let 1 and 2 represent heads and tails, respectively; and let the outcomes of the first coin be shown on the X axis and the outcomes of the second coin be shown on the Y axis. Show the sample space in a graphic form.

8. Two dice are rolled. Let X designate the number of spots showing up on the first die and Y designate the number of spots showing up on the second die. Use a two-dimensional plane to represent the sample space of this experiment.

3-3 rules for counting events

In the previous section, we listed all the possible outcomes or events in two experiments. In many situations, however, the job of listing all the outcomes is a difficult, if not impossible, undertaking. Even when an experiment involves a relatively small number of simple events, the task of constructing a complete and accurate list is a tedious and troublesome one. Actually, most problem situations do not require a complete listing: it is enough to know the total number of possible outcomes. In this section, we shall discuss the various rules for calculating the total number of outcomes. These rules include the multiplication rule, permutation, and combination.

the multiplication rule

The multiplication rule may be considered under two different but similar situations.

RULE 3-1   If a number (n) of acts are carried out, and each act can be performed in the same number of ways (k), then the total number of possible outcomes for n acts is

$$(k) (k) \ldots (k), \text{ or } k^n \qquad (3\text{-}1)$$

Here each outcome involves one simple event. In the familiar example of tossing three coins to see how many heads occur, there are three acts and each can be performed in two ways: heads or tails. Accordingly, there are 2^3, or 8, possible outcomes for the three acts. Observe that each of the eight sequences in example 3-1 is a simple event.

For further clarification of this concept, let us consider another familiar example. The outcome of rolling two dice is a simple event: the number of spots upmost on the first die and the number of spots upmost on the second die. The rolling of two dice involves two acts; the number of ways in which each act can be performed is six; that is, $n = 2$ and $k = 6$. The total number of simple events, or outcomes, is therefore 6^2, or 36.

RULE 3-2   If there are n acts which can be performed in $k_1, k_2, \ldots, k_n$ ways, respectively, then the total number of different possible outcomes for the n acts in succession is

$$(k_1) (k_2) \ldots (k_n) \qquad (3\text{-}2)$$

EXAMPLE 3-3   Imagine that you are going to take a trip from Los Angeles to New York by way of Denver and Chicago. Suppose further that there are four ways of transportation available between Los Angeles and Denver, three ways between Denver and Chicago, and two ways between Chicago and New York, as follows:

Los Angeles to Denver		Denver to Chicago		Chicago to New York	
Airplane	(A)	Greyhound bus	(G)	Steamship	(S)
Bicycle	(B)	Hitchhiking	(H)	Train	(T)
Car	(C)	Motorcycle	(M)		
Donkey	(D)				

The total number of ways in which you can travel from Los Angeles to New York is therefore (4) (3) (2), or 24. Figure 3-1 presents a tree diagram showing the 24 possibilities.

The tree diagram shows the four ways (four branches), namely, A, B, C, and D, you can travel from Los Angeles to Denver. From Denver to Chicago, there are three ways (G, H, and M) available. (Note the three branches growing out of each of the four branches mentioned above.) So there are (4)(3), or 12 ways (12 branches), in which to travel from Los Angeles to Chicago by way of Denver. Finally, there are the two ways (S and T) available from Chicago to New York, or two branches growing out of each of the 12 branches. In other words, there are 24 branches, or 24 different possible outcomes, in this experiment.

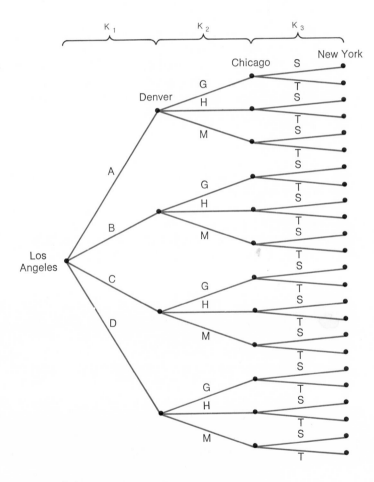

FIGURE 3-1
A tree diagram showing the total number of sequences for $k_1 = 4$, $k_2 = 3$, and $k_3 = 2$

permutations

A permutation is an arrangement in order; it refers to any of the ways in which objects are arranged. If there is a set of n objects, the number of ordered arrangements of objects will depend on r, the number of objects to be selected and arranged. Permutation rules may be applied under two circumstances: $n > r$ and $n = r$. You need not be concerned about problems with n smaller than r; they do not exist.

RULE 3-3 $n > r$ The number of permutations of r objects taken from a set of n objects (P_r^n) is expressed as follows:

$$P_r^n = n(n-1)(n-2) \ldots (n-r+1) \tag{3-3}$$

The rationale of this formula is as follows: the first object in an ordered arrangement is selected from the entire set of n objects; thus there are n ways to select the first object. After the first selection $(n-1)$ objects remain; therefore there are $(n-1)$ ways to select the second object. After the second selection there are $(n-2)$ ways to select the third object, and so forth. The rth object is selected from the

$$n - (r-1) = (n-r+1)$$

objects that remain after the first $(r-1)$ objects have been selected. A direct application of rule 3-2 $[(k_1)(k_2) \ldots (k_n)]$ will then lead to the equation expressed in rule 3-3 above. Observe that the number of permutations of r objects taken from a set of n objects is always the product of r factors.

Frequently, the number of permutations is expressed in terms of factorials. The symbol $n!$ is called the *n factorial;* the n factorial is the product of

$$n(n-1)(n-2)(n-3) \ldots (4)(3)(2)(1)$$

with $0!$ defined as 1. For instance, $6! = (6)(5)(4)(3)(2)(1) = 720$, and $6! = (6)(5)(4)(3!)$. In general, $n! = n(n-1)(n-2)!$ and $n! = n(n-1)(n-2)(n-3)!$. However, $(n-k)! \neq n! - k!$; for example, $(5-2)! \neq 5! - 2!$. Since the last factor in the equation under rule 3-3 is $(n-r+1)$, we need to multiply the right-hand side by $(n-r)!$ in order to make it $n!$. Keeping the quantity unchanged, we must divide $n!$ by $(n-r)!$. Thus equation 3-3 can also be expressed as follows:

$$P_r^n = \frac{n!}{(n-r)!} \tag{3-4}$$

EXAMPLE 3-4   Suppose that there are 10 candidates for five positions, namely, president, vice-president, secretary, treasurer, and public relations director. How many ways can these five positions be filled?

In this problem $n = 10$ and $r = 5$. There are obviously 10 ways to fill the first position. Once this is done, there are nine candidates left; therefore there are nine ways to fill the second position. Similarly, there are eight ways to fill the third position, seven ways to fill the fourth position, and six ways to fill the last position. Thus, the total number of ways, or permutations, to fill five positions from 10 candidates is

$$P_5^{10} = (10)\,(9)\,(8)\,(7)\,(6) = 30,240$$

which is the product of five factors. The same answer is obtained if the alternate equation is used.

$$P_5^{10} = \frac{10!}{(10-5)!} = \frac{(10)\,(9)\,(8)\,(7)\,(6)\,(5!)}{5!}$$

$$= (10)\,(9)\,(8)\,(7)\,(6) = 30,240$$

RULE 3-4 $n = r$ The number of permutations of n objects taken all together is

$$P_n^n = n! \tag{3-5}$$

Observe that this is just a special case of rule 3-3. When $r = n$, the last factor of the permutation equation becomes $(n - r + 1) = (n - n + 1) = 1$. Similarly, in the alternate equation, the denominator becomes $(n - r)! = (n - n)! = 0! = 1$, and the right-hand side of the equation becomes $n!$.

EXAMPLE 3-5   Suppose that five different scholarships are awarded to the five top students of a graduating class. How many ways can the scholarships be awarded? It is evident that any of the five students could be awarded the first scholarship, giving five ways for awarding scholarship 1. After scholarship 1 is awarded, there are only four students for scholarship 2; after scholarship 2 is awarded, there are only three students for scholarship 3; and so forth. Consequently, there are $5! = (5)\,(4)\,(3)\,(2)\,(1) = 120$ ways to award the five scholarships to the five students.

As another example, the number of ways to seat nine judges in nine chairs is $9! = (9)\,(8)\,(7)\,(6)\,(5)\,(4)\,(3)\,(2)\,(1) = 362,880$. Three candidates (X, Y, and Z) for the three positions of president (P), vice-president (V), and secretary-treasurer (S) can be arranged in six different ways; that is, $3! = (3)\,(2)\,(1)$.

$$
\begin{array}{lcccccc}
\text{P:} & \text{X} & \text{X} & \text{Y} & \text{Y} & \text{Z} & \text{Z} \\
\text{V:} & \text{Y} & \text{Z} & \text{X} & \text{Z} & \text{X} & \text{Y} \\
\text{S:} & \text{Z} & \text{Y} & \text{Z} & \text{X} & \text{Y} & \text{X}
\end{array}
$$

Observe that each permutation is an outcome or simple event. The total number of permutations is the total of possibilities in a particular problem situation.

combinations

In many probability problems, we are interested not in ordered arrangements but in the number of ways that r objects can be selected from n objects. The total number of ways that r objects can be selected from a set of n objects, irrespective of order, is called the number of combinations of r objects taken from n objects. Thus, combinations differ from permutations in that the latter take the order in which objects are selected into consideration, and the former do not.

The concept of a combination, however, is closely related to that of a permutation. Recall that the number of permutations of r objects taken all together is $r!$, and that the number of ways in which the three positions mentioned above can be filled by the three candidates is $3!$, or 6. The six permutations constitute only one combination because the order in which each is selected is disregarded.

RULE 3-5   Let the symbol $\binom{n}{r}$ be the number of combinations of r objects selected from a set of n objects. Since each combination of r objects has $r!$ permutations, the total number of permutations of r objects selected from n objects becomes

$$\binom{n}{r} r! = \frac{n!}{(n-r)!}$$

Dividing each side of this equation by $r!$, we obtain the formula for the number of ways in which a subset of r objects can be taken from a set of n objects irrespective of order.

$$\binom{n}{r} = \frac{n!}{(n-r)!\,r!} = \frac{n(n-1)(n-2)\ldots(n-r+1)}{r!} \tag{3-6}$$

EXAMPLE 3-6   Suppose that 10 people are candidates for the board of trustees of a certain school district. Three trustees are to be elected. How many ways can three people be selected from the 10 candidates?

In this problem, $n = 10$, $r = 3$. Accordingly, we have

$$\binom{10}{3} = \frac{10(9)(8)}{3!} = \frac{720}{6} = 120$$

That is, the total number of ways of selecting three trustees from 10 candidates is 120. Note that the order in which the three trustees are selected is not considered.

The foregoing discussion should have made clear the distinction between permutation and combination. Many students, however, are not always sure whether to apply the permutation or the combination rule when faced with counting problems. For further clarification of the distinction between these two rules, let us consider another example. Suppose that a statistics class consists of 20 students. If a committee of three students is selected from this class, in how many ways can this be done? Since the order of arrangement of the three students is not a matter of interest, this is a problem of combination. The question involved here is: How many combinations of three students can be taken from a class of 20 students? Thus the total number of ways, or combinations, is

$$\binom{20}{3} = \frac{20(19)(18)}{3(2)(1)} = 1140$$

On the other hand, if three students are selected from the class and are to be awarded three different prizes, the question of how many ways this can be done will be a problem of permutation, since the order of arrangement is now a matter of concern. There are 20 ways to award the first prize, 19 ways to award the second prize, and 18 ways to award the third prize. Consequently, the total number of ways, or permutations, to award the three prizes to three of the 20 students is

$$P_3^{20} = 20(19)(18) = 6840$$

A committee of three students constitutes only *one* combination. As indicated in rule 3-4, a combination of three individuals has

$$3! = 3(2)(1) = 6 \text{ permutations}$$

Since the total number of ways, or combinations, a committee of three students can be selected from a class of 20 students is 1140, and since for each combination of three students there are six permutations, or ways, in which the three prizes can be awarded, the total number of ways in which the three prizes can be awarded to three of the 20 students is

$$1140(6) = 6840$$

which is the same as obtained above.

For a problem involving the rule of multiplication, the set of all possible outcomes constitutes the sample space. For instance, in the travel problem discussed in example 3-3, the 24 possible outcomes form the sample space. If we are interested in the order in which a set of objects can be selected, then the sample space of the problem is made up of all the possible permutations. Thus in example 3-4, the 30,240 permutations involved in filling five positions with 10 candidates constitute the sample space. On the other hand, if we are interested only in the number of subsets of r objects selected from n objects regardless of the order in which each of the r objects appears, all the possible combinations will form the sample space. Accordingly, the 120 combinations in example 3-6 become the sample space of that problem. Each outcome, each permutation, or each combination is a simple event, a distinct possible result of the experiment in a particular problem situation.

3-3 exercises

1. Find the value for each of the following.

 a) $6!$ b) $(n - r + 1)!$ for $n = 5$ and $r = 3$ c) P_3^7

 d) P_4^4 e) $\binom{6}{3}$ f) $\binom{10}{10}$ g) $\binom{5}{0}$

2. List all the possible ways in which the letters A, B, and C can be ordered.

3. By using $n = 7$ and $r = 2$, verify that

 a) $\binom{n}{r} r! = P_r^n$ b) $\binom{n}{r} = \binom{n}{n - r}$

4. Suppose that four cities, New York, Miami, Los Angeles, and Seattle, have asked the National Football League for permission to hold the Superbowl in 1976 and 1977. Assume that each city may host the Superbowl in either 1976 or 1977 but not in both years. Draw a tree diagram to show the total number of sequences in which the two Superbowls can be held.

5. A building contractor offers houses with five different floor plans, three types of roofs, and two kinds of carpeting. How many ways can a homebuyer make his selection? Show the total number of selections by the use of a tree diagram.

6. Six dice are rolled. How many ways can they show up?

7. Let A be a set of five elements, a, b, c, d, and e. How many permutations are there if two elements of A are selected at a time?

8. The auto license plates issued by a certain state have two letters followed by three digits. How many different plates can be issued?

9. A quiz consists of 10 true-false questions. How many possible ways can the quiz be marked?

10. A class consists of 10 students. How many ways can a committee of three students be selected?

11. Three different prizes are awarded, and there are six contestants. How many ways can the prizes be awarded? How many ways can six prizes be awarded to the contestants?

12. In a class of 30 students there are 20 men and 10 women. (a) How many ways can a committee of three men and two women be selected? (b) How many ways can a committee of five students be

selected? (c) How many ways can a committee of five students be selected if all five are of the same sex?

13. A club consists of 20 members. How many ways can the three officers, president, vice-president, and secretary, be selected?

14. The board of directors of the ABC Company has 11 members. A committee of three members is to be selected. In how many ways can this be done? How many ways can five directors be selected from the board to fill five different positions?

15. A club contains 30 members: 15 whites, 10 blacks, and 5 of other races. A committee of six members is to be formed. If all the three groups are proportionally equally represented, how many ways can this be done?

16. A poker hand consists of five cards dealt from an ordinary deck of 52 playing cards. How many different hands can be dealt from a deck?

3-4
computing
probability

what is probability?

The word "probability" is often used in connection with an event. The *probability of an event* is the percentage of time that the event will occur in the long run. Thus, when we say that there is a probability of 0.08 that a part produced by an automatic machine will be defective, we mean that in the long run 8 percent of the parts produced by the machine will be defective. We cannot guarantee that in any batch of 100 parts there will be exactly eight defective parts; but if we examine a sufficiently large number of batches, we can expect to find an average of eight defective parts per batch.

This common usage of probability may also apply to an event which can happen only once. For instance, we may wish to know the probability that a person's lung cancer will heal perfectly. For this purpose we simply examine the medical records showing the proportion of patients who have had complete recovery from such a disease. If the records show that out of a large number of such cases five out of every 100 patients recovered completely, we may say that the probability of this person's recovery is 0.05.

When the sample space is made up of events that are *equiprobable,* which means that each event has the same probability of occurring as any other event, the computation of probability will be a simple and straightforward one. In general, if there are N equiprobable distinct outcomes in an experiment, and there are n possible outcomes for a particular compound event (A), then the probability that the compound event A will occur is simply n/N. Letting $P(A)$ designate the probability of A, we have

$$P(A) = \frac{n}{N} \qquad (3\text{-}7)$$
$$= \frac{\text{The number of possible outcomes in } A}{\text{The total number of possible outcomes in the sample space}}$$

That is, the probability of some particular event is just the ratio of the number of outcomes contained in the event to the total number of possible outcomes in the sample space as long as the outcomes are all equally probable.

probability computations

Applying the above formula and the various rules of counting discussed in the last section, we can compute the probability of any event in a particular experiment. Keep in mind that the computation of probabilities always involves the counting of events.

EXAMPLE 3-7   The automobile license plates issued by a certain state have three digits followed by three letters. An auto manufacturing company is conducting a four-week promotion to attract potential customers to its dealers in that state. A license number is selected at random, and the number posted in each dealer's showroom. Anyone whose license plate number matches the number drawn will win a new car. What is the probability that you will win the new car if you already own three automobiles registered in that state? (It is assumed that all the possible license plate numbers have been issued.)

There are 10 digits and 26 letters. The total number of outcomes of six objects, namely, three digits followed by three letters, is

$$(10)(10)(10)(26)(26)(26) = 17{,}576{,}000$$

You have three plates on your three cars. Accordingly, the probability of your winning a car is

$$\frac{3}{17{,}576{,}000}$$

EXAMPLE 3-8   In lotteries it is customary for the person whose number is drawn first to receive the largest prize, the person whose number is drawn next to receive a somewhat smaller prize, and so on, until r numbers are drawn and r different prizes are awarded. Suppose that in a lottery 30 tickets were sold to 30 persons and only three are to be drawn to determine the first three prize winners. Suppose further that you, your brother, and your sister each bought a ticket. What is the probability that you will win the first prize, your brother the second prize, and your sister the third prize?

Here $n = 30$ and $r = 3$. By the rule of permutation, the number of ordered arrangements of three different prizes to three persons out of a total of 30 persons is

$$\frac{30!}{(30-3)!} = \frac{(30)(29)(28)(27)!}{(27)!} = (30)(29)(28) = 24,360$$

The probability that you three will win the three prizes as specified is therefore

$$\frac{1}{24,360}$$

since there is only one arrangement out of a total of 24,360 by which you win the first prize, your brother the second, and your sister the third.

If you want to know the probability of the three of you winning the three prizes without regard to the order in which they are won, you need only to recall that the number of permutations of three objects taken all together is $3! = 6$. Consequently, the probability that all the three prizes would be awarded to you three is

$$\frac{6}{24,360}$$

EXAMPLE 3-9   From a class of 40 students a five-student committee is to be selected at random. What is the probability that a group of five close friends in the class will be chosen for that committee?

Since this problem does not involve any arrangement in order, it can be solved by the rule of combination. (See rule 3-5 on page 44.) The total number of combinations of five students from a set of 40 students is

$$\binom{40}{5} = \frac{(40)(39)(38)(37)(36)}{(5)(4)(3)(2)(1)} = 658,008$$

The group of five close friends can form only one combination. Thus, the probability that those five students will be chosen is

$$\frac{1}{658,008}$$

EXAMPLE 3-10   A box contains 20 units of a certain electronic product; 4 of them are defective, and 16 of them are good. Four units are randomly selected and sold. Find (a) the probability that all the four units sold are defective; (b) the probability that among the four units sold two are good and two are defective; and (c) the probability that at least three defective units are sold.

a) The total number of combinations of four units taken from 20 units is

$$\binom{20}{4} = \frac{20!}{4!\,16!} = \frac{20(19)\,(18)\,(17)}{4(3)\,(2)\,(1)} = 4845$$

The four defective units constitute only one of these 4845 combinations, since

$$\binom{4}{4} = \frac{4!}{0!\,4!} = 1$$

Thus the probability that all the four units sold are defective is

$$\frac{\binom{4}{4}}{\binom{20}{4}} = \frac{1}{4845}$$

b) The number of combinations of two units taken from the 16 good units is

$$\binom{16}{2} = \frac{16(15)}{2} = 120$$

The number of combinations of two units taken from the four defective units is

$$\binom{4}{2} = \frac{4(3)}{2} = 6$$

Thus by the rule of multiplication the total number of combinations of two good and two defective units is

$$\binom{16}{2}\binom{4}{2} = 120(6) = 720$$

As a result, the probability that two good and two defective units are sold is

$$\frac{\binom{16}{2}\binom{4}{2}}{\binom{20}{4}} = \frac{720}{4845} = \frac{144}{969}$$

c) From (a) above, we know that there is only one combination of four units from the four defective units. The total number of combinations of one good and three defective units is

$$\binom{16}{1}\binom{4}{3} = \frac{16!}{1!\,15!} \cdot \frac{4!}{1!\,3!} = 16(4) = 64$$

Consequently, the probability that at least three defective units are sold is

$$\frac{\binom{4}{4} + \binom{16}{1}\binom{4}{3}}{\binom{20}{4}} = \frac{1 + 64}{4845} = \frac{65}{4845} = \frac{13}{969}$$

3-4 exercises

1. In a poll to ascertain the popularity of a certain candidate for the presidency of the United States, it is found that 495 of the 1000 voters interviewed are in favor of the candidate. What is the probability that any one voter will favor this candidate?

2. Suppose that statistics compiled by the weather bureau in Los Angeles show that it has rained during the Rose Parade in Pasadena 14 times in the last 80 years.
   a) What is the probability that it will rain during the Rose Parade in Pasadena on the next New Year's day?
   b) What is the probability that it will not rain there then?

3. Suppose that in a lottery 20 tickets were sold and two prizes will be given. Assume that you have bought two tickets. What is the probability that you will win both the prizes?

4. Look again at example 3-7. If you own a car registered in that state, what is the probability that you will win a new car?

5. An auto dealer has just received a shipment of 20 new cars, 15 of which are sedans and 5 of which are convertibles. If two cars are sold, what is the probability that the two cars sold will be of the same model?

6. A club has 30 members: 25 men and 5 women. A committee of five members is to be formed. What is the probability that all five women will be included in the committee if the committee membership is selected at random?

7. A stereo shop has received a shipment of ten new sets, seven of model X and three of model Y. If four sets are sold at random, what is the probability that two of each model will be sold?

8. A committee of three persons is to be selected from the board of directors of a company. The board consists of 15 members, one-third of whom are women and two-thirds men. What is the prob-

ability that all three persons in the committee will be of the same sex?

9. A poker hand consists of five cards. What is the probability that the five cards are all of the same suit?

10. Assume that in the experiment of tossing three coins to see how many heads (H) or tails (T) will occur, that each of the eight outcomes is equiprobable. (See example 3-1.) Find

    a) the probability that two heads will occur
    b) the probability that one or two heads will occur
    c) the probability that three tails or three heads will occur

11. In the experiment of rolling two dice to see what number of spots will show up, assume that the 36 simple random events are equiprobable. (See example 3-2.)

    a) What is the probability that a sum of seven spots will occur?
    b) What is the probability that a sum of two or three will occur?
    c) What is the probability that a sum of less than five will occur?
    d) What is the probability that a sum of at least 10 will occur?

rules of probability ④

and probability functions

4-1
introduction

In the preceding chapter, our discussion of probability was centered on the counting of events. Counting is sufficient for simple probability computations. When a problem is complex, however, we need various rules to help us determine probabilities. In this chapter, we shall first summarize the most fundamental axioms of probability, the axioms on which the probability rules are based. Next, we shall consider certain basic rules about the *union* of two or more events and the *intersection* of events. (The latter is discussed under the headings of *independent* and *dependent* events.) Finally, the notions of a *random variable* and a *probability function* will be explored. These notions are closely related to the concept of a random event discussed in the last chapter and to the basic rules of probability to be considered in this chapter.

In the last section of this chapter the *binomial function* will be reviewed. The mathematical expression of the binomial function involves some of the counting rules discussed in chapter 3. The binomial function is known as a discrete probability distribution. By discrete we mean that the variable involved can take on values that differ from one another by finite amounts. Usually such a variable is associated with a sample space that is finite in nature. In statistics the antonym for "discrete" is "continuous." An example of the continuous probability function is the normal distribution, which will be discussed in a later chapter. At present it is only necessary to point out that a *continuous probability distribution* involves a variable that can take on values differing from one another by infinitesimal amounts and is associated with an infinite number of elements in its sample space.

4-2
the
basic
rules of
probability

Let E be any event for which we wish to compute the probability. Then the following axioms and rules hold.

axioms

1) $P(E) \geq 0$ The probability of any event must be a positive value or

zero. If the probability is zero, it means that the event under consideration is certain not to appear.

2) $P(U) = 1$ Since U stands for the sample space, which includes all possible outcomes or events, the probability of U must be 1. This is the case when an event is certain to occur.

3) $P(E) \leq 1$ The probability of an event can never be larger than 1. It may be equal to 1 if E is identical with the sample space.

addition rules

The above three axioms are regarded as the most basic rules of probability. The rules to be examined here involve the additions of probabilities, which will be frequently referred to in subsequent discussions.

RULE 4-1 $$P(E) + P(E') = 1 \qquad (4\text{-}1)$$

For any event (E), there is an event called the "complement of E" (E'); the union of E and E' (or $E \cup E'$) must be equal to the universal set. Thus, the sum of the probabilities of event E and its complement E' must be equal to 1. Accordingly,

$$P(E) = 1 - P(E') \text{ and } P(E') = 1 - P(E)$$

RULE 4-2 *Special Addition Rule* If two events E_1 and E_2 are *mutually exclusive* $(E_1 \cap E_2 = \emptyset)$, then

$$P(E_1 \cup E_2) = P(E_1) + P(E_2) \qquad (4\text{-}2)$$

EXAMPLE 4-1 Suppose that a bag contains 10 balls marked 1, 2, . . . , 10. Let E_1 be the event of drawing a ball marked 3 or less, and E_2 be the event of drawing a ball marked 6 or more. Then, the probability of E_1 or E_2 (that is, $E_1 \cup E_2$) to occur is the sum of the respective probabilities. That is,

$$P(E_1 \cup E_2) = P(E_1) + P(E_2) = \frac{3}{10} + \frac{5}{10} = \frac{8}{10} = \frac{4}{5}$$

The special addition rule is applied in this case because E_1 and E_2 are mutually exclusive.

RULE 4-3 *General Addition Rule*

$$P(E_1 \cup E_2) = P(E_1) + P(E_2) - P(E_1 \cap E_2) \qquad (4\text{-}3)$$

This rule is called the *general addition rule* because it applies to any two events whether mutually exclusive or not. If they are mutually exclusive, then $E_1 \cap E_2 = \emptyset$ and the probability of $\emptyset$ is zero. Accordingly, the equation for the general addition rule will be identical with that for

54

the special addition rule. If E_1 and E_2 are overlapping such as A and B in figure 2-5, the intersection of E_1 and E_2 is counted twice, once in E_1 and then again in E_2. For this reason, the probability of the intersection must be subtracted from the sum of the probabilities of E_1 and E_2.

EXAMPLE 4-2   Given 10 balls as specified in example 4-1 above, let E_1 be the event of drawing a ball marked with an even number, and E_2 be the event of drawing a ball marked 5 or less. The probability that E_1 or E_2 will occur then becomes $P(E_1 \cup E_2) = P(E_1) + P(E_2) - P(E_1 \cap E_2) = \frac{1}{2} + \frac{1}{2} - \frac{2}{10} = \frac{4}{5}$. The probability of $E_1 \cap E_2$ is $\frac{2}{10}$ because two balls, marked 2 and 4, belong to both events.

EXAMPLE 4-3   Suppose that one card is drawn from an ordinary deck of 52 playing cards. Find the probability of drawing a face (F) card (king, queen, and jack) or a spade (S).

There are 12 face cards in the deck; the probability of drawing a face card is therefore $P(F) = \frac{12}{52}$.

There are 13 spades; the probability of drawing a spade is $P(S) = \frac{13}{52}$.

However, the three face cards of spade are counted twice in figuring these two probabilities. They should be subtracted once when determining the probability of drawing a face card or a spade. Accordingly,

$$P(F \cup S) = P(F) + P(S) - P(F \cap S)$$
$$= \frac{12}{52} + \frac{13}{52} - \frac{3}{52} = \frac{22}{52} = \frac{11}{26}$$

RULE 4-4   If there are n events $E_1, E_2, \ldots, E_n$ which are mutually exclusive, then the probability of any of these events happening is the sum of all their individual probabilities. That is,

$$P(E_1 \cup E_2 \cup \ldots \cup E_n) = P(E_1) + P(E_2) + \ldots + P(E_n) \quad (4\text{-}4)$$

EXAMPLE 4-4   Suppose that Mr. Smith is planning to go out next Saturday evening. The probability that he will see a basketball game (G), a movie (M), or a horse race (H) are 0.35, 0.30, and 0.20, respectively. The probability that he will do one of these three things is

$$P(G \cup M \cup H) = P(G) + P(M) + P(H) = 0.35 + 0.30 + 0.20 = 0.85$$

For further clarification of the rules presented in this section, let us consider the following example: Suppose that a statistics class consists of 40 students: 5 freshmen, 25 sophomores, 6 juniors, and 4 seniors. Among the 40 students, 2 freshmen, 12 sophomores, 3 juniors, and 3 seniors are male students. Then, if a student is selected at random,

a) The probability that the student will be either male or female is
$$P(\text{male} \cup \text{female}) = P(\text{male}) + P(\text{female}) = \frac{20}{40} + \frac{20}{40} = \frac{40}{40} = 1$$

b) The probability that the student will be either a freshman or a sophomore is

$P(\text{freshman} \cup \text{sophomore}) = P(\text{freshman}) + P(\text{sophomore}) = \frac{5}{40} + \frac{25}{40} = 0.75$

c) The probability that the student will be either a female student or a sophomore is

$P(\text{female} \cup \text{sophomore})$

$= P(\text{female}) + P(\text{sophomore}) - P(\text{female} \cap \text{sophomore})$

$= \frac{20}{40} + \frac{25}{40} - \frac{13}{40} = \frac{45}{40} - \frac{13}{40} = \frac{32}{40} = 0.80$

d) The probability that the student will be either a sophomore, or a junior, or a senior is

$P(\text{sophomore} \cup \text{junior} \cup \text{senior}) = P(\text{sophomore}) + P(\text{junior}) + P(\text{senior}) = \frac{25}{40} + \frac{6}{40} + \frac{4}{40} = \frac{35}{40} = 0.875$

4-2 exercises

1. In the experiment of tossing three coins, it is assumed that the eight possible outcomes (see example 3-1) are equiprobable. If E_1 designates the event that two heads will occur and E_2 designates the event that three heads will occur, what is the probability that either E_1 or E_2 will occur? That is, what is $P(E_1 \cup E_2)$? Use a Venn diagram to show the union of the two events.

2. In the preceding problem, if A denotes the event that two or more heads will occur and B denotes the event that two or fewer heads will occur, what is the probability that either A or B will occur. That is, what is $P(A \cup B)$?

3. In the experiment of rolling two dice, let us assume that the 36 possible outcomes are equiprobable (see example 3-2). Let A denote the event that four or fewer spots will occur, and B denote the event that 10 or more spots will occur. Are A and B mutually exclusive? Find the probability that A or B will occur.

4. In problem 3 above, if A denotes the event that an odd number of spots will occur, and B denotes the event that more than seven spots will occur, what is $P(A \cup B)$? Use a Venn diagram to show the union of A and B.

5. See problem 3 above. Find the probability that one spot or four spots or seven spots or nine spots will occur.

6. Suppose that a bag contains 10 balls marked 1, 2, 3, . . . , 10. Let E be the event of drawing a ball marked with an even number, and

let F be the event of drawing a ball marked with a 5 or a higher number. Are E and F mutually exclusive? Find $P(E \cup F)$.

7. If a card is drawn at random from an ordinary deck of 52 well-shuffled cards, what is the probability of drawing either a spade or a heart or a diamond? What is the probability of drawing a diamond or an ace?

8. In the above problem, what is the probability of drawing
   - a) a red king or a black card?
   - b) a red king or a red card?
   - c) a red or a black queen?

9. Mr. Jones earned $3000 last summer, and he is considering spending the money for one of three things: a new car, a vacation trip, or a new set of furniture for his living room. The probabilities that he will buy a new car, take the vacation trip, or buy furniture, are respectively, 0.30, 0.25, and 0.33. What is the probability that he will do one of these three things? What is the probability that he will forget these three possibilities and use the money for something else?

10. Suppose that 80 percent of all Americans vacationing in the Far East visit Tokyo, 80 percent of them visit Hong Kong, and 70 percent of them visit both Tokyo and Hong Kong. What is the probability that an American tourist vacationing in the Far East will visit either Tokyo or Hong Kong? What is the probability that he will visit neither city?

11. The probability that Mr. Smith will invest in common stock A is 0.20, in common stock B 0.30, and in both A and B 0.10. What is the probability that he will invest in neither A nor B?

12. The probabilities that an auto dealer will sell, in one week, zero, one, two, three, four, or five or more cars are 0.05, 0.10, 0.18, 0.25, 0.20, and 0.22, respectively. Find the probabilities of
   - a) selling three or more cars in one week
   - b) selling three or fewer cars in one week

4-3 independent events

The concepts of dependence and independence are integral parts of the theory of probability. In this section we are concerned with the probabilities of independent events.

Suppose that there are two events: A and B. If the occurrence of A (or B) does not affect the probability that B (or A) will occur, then A and B are said to be *independent*. They are called independent because the fact that one event has happened will in no way influence the probability of occurrence or nonoccurrence in the other.

RULE 4-5 *Special Multiplication Rule* If A and B are two independent events, then the probability that both A and B will occur is equal to the product of their respective probabilities. Symbolically,

$$P(A \cap B) = P(A) \cdot P(B) \tag{4-5}$$

where A intersection B, or $A \cap B$, means that A and B will both occur. Two events form a *joint event* if both of them occur. The probability of a joint event is called *joint probability*. Thus $P(A \cap B)$ is called a joint probability. For instance, if A is the event that heads will show up in the first toss of a balanced coin and B is the event that heads will show up in the second toss of the coin, then the joint event is the event that heads will show up in *both* tosses. Since A and B are independent—the occurrence of heads in the first toss in no way affects the probability that heads will occur in the second toss, the probability of this joint event, or *joint probability*, is

$$P(A \cap B) = P(A) \cdot P(B) = (0.5)(0.5) = 0.25$$

EXAMPLE 4-5 Suppose that two cards are drawn in succession from an ordinary deck of 52 playing cards with the first card being replaced before the second is drawn. The probability of drawing two kings is simply $\left(\frac{4}{52}\right)\left(\frac{4}{52}\right) = \left(\frac{1}{13}\right)\left(\frac{1}{13}\right) = \frac{1}{169}$. Since the first card is replaced, the probability that the second draw will be a king remains $\frac{4}{52}$ even though a king may have been selected in the first draw.

In general, the event of drawing an object from a set of objects and the event of a subsequent draw are independent, provided that the object that has been drawn is replaced.

EXAMPLE 4-6 A balanced die is rolled three times. What is the probability that six spots will show up on all three rolls?

Let A, B, and C be the events that six spots will show up on the first, second, and third rolls. Since the outcome of one roll is independent of the outcome of any other roll, the joint probability of these three events is equal to the product of their respective probabilities. Because the die is a balanced one, the probability that each side will show up on a single roll is $\frac{1}{6}$. Thus, we have

$$P(A \cap B \cap C) = P(A) \cdot P(B) \cdot P(C) = \left(\frac{1}{6}\right)\left(\frac{1}{6}\right)\left(\frac{1}{6}\right) = \frac{1}{216}$$

EXAMPLE 4-7 Suppose that in a large company both people with and without college training are hired to do the same kind of work. After a period of time they are rated by their supervisors. Table 4-1 shows the joint probabilities of performance rating for the two types of employees.

Table 4-1 Joint probabilities of independent events—training and performance rating of employees

Performance rating	With college training (C)	Without college training (C')	Total
Good (R)	0.12	0.18	0.30
Poor (R')	0.28	0.42	0.70
Total	0.40	0.60	1.00

This example is hypothetical and the table entries are deliberately designed to show that college training (C) or no college training (C') for employees has nothing to do with performance rating which is classified into two categories: Good (R) and Poor (R').

As shown, $P(C) = 0.40$ and $P(C') = 0.60$, which are presented in the bottom row. Similarly, $P(R) = 0.30$, $P(R') = 0.70$, which are presented in the last column. These probabilities are called *marginal probabilities* for the obvious reason that they appear in the margins of the table. All the events appearing along any particular margin of the table must be mutually exclusive and exhaustive. Each cell entry is the *joint probability* for the two events represented by the captions of the corresponding row and column. For instance, the joint probability $P(R \cap C) = 0.12$, $P(R' \cap C) = 0.28$. Note that each marginal probability is equal to the sum of all the joint probabilities in a particular row or column of the table.

Table 4-1 serves to illustrate the equation $P(A \cap B) = P(A) \cdot P(B)$ as follows:

$$P(R \cap C) = P(R) \cdot P(C) = (0.30)(0.4) = 0.12$$
$$P(R' \cap C) = P(R') \cdot P(C) = (0.70)(0.40) = 0.28$$
$$P(R \cap C') = P(R) \cdot P(C') = (0.30)(0.60) = 0.18$$
$$P(R' \cap C') = P(R') \cdot P(C') = (0.70)(0.60) = 0.42$$

Observe that the joint probability appearing in each cell of the table is equal to the product of the corresponding marginal probabilities. Thus the test for determining whether a joint probability table involves independence between two sets of events is to see whether the product of the corresponding marginal probabilities equals the entry of the cell representing the joint probability.

College training, or the absence of it, is independent of a good or poor performance rating because one does not affect the probability that the other will occur. Any employee taken at random will have the probability of 0.30 of being rated R, regardless of whether he has attended college or not. If we are told that an employee has attended college, the probability that he is rated R is still 0.30; for we are now concerned only with the column headed C and the probability of R on the condition that C has occurred is $0.12/0.40 = 0.30$. Similarly, if we are told that an employee has not attended college, the probability that this employee is rated R is still 0.30; for our interest is limited to the column headed C'. The probability of R on the condition that C' has occurred is $0.18/0.60 = 0.30$. In short, the occurrence of one event does not affect the probability that the other event will occur, which is the definition of independence between two events discussed at the beginning of this section.

The probability of an event occurring on the condition that another event has occurred is called conditional probability. Thus the probability of R given C, written $P(R|C)$, is a conditional probability. *When two events are independent, the conditional probability is equal to the marginal probability.* For instance, given the occurrence of C, the probability of R is

$$P(R|C) = 0.12/0.40 = 0.30$$

because 40 out of 100 are classified as C, and 12 out of the 40 are classified as R. Observe that

$$P(R|C) = P(R) = 0.30$$

Similarly,

$$P(C|R) = 0.12/0.30 = 0.40 = P(C)$$

and

$$P(C'|R) = 0.18/0.30 = 0.60 = P(C')$$

Likewise, in example 4-5, the probability that the second draw will be a king on the condition that the first draw is also a king is a conditional probability. Since the second draw is independent of the first draw, the probability that the second draw will be a king is not affected by the fact that the first draw was or was not a king. That is,

$$P(\text{the second draw is a king}|\text{the first draw is a king})$$
$$= P(\text{drawing a king}) = \tfrac{4}{52} = \tfrac{1}{13}$$

Furthermore, in example 4-6, the probability that six spots will turn up on the third roll on the condition that six spots has already turned up on the first and second rolls is a conditional probability. Since the outcome of each roll is independent of the outcomes of previous rolls, this conditional probability is not affected by the earlier rolls. That is,

$$P(\text{six on the third roll}|\text{six has shown up on first and second rolls})$$
$$= P(\text{six spots will show up}) = \tfrac{1}{6}$$

In general, if two events (A and B) are independent, then the probability that B will happen on the condition that A has happened is simply the probability that B will happen regardless of whether A has happened. Symbolically,

$$P(B|A) = P(B)$$

or

$$P(A|B) = P(A)$$

4-3 exercises

1. Which of the following pairs of events are independent?
   a) Getting heads in two successive tosses of a balanced coin
   b) Being a corporation president and having blue eyes
   c) Producing a second child of the same sex as the first
   d) Being intoxicated while driving a car and having a fatal accident
   e) Drawing a spade from a deck of playing cards, replacing the spade, and drawing another spade
   f) Drawing a king from a deck of playing cards and drawing a queen without replacing the first card drawn

2. Use your own words to distinguish between a joint probability and a conditional probability.

3. A box contains 10 balls. Five of them are white, three are red, and two are black. A ball is selected at random with replacement.
   a) What is the probability of drawing two white balls in a row?
   b) What is the probability of drawing a red ball and then a black ball?
   c) What is the probability of drawing three red balls in a row?
   d) What is the probability of drawing a black ball, then a red ball, and finally a white ball?

4. A card is drawn from an ordinary deck of playing cards. If the card that has been drawn is replaced before the next one is drawn, what is the probability of drawing
   a) four aces and any one of the other cards?
   b) three aces and two kings?
   c) five cards of the same suit?

5. Four balanced coins are tossed. What is the probability of obtaining
   a) four tails?
   b) four heads?
   c) one head and three tails?

6. Two balanced dice are rolled. What is the probability that an even number of spots will appear on the first die *and* an odd number of spots will appear on the second die?

7. Mr. Hanson and his wife are, respectively, 55 and 50 years old. If the probability that a man aged 55 will live at least another 15 years is 0.70, and the probability that a woman aged 50 will live at least another 15 years is 0.85, what is the probability that both Mr. Hanson and his wife will still be alive 15 years hence? Assume that the longevity of the husband and wife are independent.

8. Assume that the probability of a coin falling heads is 0.6. If a coin is tossed three times, what is the probability of obtaining

   a) all tails? c) at least two heads?
   b) at most two heads? d) at least two tails?

9. Two fire engines are available for emergencies. The probability that any one of the two engines will be ready when needed is 90 percent. It is assumed that the availability of one engine is independent of the other.

   a) In the event of a fire alarm, what is the probability that both engines will be ready?
   b) What is the probability that both engines will not be ready?
   c) What is the probability that one engine will be ready?

10. A joint probability table of the sex and marital status of employees in a large institution is given as follows:

Marital status	Female F	Male F'	Total
Married (M)	0.42	0.18	0.60
Not married (M')	0.28	0.12	0.40
Total	0.70	0.30	1.00

   a) Are sex and marital status independent? Why or why not?
   b) Find $P(M|F)$, $P(M|F')$, and $P(M)$.
   c) Find $P(F|M)$, $P(F|M')$ and $P(F)$.
   d) Find $P(M'|F')$, $P(M'|F)$, and $P(M')$.
   e) Find $P(F'|M)$, $P(F'|M')$, and $P(F')$.

4-4 dependent events

Recall that if two events are independent, the occurrence of one will not affect the probability that the other will or will not occur. *Conversely, two events (A and B) are said to be dependent if the occurrence of A (or B) affects the probability that B (or A) will or will not occur.* Thus the conditional probability is no longer equal to the marginal probability, because the fact that one event has occurred will influence the probability that the other will occur, and vice versa. Consequently, the joint probability of two events A and B is no longer equal to the product of their respective probabilities.

RULE 4-6 *General Multiplication Rule* If A and B are two dependent events, then the probability that both A and B will occur is equal to the product of the probability of A (or B) multiplied by the conditional probability of B given A (or A given B). Symbolically,

$$P(A \cap B) = P(A) \cdot P(B|A) \qquad (4\text{-}6)$$

and

$$P(A \cap B) = P(B) \cdot P(A|B) \qquad (4\text{-}7)$$

That is, for dependent events,

$$P(A) \neq P(A|B) \text{ and } P(B) \neq P(B|A)$$

Rule 4-6 is referred to as general because it applies to both dependent and independent events. When A and B are independent, $P(B|A) = P(B)$ and $P(A|B) = P(A)$, and equations 4-6 and 4-7 will be identical with equation 4-5.

EXAMPLE 4-8 If two cards are drawn from an ordinary deck of 52 playing cards *without* replacing the first card before the second card is drawn, then the outcomes of the two draws are dependent. The probability of obtaining a king in the first draw is $\frac{4}{52}$. If the first draw is a king, the probability of drawing a king in the next draw is $\frac{3}{51}$. Consequently, the joint probability of drawing two kings in the two trials is $\frac{4}{52} \cdot \frac{3}{51} = \frac{1}{221}$ In general, the event of drawing an object from a finite set of objects and the event of a subsequent draw are *dependent* if the object that has been previously drawn is *not replaced.*

EXAMPLE 4-9 Two groups of machines are purchased from two different suppliers by a manufacturing company. Group I consists of four machines, and group II of six machines. The 10 machines, which all have the same capacity, are used to produce identical parts. Five percent of the parts produced by group I machines are defective (D), while 90 percent of those produced by group II machines are nondefective (D'). Suppose that a machine is selected at random, and then one of the parts produced by this machine is drawn and examined. Determine each of the four possible joint probabilities.

Here we have $P(I) = 0.4$ and $P(II) = 0.6$.

$$P(D|I) = 0.05 \qquad P(D'|I) = 0.95$$
$$P(D|II) = 0.10 \qquad P(D'|II) = 0.90$$

Thus,

1) $P(I \cap D) = P(I) \cdot P(D|I) = 0.4(0.05) = 0.02$
2) $P(II \cap D) = P(II) \cdot P(D|II) = 0.6(0.10) = 0.06$
3) $P(I \cap D') = P(I) \cdot P(D'|I) = 0.4(0.95) = 0.38$

and

4) $P(II \cap D') = P(II) \cdot P(D'|II) = 0.6(0.90) = 0.54$

The various probabilities involved in this example are shown in the tree diagram in figure 4-1. The probabilities for the two groups of machines are presented as two main branches of the tree. The conditional probabilities for each group are shown as branches growing out of the main branch. Each of the joint probabilities is the product of the two probabilities along a sequence of two branches.

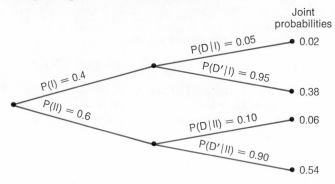

FIGURE 4-1
A tree diagram showing the
computation of joint probabilities

<table>
<tr><td></td><td>Joint probabilities</td></tr>
</table>

4-4
exercises

1. From the following pairs of events, identify those that are dependent.

    a) Mr. Smith received a large salary increase, and his wife bought a new dress.

    b) Mr. Jones drives on a slippery road, and Mr. Jones has an accident.

    c) Mr. Andrews is a Protestant, and his son is a Protestant.

    d) The probability of a coin falling heads is assumed to be 0.6; its first toss results in heads, and its second toss also results in heads.

    e) Two people randomly selected have the same blood type.

    f) Two people in the same family have the same blood type.

    g) A mother has an IQ measured at 120 and her child has an IQ measured at 132.

    h) The rate of cigarette smoking (packs per day) and the incidence of lung cancer.

2. Let $P(A) = 0.6$, $P(B) = 0.4$, and $P(A \cap B) = 0.18$. Find
   a) $P(B|A)$
   b) $P(A|B)$

3. A box contains 10 balls. Five of them are white, three are red, and two are black. A ball is drawn at random *without* replacement. What is the probability of drawing

   a) two white balls in a row?
   b) a red ball and then a black ball?
   c) three red balls in a row?
   d) a black ball, then a red ball, and finally a white ball?

4. Cards are drawn from an ordinary deck of playing cards. If the cards that have been drawn are not replaced before the next card is drawn, what is the probability of drawing

   a) four aces and any one of the other cards?
   b) three aces and two kings?
   c) five cards of the same suit?

5. A drawer contains sixteen socks; eight of them are brown, six are green, and two are yellow. Two socks are drawn in succession without replacement. What is the probability that the two socks will match in color?

6. A carton contains 20 eggs, five of which are spoiled. If we randomly select three eggs without replacement, what is the probability that all three eggs will be spoiled?

7. The owner of a resort in Newport Beach, California, has been informed that the probability of having a hot summer is 0.70. If the summer is hot, then the probability that he will make more than his normal profit is 0.90. What is the probability that there will be a hot summer and that he will make an above-normal profit?

8. In a certain large company, 80 percent of the employees are white (W), and 20 percent are nonwhite (W'). Of the whites, 20 percent have graduate training (G), 50 percent have college training (C), and 30 percent have high school training (H). Of the nonwhites, 30 percent have graduate training, 50 percent have college training, and 20 percent have only high school training. If an employee is selected at random from the company, what is the probability that the employee

   a) is nonwhite and has graduate training?
   b) has only high school training?

   Use a tree diagram to show the answers.

9. Suppose that in a dark room there are six white boxes, each containing three green and five yellow balls; and two black boxes, each containing two green and four yellow balls. If you went to the room and selected a box at random and then picked up a ball from that box, what is the probability that you selected a green ball from a white box? Use a tree diagram to show your answer.

10. Look again at problem 10 at the end of section 4-3. Suppose that the marginal probabilities remain unchanged, but the joint probabilities have changed as shown below.

Marital status	Female (F)	Male (F')	Total
Married (M)	0.40	0.20	0.60
Not married (M')	0.30	0.10	0.40
Total	0.70	0.30	1.00

a) Are sex and marital status independent? Why or why not?
b) Find $P(M|F)$; $P(M'|F)$; $P(M|F')$; and $P(M'|F')$.

11. Suppose that it is the policy of an insurance company to have its salesmen make house-to-house calls. According to past experiences, 20 percent of the calls resulted in sales (S), or $P(S) = 0.20$; and 80 percent of the calls did not (S'), or $P(S') = 0.80$. Of those families that bought insurance policies, 30 percent lived in single-unit two-story houses (T), or $P(T|S) = 0.30$. Seventy percent of the buyers lived in other types of buildings (T'), or $P(T'|S) = 0.70$. Of those families that did not buy any policy, 60 percent lived in single-unit two-story houses, or $P(T|S') = 0.60$; and 40 percent lived in other types of housing, or $P(T'|S') = 0.40$. What is the probability that

a) the next call will result in a sale if the family lives in a single-unit two-story house? Or, what is $P(S|T)$?
b) the next call will not result in a sale if the family lives in any other type of building? Or, what is $P(S'|T')$?

(Hint: Compute the joint probabilities and then set up a joint probability table such as table 4-1.)

12. In a survey of graduating students at a community college, it is found that 40 percent of the students will continue their training in a four-year college or university (T) and 60 percent will not (T'). The proportions of students in these two categories earning average grades of A, B, and C or lower are shown below.

Students	Average grades			
	A	B	C or lower	Total
T	0.10	0.30	0.60	1.00
T'	0.05	0.40	0.55	1.00

a) A student is selected at random and his average grade is A. What is the probability that he will continue his education?

b) What is the probability that he will not continue his education if his average grade is B?

(Hint: Use the same approach suggested in problem 11 above.)

4-5 probability functions

Virtually all the theoretical functions in statistics are probability functions. As pointed out in section 2-4, a function has its domain and range. The domain of a probability function includes all sets of values for the random variable, whereas its range is the set of values called probabilities. The term "random variable" is defined in the following section.

random variable

A symbol such as the letter X can be called a random variable provided that it meets certain requirements.

> If the values which a symbol takes on are associated with the simple random events in a given experiment, and hence are dependent on chance occurrences, we refer to the symbol as a random variable.

It is easy to imagine that a single value may be assigned to each and every possible simple random event in the sample space of a given experiment. Let the symbol X designate a single value assigned to any of all the distinct random events of the sample space of a particular experiment. Then, corresponding to each simple event there is one value designated by X. Thus the event of a "yes" answer to a question in a survey may be assigned the value 1, while a "no" answer is assigned the value 0. The height of a person may be measured at 70 inches, and the defective parts from a batch of 100 parts may be numbered at 8. Each of the values mentioned here is not predetermined but dependent on chance. It is a value for X, the random variable for the particular experiment. Let us use the coin-tossing experiment for further clarification of this concept.

EXAMPLE 4-10   Let the symbol X designate the number of heads obtained in the toss of three coins. Each of the simple random events in the sample space is paired with a value of the random variable X, as shown in table 4-2. Although there are eight values for X corresponding to the eight simple random events, there are only four distinct values for X.

Table 4-2   The random variable X as the number of heads in the toss of three coins

Sample space U	Value of the random variable X
TTT	0
TTH	1
THT	1
HTT	1
HHT	2
HTH	2
THH	2
HHH	3

It is hardly necessary to point out that any value of X is also an event because there will be some simple random event or compound event to which this value is assigned. Not only is any single value of X an event, but also two or more values for X constitute an event. Thus, each of the following can be called an event.

$$X = 0 \qquad X \geq 1 \qquad 1 \leq X \leq 3$$
$$X \leq 2 \qquad X = 2 \qquad 0 \leq X \leq 2$$

where $X = 0$ is a simple event, and all the others are compound events.

EXAMPLE 4-11   In the experiment of rolling two dice, the random variable X is the number of spots uppermost on the two dice after being rolled. There are 36 distinct outcomes in the experiment: For $X = 2$, there is only one outcome, namely, $1 + 1$; for $X = 3$, two outcomes; for $X = 4$, three outcomes; . . . ; for $X = 11$, two outcomes; and for $X = 12$, one outcome. Again, each of these values for X represents an event. The two events $X = 2$ and $X = 12$ are simple events, whereas all the others are compound events.

probability as a function

It is interesting to note that the probability of a value has a functional relationship with the possible values of the random variable. Suppose that a sample space of distinct outcomes, or simple events, is generated by a certain experiment. We can then divide the sample space into a set of *mutually exclusive* and *exhaustive* subsets, which are simple or compound events. Corresponding to each subset, there is a value for the random variable X. Each of the distinct values of X is associated with one and only one real number called probability. Thus we say that

> A probability is the value of a function which has as its domain a set of distinct values each corresponding to one of the mutually exclusive and exhaustive events, simple or compound, in the sample space of a given experiment.

EXAMPLE 4-12 Referring to example 4-10, we find that the distinct values for X are 0, 1, 2, and 3. The probability function of X is shown in the last column of table 4-3.

Table 4-3 Probability function for X, the number of heads obtained in the toss of three ideal coins

Sample space U	Distinct values for X	Probability function $P(X)$
TTT	0	$\frac{1}{8}$
TTH THT HTT	1	$\frac{3}{8}$
HHT HTH THH	2	$\frac{3}{8}$
HHH	3	$\frac{1}{8}$
Total		1.0

Observe that each distinct value of X is associated with one and only one value called probability, and that the sum of all the probabilities is equal to 1.

EXAMPLE 4-13 Let the random variable X stand for the sum of spots obtained in the roll of two dice. The set of distinct values for X is $S = \{2, 3, 4, 5, 6, 7, 8, 9, 10, 11, 12\}$, which is shown in the first column of table 4-4. Each of the distinct values in S represents an event, simple or compound, of the experiment; the second column shows the number of simple events represented by each distinct value of X (see table 3-1); and the last column shows the probability for each distinct value, which is expressed as a fraction with the number of simple events for each value as the numerator and the total number of simple events in the sample space, namely, 36, as the denominator.

Table 4-4 Probability function for X, the sum of spots shown on the faces of a pair of dice

Distinct values for X	Number of simple events	Probability function $P(X)$
2	1	$\frac{1}{36}$
3	2	$\frac{2}{36}$
4	3	$\frac{3}{36}$
5	4	$\frac{4}{36}$
6	5	$\frac{5}{36}$
7	6	$\frac{6}{36}$
8	5	$\frac{5}{36}$
9	4	$\frac{4}{36}$
10	3	$\frac{3}{36}$
11	2	$\frac{2}{36}$
12	1	$\frac{1}{36}$
Total		1.0

From the above examples, we can see that a probability is a function because one, and only one, probability value corresponds to each of the possible values of X.

Probability functions such as shown in the last columns of tables 4-3 and 4-4 are also known as *probability distributions*, since the total probability of 100 percent is *distributed* over and among all the mutually exclusive and exhaustive events of the sample space.

4-5 exercises

1. Four balls marked 2, 4, 6, and 8 are put in a box and are thoroughly mixed before a ball is drawn. Let X be the random variable designating the number on the ball that is drawn with replacement. Find the probability distribution of X.

2. A balanced die is rolled to see which side will show up.
   a) What does the random variable X designate?
   b) What is the probability function of X?

3. Let X designate the number of heads obtained in the toss of four balanced coins. Find
   a) $P(X \leq 2)$ d) $P(1 \leq X \leq 3)$
   b) $P(X \geq 2)$ e) $P(0 < X \leq 4)$
   c) $P(X \leq 3)$ f) $P(1 < X < 3)$

4. A nonbalanced coin is tossed three times. Assuming that the probability that heads will occur is 0.6, find the probability distribution of X, the number of heads obtained in three tosses.

5. A retailer sells two different models of stereo receivers, H and T. Assume that the two models are equally popular: 50 percent of all potential buyers prefer model H, and 50 percent prefer model T. Assume, too, that the retailer stocks three receivers of each model, and that three receivers are sold on a single day.

    a) What is the probability that three receivers of the same model are sold on a given day?

    b) Define the random variable in this experiment.

    c) What are the simple random events and their corresponding values of the random variable?

    d) What is the probability distribution of the random variable?

6. Suppose that a balanced coin is tossed three times. If all three tosses end up in tails, you win nothing. If heads turns up only once, you win $1. If two of the tosses result in heads, you win $4. If all three tosses result in heads, you win $9. In other words, if X designates the number of heads obtained, and if Y designates the amount of money you win, $Y = X^2$. Find the probability distribution of Y.

7. Let the random variable X be the number of heads obtained in the toss of four balanced coins. Find the probability distribution of X.

8. A die is biased in such a way that the probability of each side of the die showing up is proportional to the number of spots on it. That is, a 2 is twice as likely to show up as a 1, a 3 is three times as likely as a 1, . . . , and a 6 is six times as likely as a 1.

    a) What does the random variable X designate?

    b) What is the probability function of X in this experiment?

9. In a survey of consumers' attitudes toward a new product, the following question is asked: "Do you like the new product?" To this question there are two and only two possible answers, Yes and No, which are assigned the values of 1 and 0, respectively. Let p be the probability that the event of a Yes answer will occur. What is the probability distribution of W, the random variable in this experiment?

**4-6
binomial
distributions**

Table 4-3 provides an illustration of the binomial distribution. It is a rather simple probability distribution. In this section, we shall try to derive a binomial function by applying the rule of combination discussed in section 3-3 for more elaborate probability computations. First of all, we shall discuss the Bernoulli process, the process on which the binomial distribution is based.

bernoulli trials

The simplest probability model is one which has only two simple events in the sample space. For instance, the toss of a coin can result in one of two possible events: heads or tails. If you ask a person whether he watched a certain TV program last night, you will get one of two possible answers: Yes or No. The inspection of a machine part drawn from a production process can yield one of two possible outcomes: defective or good. Experiments that can lead to only one of two possible outcomes are commonly referred to as Bernoulli trials. (Jacques Bernoulli [1654–1705] is the Swiss mathematician who first described these experiments in the seventeenth century.)

Statisticians usually call one of the two possible outcomes of a Bernoulli trial a success and the other a failure. These two terms are used only to identify the outcomes, and not to indicate that one outcome is actually more desirable than the other. Thus, we may call one of two possible outcomes a success and the other a failure as long as we are consistent throughout our discussion.

Customarily, the event called a success is assigned the value 1, and the event called a failure assigned the value 0. Let W be a Bernoulli random variable which has the possible values 0 and 1. Further, let p be the probability of a success and $(1-p)$ the probability of a failure. Then we shall have the probability model of W, as shown in table 4-5.

Table 4-5   Bernoulli probability model

W	$P(W)$
0	$1-p$
1	p

If we call the event of obtaining six spots in the roll of a die a success and any other outcome a failure, then $P(0) = 1 - \frac{1}{6} = \frac{5}{6}$, and $P(1) = \frac{1}{6}$. Similarly, in the toss of a coin, we may call heads a success and tails a failure. Then the random variable W will have the probability distribution: $P(0) = \frac{1}{2}$ and $P(1) = \frac{1}{2}$.

binomial random variable

Suppose that there exists a Bernoulli process for W which has a sample space of only two events with probabilities $(1-p)$ and p. Suppose further that a sample is taken either with replacement or from an infinite set of simple events so that the probability p is unchanged during the process of sampling. Let us make n trials for W, so the sample consists of n observations in the following sequence.

$$W_1, W_2, W_3, \ldots, W_n$$

Let X be the *binomial random variable*. Then X can be defined as follows:

$$X = W_1 + W_2 + W_3 + \ldots + W_n = \sum_{i=1}^{n} W \qquad (4\text{-}8)$$

That is, the random variable X is the number of successes in n observations taken by a Bernoulli process.

The fact that the binomial random variable X, which is defined as the number of successes in n trials, is the sum of the $W's$ can be easily verified by a simple example.

EXAMPLE 4-14   Suppose that a coin is tossed 10 times. When heads occur (success), we assign the value 1 to the observation; when tails occur (failure), we assign the value 0. Suppose further that the outcomes of the 10 tosses are as follows:

W_1	W_2	W_3	W_4	W_5	W_6	W_7	W_8	W_9	W_{10}
1	0	0	1	1	1	0	1	1	0

Since there are six successes, this sample of 10 observations yields a value of 6 for X. That is,

$$X = \sum W = 1 + 0 + 0 + 1 + 1 + 1 + 0 + 1 + 1 + 0 = 6$$

Observe that in a sample of n observations, the random variable X may assume any one of the possible $(n + 1)$ values: $0, 1, 2, 3, \ldots, n$. Note also that X is a *discrete* random variable, since the possible values it may take on are counts of successes, which are integers and differ from each other by a finite value 1.

binomial function

Table 4-5 shows the probability function (Bernoulli model) of an experiment which involves a single trial that has only one of two possible outcomes success or failure. A binomial experiment, on the other hand, involves *two or more* trials. That is, a sample taken for the binomial random variable X must contain at least two observations ($n \geq 2$). The expression of the probability function for X will be easier for the student to understand if a simple illustration is provided at the outset.

EXAMPLE 4-15   Four coins are tossed simultaneously. The outcomes of the tosses are independent; this is tantamount to sampling from an infinite population, or from a finite population with replacement, because the probability of obtaining heads or tails remains the same for all the coins. Because there are two ways in which each coin can perform ($k = 2$), and because there are four coins ($n = 4$), there will be 2^4, or 16, different possible outcomes in this experiment. The 16 outcomes are listed in the left-hand column of table 4-6 with the possible results of the first coin represented by the first letter from the left, results of the second coin by the second letter, and so on. The probability of any particular outcome of four observations will depend on p, the probability that heads will occur.

Table 4-6   Probability distribution of X, the number of heads obtained in the toss of four balanced coins

Outcomes	X	Probability distribution $P(X)$	
$TTTT$	0	$(1-p)^4 = \left(\frac{1}{2}\right)^4$	$= \frac{1}{16}$
$HTTT$ $THTT$ $TTHT$ $TTTH$	1	$4(p)(1-p)^3 = 4\left(\frac{1}{2}\right)\left(\frac{1}{2}\right)^3$	$= \frac{4}{16}$
$HHTT$ $HTHT$ $HTTH$ $THTH$ $THHT$ $TTHH$	2	$6(p)^2(1-p)^2 = 6\left(\frac{1}{2}\right)^2\left(\frac{1}{2}\right)^2$	$= \frac{6}{16}$
$HHHT$ $HHTH$ $HTHH$ $THHH$	3	$4(p)^3(1-p) = 4\left(\frac{1}{2}\right)^3\left(\frac{1}{2}\right)$	$= \frac{4}{16}$
$HHHH$	4	$(p)^4 = \left(\frac{1}{2}\right)^4$	$= \frac{1}{16}$
Total			$\frac{16}{16} = 1$

The variable X here designates the number of heads turning up in a total of four observations; that is, $X = W_1 + W_2 + W_3 + W_4$. The set of possible values for X is therefore $\{0, 1, 2, 3, 4\}$. On the assumption that the coins are balanced ($p = \frac{1}{2}$), and by the use of equations 4-4 and 4-5, it is possible to find the probability function for X, as expressed in table 4-6.

Let us consider $X = 0$, which signifies that all four coins will be tails. There is only one possible outcome here: $TTTT$. The probability that tails will turn up is $1 - p$. Thus, we have

$$P(X = 0) = P(TTTT)$$
$$= P(W_1 = 0 \text{ and } W_2 = 0 \text{ and } W_3 = 0 \text{ and } W_4 = 0)$$
$$= P(W_1 = 0) \cdot P(W_2 = 0) \cdot P(W_3 = 0) \cdot P(W_4 = 0)$$
$$= (1 - p) \cdot (1 - p) \cdot (1 - p) \cdot (1 - p) = (1 - p)^4 = (\tfrac{1}{2})^4 = \tfrac{1}{16}$$

For $X = 1$, there are four possible outcomes: $HTTT$, $THTT$, $TTHT$, and $TTTH$. By the special addition rule (rule 4-2), the probabilities for the four outcomes are added to obtain the probability of $X = 1$.

$$P(HTTT) = (p)(1 - p)(1 - p)(1 - p) = (p)(1 - p)^3 = \tfrac{1}{16}$$
$$P(THTT) = (1 - p)(p)(1 - p)(1 - p) = (p)(1 - p)^3 = \tfrac{1}{16}$$
$$P(TTHT) = (1 - p)(1 - p)(p)(1 - p) = (p)(1 - p)^3 = \tfrac{1}{16}$$
$$P(TTTH) = (1 - p)(1 - p)(1 - p)(p) = (p)(1 - p)^3 = \tfrac{1}{16}$$

Consequently,
$$P(X = 1) = \tfrac{1}{16} + \tfrac{1}{16} + \tfrac{1}{16} + \tfrac{1}{16} = \tfrac{4}{16}$$

For $X = 2$, there are six possible outcomes: $HHTT$, $HTHT$, $HTTH$, $THTH$, $THHT$, and $TTHH$. The probability of obtaining two heads is the sum of the probabilities for the six outcomes.

$$P(HHTT) = (p)(p)(1 - p)(1 - p) = p^2(1 - p)^2 = \tfrac{1}{16}$$
$$P(HTHT) = (p)(1 - p)(p)(1 - p) = p^2(1 - p)^2 = \tfrac{1}{16}$$
$$P(HTTH) = (p)(1 - p)(1 - p)(p) = p^2(1 - p)^2 = \tfrac{1}{16}$$
$$P(THTH) = (1 - p)(p)(1 - p)(p) = p^2(1 - p)^2 = \tfrac{1}{16}$$
$$P(THHT) = (1 - p)(p)(p)(1 - p) = p^2(1 - p)^2 = \tfrac{1}{16}$$
$$P(TTHH) = (1 - p)(1 - p)(p)(p) = p^2(1 - p)^2 = \tfrac{1}{16}$$

Thus,
$$P(X = 2) = 6\left(\tfrac{1}{16}\right) = \tfrac{6}{16}$$

There are four possible outcomes for $X = 3$: $HHHT$, $HHTH$, $HTHH$, and $THHH$, and one possible outcome for $X = 4$, $HHHH$. The probabilities for $X = 3$ and $X = 4$ can be similarly derived, and the results are shown below.

$$P(X = 3) = 4(p)^3(1 - p) = \tfrac{4}{16}$$
and
$$P(X = 4) = p^4 = \tfrac{1}{16}$$

The probability distribution of X is summarized in table 4-6.

Table 4-6 listed all the possible outcomes of the above experiment. In most cases, however, the size of the sample makes it difficult or impractical to list all possible outcomes. For instance, if $n = 10$, the total number of outcomes will be $2^{10} = 1024$. Fortunately, the rule of combination can be applied for determining the number of all possible outcomes for each value of X whatever the sample size.

It is interesting to note that each of the outcomes listed in table 4-6 is a *combination,* and *not a permutation.* Let the numbers 1, 2, 3, and 4 identify the four coins and be used as subscripts of the letters T and H. Thus T_1 designates the fact that the first coin turns up tails, H_2 designates the fact that the second coin turns up heads, and so on. Accordingly,

$$T_1 T_2 T_3 T_4$$

designates the outcome that all four coins turn up tails, and

$$T_1 T_2 H_3 H_4$$

designates the outcome that the first and second coins turn up tails and the third and fourth coins turn up heads. Each of these outcomes is a combination because the order of arrangement is not a matter of concern here. If the combination $T_1 T_2 T_3 T_4$ is arranged with the order taken into consideration, we would have $P_4^4 = (4)(3)(2)(1) = 24$ permutations as follows:

$$
\begin{array}{llll}
T_1 T_2 T_3 T_4 & T_2 T_1 T_3 T_4 & T_3 T_1 T_2 T_4 & T_4 T_1 T_2 T_3 \\
T_1 T_2 T_4 T_3 & T_2 T_1 T_4 T_3 & T_3 T_1 T_4 T_2 & T_4 T_1 T_3 T_2 \\
T_1 T_3 T_2 T_4 & T_2 T_3 T_1 T_4 & T_3 T_2 T_1 T_4 & T_4 T_2 T_1 T_3 \\
T_1 T_3 T_4 T_2 & T_2 T_3 T_4 T_1 & T_3 T_2 T_4 T_1 & T_4 T_2 T_3 T_1 \\
T_1 T_4 T_2 T_3 & T_2 T_4 T_1 T_3 & T_3 T_4 T_1 T_2 & T_4 T_3 T_1 T_2 \\
T_1 T_4 T_3 T_2 & T_2 T_4 T_3 T_1 & T_3 T_4 T_2 T_1 & T_4 T_3 T_2 T_1
\end{array}
$$

These 24 permutations constitute only one combination. Similarly, $T_1 T_2 H_3 H_4$, as well as all the other outcomes listed in the left-hand column of table 4-6, forms only one combination.

The number of outcomes for each distinct value of X from a sample of four observations equals the number of combinations of X objects taken from four objects. Table 4-7 shows the computations.

Table 4-7 Combinations of X heads from a toss of four coins

X	Combinations
0	$\binom{4}{0} = \dfrac{4!}{(4-0)!\,0!} = \dfrac{4!}{4!\,0!} = 1$
1	$\binom{4}{1} = \dfrac{4!}{(4-1)!\,1!} = \dfrac{4!}{3!\,1!} = 4$
2	$\binom{4}{2} = \dfrac{4!}{(4-2)!\,2!} = \dfrac{4!}{2!\,2!} = 6$
3	$\binom{4}{3} = \dfrac{4!}{(4-3)!\,3!} = \dfrac{4!}{1!\,3!} = 4$
4	$\binom{4}{4} = \dfrac{4!}{(4-4)!\,4!} = \dfrac{4!}{0!\,4!} = 1$

The values 1, 4, 6, 4, and 1 are identical with the first factors in the right-hand column of table 4-6. They are called *binomial coefficients*.

Observe, furthermore, that for each value of X the probability of a success p is raised to the Xth power and $(1-p)$ to the $(n-X)$th power. Accordingly, a generalization can be made to express the binomial probability function as follows:

$$P(X) = \binom{n}{X} (p)^X (1-p)^{n-X} \text{ for } X = 0, 1, 2, \ldots, \text{ or } n \quad (4\text{-}9)$$

Although example 4-15 seems of interest only to people who are addicted to the tossing of coins, equation 4-9 can be applied to problems of an entirely different nature. The following is another example illustrating the application of the binomial distribution.

EXAMPLE 4-16 Consider five sequoia trees planted in the California desert. If the probability for each tree to survive the hot summer is 0.5, what is the probability for three of the five trees to survive? Four trees? All five trees?

Using equation 4-9, we can compute the probabilities of survival for three, four, and five trees, respectively.

$$P(X = 3) = \binom{5}{3} \left(\tfrac{1}{2}\right)^3 \left(\tfrac{1}{2}\right)^2 = 10\left(\tfrac{1}{2}\right)^5 = \tfrac{10}{32} = 0.3125$$

$$P(X = 4) = \binom{5}{4} \left(\tfrac{1}{2}\right)^4 \left(\tfrac{1}{2}\right)^1 = 5\left(\tfrac{1}{2}\right)^5 = \tfrac{5}{32} = 0.15625$$

and

$$P(X = 5) = \binom{5}{5} \left(\tfrac{1}{2}\right)^5 \left(\tfrac{1}{2}\right)^0 = 1\left(\tfrac{1}{2}\right)^5 = \tfrac{1}{32} = 0.03125$$

Appendix A provides binomial probability distributions for $n = 5$, 10, 15, 20, and 30 with $p = 0.1, 0.2, \ldots, 0.9$. The table entries are probabilities computed by the use of equation 4-9. In the above example, the probability for $X = 3$ is found in the cell corresponding to row 3 and column 0.5 in the table for $n = 5$. The same number 0.3125 is found in the table.

EXAMPLE 4-17 Assume that 70 percent of all the patients who have taken a certain medicine can be cured. What is the probability that in a sample of 30 patients who have taken the medicine 20 will be cured?

From appendix A, we find that for $p = 0.7$ and $n = 30$ the probability of $X = 20$ is $P(X = 20) = 0.1416$.

cumulative distribution

In many problems, we are interested in the probability that the variable X is *equal to or smaller than* a certain value. A function showing the probability equal to or smaller than each of the possible values is known as the cumulative probability distribution. If each of the five sequoia trees has the probability of 0.5 to survive, then the probability that no tree will survive is

$$P(X = 0) = \binom{5}{0} \left(\frac{1}{2}\right)^0 \left(\frac{1}{2}\right)^5 = \frac{1}{32} = 0.03125$$

The probabilities that one or two trees will survive are, respectively,

$$P(X = 1) = \binom{5}{1} \left(\frac{1}{2}\right)^1 \left(\frac{1}{2}\right)^4 = 5\left(\frac{1}{32}\right) = \frac{5}{32} = 0.15625$$

and

$$P(X = 2) = \binom{5}{2} \left(\frac{1}{2}\right)^2 \left(\frac{1}{2}\right)^3 = 10\left(\frac{1}{32}\right) = \frac{10}{32} = 0.3125$$

Thus, the probability that two or fewer trees will survive, or $P(X \leq 2)$, is

$$P(X \leq 2) = P(X = 0) + P(X = 1) + P(X = 2)$$
$$= 0.03125 + 0.15625 + 0.3125 = 0.5000$$

That is, the cumulative probability is obtained by adding the probabilities of 0, 1, 2, and so on, until reaching the value for which cumulative probability is desired. Table 4-8 shows the cumulative probability distribution for $n = 5$ and $p = 0.5$.

Table 4-8 Cumulative probability distribution for $n = 5$ and $p = 0.5$

Possible values of X	$P(X)$	Cumulative probability distribution
0	$\frac{1}{32}$	$\frac{1}{32}$
1	$\frac{5}{32}$	$\frac{6}{32}$
2	$\frac{10}{32}$	$\frac{16}{32}$
3	$\frac{10}{32}$	$\frac{26}{32}$
4	$\frac{5}{32}$	$\frac{31}{32}$
5	$\frac{1}{32}$	$\frac{32}{32}$

EXAMPLE 4-18 From experience it is known that 30 percent of all housewives who let a door-to-door salesman into their homes will end up buying an accident insurance policy. Assuming that 10 housewives have let such a salesman into their homes, find the probability that (1) *at most* four of them will buy the policy and (2) *at least* four of them will buy the policy.

1) "At most four" refers to the values 0, 1, 2, 3, or 4. The probability that at most four housewives will buy the insurance policy, given $n = 10$ and $p = 0.3$, is designated by $P(X \leq 4 \mid n = 10, p = 0.3)$.

Appendix B provides cumulative probability distributions for specified values of n and p. The desired probability is found in the cell corresponding to row 4 and column 0.3 in the table for $n = 10$. It is found to be 0.8497.

2) "At least four" here refers to the values 4, 5, 6, 7, 8, 9, and 10. It excludes the values 0, 1, 2, and 3. The cumulative probability for $X \leq 3$ is 0.6496, which is found in the cell corresponding to row 3 and column 0.3 in the table for $n = 10$. Consequently, the probability that at least four housewives will buy the policy is

$$P(X \geq 4 \mid n = 10, p = 0.3) = P(X > 3 \mid n = 10, p = 0.3)$$
$$= 1 - P(X \leq 3 \mid n = 10, p = 0.3) = 1 - 0.6496 = 0.3504$$

since for each value of n the sum of the probabilities in each column is 1.

4-6 exercises

1. Indicate which of the following trials are Bernoulli and which are binomial.

   a) Answering a single true-false question.
   b) Making a telephone call to find out whether the respondent is going to vote in the next election
   c) Finding the number of correct answers to 10 true-false questions
   d) Interviewing a family of five people to determine how many of them like a particular product

2. Find the values for the following expressions.

   a) $\binom{3}{1} (0.5)^1 (0.5)^2$ c) $\binom{5}{3} (0.4)^3 (0.6)^2$

   b) $\binom{4}{3} (0.5)^3 (0.5)^1$ d) $P(X = 3 \mid n = 5$ and $p = 0.4)$

3. Verify the following and then generalize.

   a) $(0.5)^1 (0.5)^2 = (0.5)^3$
   b) $(0.5)^3 (0.5)^1 = (0.5)^4$
   c) $(0.4)^1 (0.6)^2 \neq (0.5)^3$

4. Find the values for the following expressions.

   a) $\sum_{X=0}^{1} \binom{3}{X} (0.5)^X (0.5)^{3-X}$

   b) $\sum_{X=0}^{2} \binom{5}{X} (0.5)^X (0.5)^{5-X}$

   c) $P(X \leq 2 \mid n = 5$ and $p = 0.5)$

Use appendix B to find the answer for (c); is it the same as the answer for (b)?

5. A test contains 20 true-false questions. If a student answers the questions by guessing, what is the probability that
   a) he will guess 10 questions correctly?
   b) he will guess five or fewer questions correctly?
   c) he will guess seven or more questions correctly?

6. Suppose that 10 multiple-choice questions are included in a test, and that one of the five answers given for each question is the correct one. If a student answers the questions simply by guessing, what is the probability that
   a) he will guess five questions correctly?
   b) he will guess three or fewer questions correctly?
   c) he will guess five or more questions correctly?

7. A certain medicine is 50 percent effective; that is, out of every 100 patients who take it, 50 are cured. Let X be the number of patients cured in a sample of 30 patients. Find the probability that
   a) twenty or fewer patients are cured, or $P(X \leq 20)$
   b) eighteen or more patients are cured, or $P(X \geq 18)$
   c) more than 12 but fewer than 22 patients are cured, or
      $P(12 < X < 22)$

8. Suppose that 10 radar sets are operating independently of each other and that the probability for a single set to detect an enemy rocket is 0.80. What is the probability that nine radar sets will detect the rocket?

9. If it is known that 90 percent of the students taking an elementary economics course pass, what is the probability that at least three students in a class of 15 will fail the course?

10. Suppose that 10 percent of a certain brand of television tubes will burn out before their warranty expires. Find the probability that out of 30 tubes sold, five or more will burn out before their warranty expires.

11. Assume that the probability of a biased coin to turn up heads is 0.7. Let X be the number of heads obtained in 30 tosses of this coin. Find the following probabilities.
   a) $P(X = 21)$ d) $P(14 < X < 20)$
   b) $P(X \leq 15)$ e) $P(14 \leq X \leq 20)$
   c) $P(X \geq 16)$

12. If 10 percent of the parts produced by a certain production process

are defective, what is the probability that out of 20 parts selected at random there are

    a) at least two defectives?

    b) at most three defectives?

    c) between two and five defectives inclusive?

13. Sixty percent of the senior class are girls. What is the probability that in a randomly selected group of 10 students from the senior class there are

    a) five girls?   c) at most five girls?

    b) at least five girls?   d) between four and six girls inclusive?

14. A professional football team is scheduled to play 15 games during the season. Suppose that 20 percent of the days in the area where the the games will be played are rainy days. What is the probability that

    a) three games will be played in the rain?

    b) at least three games will be played in the rain?

    c) at most three games will be played in the rain?

15. Suppose that the probability that an odd number of spots will show up on a die is 0.4. What is the probability that in five rolls of the die the number of times that an odd number of spots will show up is

    a) less than two?

    b) more than two?

    c) between two and four inclusive?

16. Suppose that the management of an automobile manufacturing company believes that three out of the 10 persons who will read its brochure about new cars will purchase a new car from one of its dealers. If five persons who have read the brochure are selected at random, what is the probability that

    a) no one will purchase a new car?

    b) all five will purchase a new car?

    c) at most three will purchase a new car?

    d) at least three will purchase a new car?

17. Suppose that 10 percent of all housewives who let a door-to-door vacuum-cleaner salesman in the house will end up buying a vacuum cleaner. If, in a certain community, 30 housewives have let such a salesman in, what is the probability that

    a) exactly 20 housewives will not buy a vacuum cleaner?

    b) at most five housewives will buy?

    c) at least five housewives will not buy?

18. In a certain section of a large city, 90 percent of all families have at least one automobile. In a randomly selected sample of 20 families, what is the probability that

a) exactly 18 families have at least one automobile?
b) eighteen or more families have at least one automobile?
c) two or fewer families have at least one automobile?

random sampling and 5

sample statistics

5-1
introduction

The collection of numerical data is an indispensable task in statistical analysis. Since modern statistical analysis involves, to a great extent, making inferences about population parameters on the basis of incomplete sample information, it is essential that any sample drawn for this purpose be representative of the population from which it is taken. The basic approach in selecting a representative sample is the technique of random sampling. In the second section of this chapter, we shall examine the concept of a random sample, and briefly discuss the various techniques commonly used in that process.

Once a random sample has been taken, the data contained in the sample must be organized and presented. Data organization and presentation are very important because unorganized raw data seldom provide any meaningful picture of the essential nature of the sample. For instance, a sales manager who wishes to know something about sales patterns would gain very little knowledge by merely looking at the monthly invoices. A simple enumeration of student test scores would reveal very little of the essential features of that mass of information.

In the third section of this chapter, we will discuss the procedure of organizing raw data into a frequency distribution and representing them graphically. The fourth section is intended to show how collected data can be summarized in one single central value around which the entire sample data are distributed. For this reason, a central value is often referred to as a *measure of central tendency*. There are many kinds of central values: the most frequently used measures are the *arithmetic mean*, the *median*, and the *mode*. *Index numbers*, a kind of average, are discussed in the exercises.

The measure of central tendency is used to locate the center of a set of observations. Very frequently, however, it is just as important to describe how observations are scattered, or spread, on each side of the center; this scattering is commonly known as the *dispersion*, or *variation*. A small dispersion signifies a high degree of uniformity in the observations; a large dispersion indicates little uniformity. If a set of observations has no dispersion, all the observations are identical. Such perfect uniformity is virtually nonexistent, however.

Two samples of observations with the same central value may still have different dispersions. Suppose, for instance, that an identical test is administered to a group of 20 men and a group of 20 women. Suppose further that the distributions of the test scores for the two groups are as follows:

Test score:	20	30	40	50	60	70	80	90	100
Men ($n = 20$):			2	4	7	6	1		
Women ($n = 20$):	1	1	2	3	6	3	2	1	1

The average scores for the 20 men and the 20 women are the same (60), but the dispersions for the two groups are obviously quite different. The 20 men show very little variability in their test scores, whereas the 20 women show much greater variation. It is the objective of the final section of this chapter to consider the commonly used techniques by which the dispersion of a set of observations is measured.

In short, this chapter is intended to deal with the collection and processing of the raw data that form the sample, and to consider certain characteristics of the sample such as central values and variability.

5-1 exercises

1. Why is it essential that a sample be representative of the population from which it is taken?

2. What does measures of central tendency mean?

3. Why is it important for statisticians to know the dispersion of a set of observations?

4. Suppose that the average age of the students in a certain class is 22 and the range from the youngest to the oldest is 30 years. The average age of another class of students is also 22, but the range from the youngest to the oldest is 10 years. What can you say about the age distribution of the first class as compared with that of the second class?

5-2 random sampling

Any set of n observations taken from a population is generally referred to as a sample of size n. However, the kind of sample that statisticians are primarily interested in is one that is truly representative of the population from which it is selected. We can expect a *randomly selected* sample to possess this quality. Before we look into the design of random sampling, we should first discuss the characteristics of a random sample.

A random sample may be taken from a finite or an infinite population. In the tossing of a coin n times to see how many heads occur, the outcomes of n tosses form a sample. The population is infinite because there is theoretically no limit to the number of tosses that can be made. In addition, the occurrence of either one of the two possible outcomes, heads or tails, in a given toss will not affect the probable outcomes of subsequent tosses. Similarly, the rolling of a die n times results in a sample. Such a sample is taken from the infinitely large number of potential observations which constitute the population. Furthermore, the outcome of any roll does not affect the outcome of any other roll. Under these and similar circumstances,

> A sample is considered a random sample as long as each observation that has been taken does not affect the probability that any other observation will be selected.

When a sample is taken from a finite population (and this is the case in most sampling situations), it is obvious that the probability for each observation of the population to be included in the sample does not remain unaffected. In fact, the probability of inclusion increases with each successive selection. Under such circumstances, a random sample is obtained as long as all the *available*, or *remaining*, potential observations have an equal probability of being included. In other words,

> A random sample is one that is drawn in such a way that every available potential observation in the population has an equal probability of being selected.

Furthermore, given that each available observation, which is usually called a *sampling unit*, has the same probability of being selected, it follows that

> Any random sample of size n drawn from any population must have the same probability of being selected.

This fact can be demonstrated by a simple example below.

EXAMPLE 5-1   Suppose that from a school of 1000 students a sample of 20 students is taken at random. Suppose further that a statistics class has 20 students. What is the probability that the 20 students in the statistics class will be included in the sample? Also, what is the probability that any other combination of 20 students will be included?

By the rule of combination (equation 3-6), the total number of ways to combine 20 objects selected from a group of 1000 objects is $\binom{1000}{20}$, with any 20 objects forming only one combination. Thus the probability that all 20 students in the statistics class will be included in the sample can be computed as follows:

$$\frac{1}{\binom{1000}{20}} = \frac{1}{\dfrac{1000!}{(20)!(1000-20)!}} = \frac{(20)(19)(18)\ldots(3)(2)(1)}{(1000)(999)(998)\ldots(982)(981)}$$

$$= \frac{20}{1000} \cdot \frac{19}{999} \cdot \frac{18}{998} \cdot \frac{17}{997} \ldots \frac{3}{983} \cdot \frac{2}{982} \cdot \frac{1}{981}$$

Similarly, any other combination of 20 students will have the same probability of being included in the sample.

We can therefore generalize that a sample is called a random sample if *any* possible sample of size n taken from a finite population of size N has the same probability of being selected as follows:

$$\frac{1}{\binom{N}{n}} = \frac{n!}{N(N-1)(N-2)\ldots(N-n+1)} \tag{5-1}$$

Observe that in example 5-1 there are 20 students in the statistics class, and that the total number of students in the school is 1000. Thus the probability that any one of the 20 students of the statistics class will be drawn first is $\frac{20}{1000}$. After the first draw (in which a statistics student was selected), there are 19 students remaining in the class and 999 students left in the school; therefore, the probability that the second selection is one from the class is $\frac{19}{999}$; and so on. After 19 students have been selected, there are 981 students remaining in the school; the probability that the last one will be selected is therefore $\frac{1}{981}$. Finally, by the rule of multiplication, we arrive at the same probability figure that could have been derived from equation 5-1 above.

sampling designs

There are at least four designs, or methods, in common use; they are *simple random sampling, systematic sampling, stratified sampling,* and *cluster sampling.* These sampling designs are briefly described below.

Simple Random Sampling Simple random sampling is the most frequently used and most basic sampling method. The "goldfish bowl" technique is one example of simple random sampling. By this technique, all available sampling units are numbered. The numbers are then written on balls, cards, or other physically identical articles. They are put in an urn or bowl, and thoroughly mixed before each drawing. Because the articles are carefully mixed before each drawing, each has an equal probability of being selected.

The procedure of drawing a simple random sample can be simplified by the use of a table of random digits. We could construct such a table by the "goldfish bowl" method with 10 cards numbered 0, 1, 2, ..., 9. After the cards are mixed, one card is randomly drawn and its number is recorded. The card is replaced before the next one is drawn, and so on. In actual practice, this is done with an electronic computer.

Appendix C contains a brief table of random digits, and can be applied as follows: If any three digits in the same row are combined, then 1000 numbers from 000 to 999 are obtained; if four digits are combined, 10,000 numbers are obtained, and so on. The number of digits to be used depends on the number of sampling units in the population from which a simple random sample is taken. Suppose that a sample of 30 units is taken from a population of 1000 units. As the first step, each of the 1000 units is assigned a number from 000 to 999. Then a number with three digits is selected at random from the table. Once such a number is chosen and is recorded, other three-digit numbers may be obtained in the same fashion, or they may be obtained by recording the number immediately below, above, or to the left or right of, the number already selected. A number is used only once. If the same number reappears, it is simply disregarded. When 30 numbers are thus selected from the table, the 30 units in the population bearing such numbers constitute the desired random sample.

The procedure of simple random sampling described above may conceivably yield a sample that is not representative of the population. This may be the case when, for instance, the units close to one another are known to be more homogeneous than those far apart. In such a case, the procedure of systematic sampling is preferable.

Systematic Sampling According to the *systematic sampling* procedure, a sample is selected by taking every kth unit of the population once the sampling units are numbered or arranged in some fashion. The letter k is the *sampling ratio*, that is, the ratio of the population size to the size of the sample. Thus, if a sample of 40 units is selected from a population of 1000 units, then $k = \frac{1000}{40}$, or 25, and the sample is obtained by taking every 25th unit of the population.

The goldfish bowl procedure may be used to determine which one of the first 25 units to start with. Suppose that the 10th unit is selected as the random start, then the sample will include the 10th, 35th, 60th, ..., 960th, and the 985th units. Thus, units close to one another would not be over or under represented in the sample.

Stratified Sampling Another frequently used sampling design is *stratified random sampling*. This procedure involves dividing the population into classes, or groups, called *strata*. Units included in each stratum are supposed to be relatively homogeneous with respect to the characteristics to be studied. A subsample is taken from each stratum by the simple random procedure. The subsamples for all strata are then combined to obtain the overall sample.

Stratified sampling is most frequently used in handling heterogeneous populations such as data on family income in a single metropolitan area. By stratification, strata are set up such that the units within each stratum are more homogeneous and the strata are different from one another. The sampling ratio may be made the same for all strata; thus the proportion of the overall sample from each stratum is not left to chance, and the representativeness of the overall sample can be assured.

Cluster Sampling   The procedure of *cluster sampling* involves, first of all, the selection at random of groups, or *clusters,* from the population; the overall sample is made up of all, or a subsample of, the units in each cluster. Cluster sampling is different from stratified sampling in that differences between clusters are usually small, and the units within each cluster are generally more heterogeneous. Each cluster, by and large, should be a miniature of the population. Although a single cluster could be a satisfactory sample, this is rarely the case in practice.

Clusters are generally known as *primary sampling units.* If all the units of the selected clusters are included in the overall sample, the procedure is referred to as *single-stage sampling.* If a subsample is taken at random from each selected cluster, and all units of each subsample included in the overall sample, the procedure is called *two-stage sampling.* If a sampling process involves three or more stages, it is called *multistage sampling.* In a survey of professors in American state universities, for instance, a certain number of state universities are randomly selected as clusters in the first stage. If all professors at the selected universities are included in the sample, it is a single-stage sampling. If within each of the selected universities a subsample of professors is taken so that the overall sample includes all the professors in each of the subsamples, it is a two-stage sampling. The third stage may involve the selection of schools within each selected university; the sample will then include all professors sampled in the selected schools of the chosen clusters. Simple random sampling is used at each stage. Although primary sampling units are selected first, the purpose of cluster sampling is to investigate the simple units contained in the clusters.

5-2
exercises

1. Are the following samples representative? Explain.
   a) The publisher of a magazine wished to predict the outcome of the next presidential election, and for this purpose 1000 subscribers are interviewed to determine their voting preferences.

b) Students in a statistics class are interviewed about their attitudes toward the legalization of marijuana; their responses will be used to predict the opinion of the college community regarding this issue.

2. Suppose that a roster of registered voters contains one million names. From this roster, a random sample of 1000 voters is to be selected. How can this be accomplished by the use of systematic sampling?

3. Should stratified sampling or cluster sampling be used, and how, to draw a random sample in each of the following situations.

   a) A study is conducted to determine the pattern of family expenditures in a certain city.
   b) A survey is conducted to determine the attitude of American college students toward capital punishment.

4. Design a random sample of 100 students to be drawn from the student body of your college.

5. Suppose that a bag contains 10 balls marked 0, 1, 2, 3, . . . , 9. Three balls are to be selected one after the other without replacement. What is the probability that the balls marked 9, 8, and 7 will be selected in that order?

6. In a certain community there are 10 supermarkets. If a sample of three supermarkets is to be selected, how many different samples can be possibly drawn? What is the probability that a random sample of three particular supermarkets will be selected?

7. In a certain city there are 12 service stations, four of which belong to the same corporation. If a sample of four stations is randomly selected, what is the probability that all four affiliated stations will be included?

8. If 10 service stations constitute a finite population, what is the probability that a combination of four particular stations will be selected as a random sample?

9. List all possible samples of two numbers that can be randomly selected from a finite population that consists of the numbers 1, 3, 5, 7, and 9. What is the probability that the digit 9 will be included in a sample that is selected at random?

10. Suppose that the 70 faculty members of a college are assigned identification numbers from 00 to 69. The governing board of the college wishes to interview 10 of the 70 faculty members at random. If a random digit table such as the one in appendix C is used for selecting the 10 faculty members, which faculty members will be selected?

(Use the last two columns of the first page beginning with the first row.)

11. A company has 849 employees, who have been numbered serially from 000 to 848. Select a random sample of 10 employees by using the first three columns of the first page of appendix C beginning with the fifth row.

5-3 data organization

Once a sample is drawn, and observations of all the units in the sample made, the task of data collection is completed. What has been collected, however, is raw data, which are seldom meaningful without organization and presentation. A hypothetical example will illustrate the usual procedure of handling unorganized information.

Let us consider a sample of wages earned on Saturdays by working college students. Suppose that 20 such students are randomly drawn, and the wages they earned on the previous Saturday obtained. Assume that an initial treatment of the data results in an *array,* which is an arrangement of the observations in the order of their magnitude, as shown in table 5-1.

Table 5-1 An array of wages (in dollars) earned by 20 college students on a Saturday

8	18	25	30
11	21	25	30
13	21	26	35
15	23	29	36
17	25	30	42

An array like this shows us that the difference between the lowest wages, $8, and the highest, $42, is $34. However, it is a rudimentary form of data organization, which would be both cumbersome and impracticable if the number of observations involved were much larger.

frequency distribution

Data organization generally involves the arrangement of observations into classes. The arrangement of data to express the frequency of occurrences of observations in each of these classes is known as a *frequency distribution.* The construction of a frequency distribution requires, first of all, the selection of *class intervals.* Although the selection of class intervals is an art and will depend on the data involved, the following rules should prove to be helpful in most instances.

1) The number of classes should not be so small (fewer than six) or so large (more than 20) that the true nature of the distribution cannot be seen.

2) The interval length of each class should be the same; the interval length should always be an odd number; and the midpoints of the classes should have the same number of digits as the raw data. This will make both the lower and the upper limits of each class have an extra decimal digit 5; none of the observed values should fall on these limits. An examination of the first two columns of table 5-2 will make this point clearer.

3) Midpoints should be values that are easy to work with.

Table 5-2   Tabulation of frequencies and cumulative frequencies

Wages $	Midpoints x	Tally	Frequency FR(x)	Cumulative frequency CF(x)
7.5-12.5	10	//	2	2
12.5-17.5	15	///	3	5
17.5-22.5	20	///	3	8
22.5-27.5	25	Ж	5	13
27.5-32.5	30	////	4	17
32.5-37.5	35	//	2	19
37.5-42.5	40	/	1	20
Total			20	

With these rules in mind, we select 10, 15, 20, 25, 30, 35, and 40 as the midpoints, with 7.5 and 12.5, 12.5 and 17.5, . . . , 37.5 and 42.5 as the lower and upper limits, respectively. The seven classes, with an interval length of 5, are presented in the first column; and the midpoints, designated by the lower-case letter x, are given in the second column of table 5-2. The third column shows the tally of the 20 observations. The tallying procedure runs as follows.

(1) On reading 8 from the array of the 20 wages, we enter a diagonal stroke in the class 7.5 to 12.5; on reading 11, we enter another stroke in the same class, and so on, until all the 20 wages are read, and the strokes entered in the appropriate columns. (2) After the tally is completed, the strokes in each class are counted, and the number for each class is entered in the fourth column as the frequency of the class. (3) Since the midpoints are designated by x, $FR(x)$ is used to stand for the frequency of the class with midpoint x, the distribution of all observations among the various classes as shown in the fourth column is called the *frequency distribution*. (4) The last column shows the *cumulative frequency* at x, designated by

$CF(x)$. It is obtained by adding successively the numbers in the fourth column; each entry signifies the total number of observations equal to or less than x, on the assumption that all observations in each class assume the value of the midpoint.

graphic representation

Both the frequency distribution and the cumulative frequency distribution can be presented graphically. The midpoints, and the lower and upper limits, of classes are represented by the X axis, and the frequencies are represented by the Y axis. For quantitative data, three types of graphs are commonly used: the histogram, the polygon, and the ogive. The pie chart is frequently used for qualitative data.

The Histogram The histogram, which is really a version of the familiar bar chart, is constructed by first marking off the class intervals along the X axis, and then drawing a rectangle to the height equal to the frequency of the class for each class interval. A histogram showing the frequency distribution for the data in table 5-2 is presented in figure 5-1. In this kind of graph, we can see at a glance the frequency pattern of the sample observations.

The Polygon The histogram can be transformed into a frequency polygon (see figure 5-2) by connecting the midpoints of the tops of the rectangles. (The rectangles themselves are not normally part of a polygon; they are used here as a visual aid.) Note that the curve begins and ends at the midpoints of the classes immediately before and after the

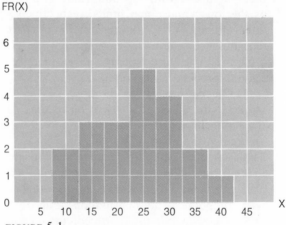

FIGURE 5-1
A histogram for the data in table
5-2

lowest and the highest classes. The curve is constructed on the assumption that the observations in each class are evenly distributed throughout the class.

The Ogive It is also possible to represent cumulative frequency distribution by a graph called an ogive. To construct an ogive, we first put a dot at the lower limit of the lowest class on the X axis to show that there is no observation of this magnitude or lower. Then, directly above the upper limit of the lowest class, we place a dot at the height equal to the frequency of the class. Then, at the upper limit of the second-to-lowest class, we place a dot at the height equal to the *cumulative frequency* (that is, the sum of the frequencies of the lowest and second-to-lowest classes). Proceed in this fashion until a dot representing the total cumulative frequency is plotted at a point directly above the upper limit of the highest class. Finally, the dots are connected by straight lines. An ogive obtained in this manner is also called a *cumulative frequency polygon*. Figure 5-3 shows the ogive for the data in table 5-2.

When data fall into qualitative categories rather than numerical classes, these categories may be represented on a *pie chart*. Note that the pie chart in figure 5-4 is divided into various sectors proportional in size to the frequencies of the various categories. To construct a pie chart, we need to convert the distribution into a *relative frequency distribution*. Relative frequency refers to the proportion of total observations that fall into each class. It is obtained by dividing the frequency of the class by the total number of observations. The relative frequencies of all the classes should, of course, add up to 1.

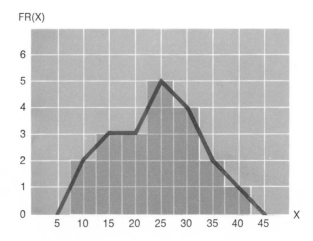

FIGURE 5-2
A frequency polygon for the data
in table 5-2

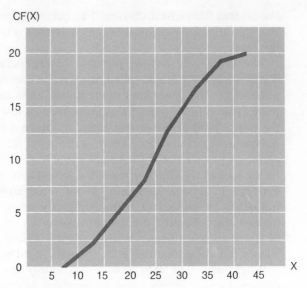

FIGURE 5-3
An ogive for the data in table 5-2

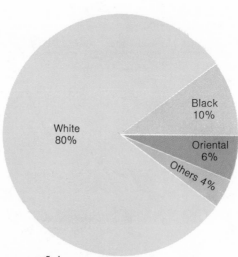

FIGURE 5-4
A pie chart showing the racial
background of 100 freshmen

If a certain class or category has a relative frequency of 10 percent, a sector of the pie chart with a central angle of 36 degrees (10 percent of $360° = 36$ degrees) will be used to represent that category. For instance, if the racial backgrounds of 100 freshmen at a certain college are White (80), Black (10), Oriental (6), and Other (4), then the pie chart is divided into four sectors as follows:

Racial background	Relative frequency	Chart area
White	0.80	$0.80(360°) = 288°$
Black	0.10	$0.10(360°) = 36°$
Oriental	0.06	$0.06(360°) = 21.6°$
Others	0.04	$0.04(360°) = 14.4°$
Total	1.00	$360°$

This distribution is shown in figure 5-4, where different shades represent different categories. (The various categories are also frequently represented by different colors.)

5-3 exercises

1. What are the advantages and disadvantages of using an array as a description of collected raw data?

2. Explain the procedures necessary to construct (a) a frequency distribution, (b) a cumulative frequency distribution, and (c) a relative frequency distribution.

3. If the lowest value of a set of observations is 25 and the highest is 94, and if we have decided to organize the sample into 10 classes, what should the length of each interval be? If we choose 9 to be the length of each interval, how many classes will we have?

4. Explain the drawbacks or advantages of using classes such as
    a) 11 to 20, 21 to 30, 31 to 40, and so on
    b) 10 to 20, 20 to 30, 20 to 40, and so on
    c) 10.5 to 19.5, 19.5 to 28.5, 28.5 to 37.5, and so on, when the raw data are all whole numbers

5. Explain why the data in table 5-1 should not be organized into as few as two or three classes or as many as 15 or 16 classes.

6. Why is it often advisable to represent collected data in graphic form?

7. In a survey of the number of customers served on a Sunday by the restaurants in a metropolitan area, the midpoints of the distribution are 15, 30, 45, 60, 75, 90, 105, and 120. What is the class interval of this distribution and what are the class limits?

8. Given the following series of numbers,
   5, 3, 4, 6, 2, 9, 2, 4, 3, 8, 4, 6, 5, 7, 5, 4, 5, 6, 7, and 5,

   a) construct a table showing the frequency, cumulative frequency, and relative frequency distributions
   b) draw a histogram of the frequency distribution
   c) draw a frequency polygon
   d) draw an ogive

9. Given the following series of numbers,
   8, 6, 7, 9, 5, 12, 5, 7, 6, 11, 7, 9, 8, 10, 8, 7, 8, 9, 10, and 8,

   a) construct a table showing the frequency, cumulative frequency, and relative frequency distributions
   b) draw a histogram of the relative frequency distribution
   c) draw a frequency polygon
   d) draw an ogive

10. Construct a pie chart to represent the marital status of 50 employees recently hired by a large corporation. The frequency distribution of these 50 employees is shown below.

Employees	Frequency
Single	25
Married	15
Divorced	7
Widowed	3
Total	50

11. A random sample of 40 students is selected from the freshman class of a large college. The high school grade-point averages of these 40 students are listed below.

2.1	2.5	2.9	3.4
2.4	3.6	3.1	3.0
2.7	2.7	3.1	3.6
3.0	3.8	3.1	2.7
3.0	2.7	3.2	3.9
3.9	2.5	2.9	3.4
3.6	3.4	2.9	2.4
3.0	2.3	2.9	3.4
3.3	3.2	2.9	2.5
3.0	2.0	2.8	3.2

   a) Organize these averages into a table similar to table 5-2. The classes are 1.85–2.15, 2.15–2.45, 2.45–2.75, 2.75–3.05, 3.05–3.35, 3.35–3.65, and 3.65–3.95.
   b) Draw a histogram of the frequency distribution.
   c) Draw a frequency polygon.
   d) Draw an ogive.

12. Many college students are gainfully employed during the summer months. A random sample of 50 students is selected and their earnings during a particular week are obtained.

$53	$57	$58	$61	$61
63	64	66	67	68
69	70	71	72	73
74	74	74	74	77
77	77	78	78	79
79	81	81	81	82
82	83	83	84	85
85	86	87	87	88
79	90	90	90	90
92	93	94	96	97

a) Organize the above data into a table similar to table 5-2. The classes are 52.5−57.5, 57.5−62.5, 62.5−67.5, . . . , and 92.5−97.5.

b) Convert the frequencies into relative frequencies, and then draw a histogram with the left-hand vertical scale representing the frequencies and the right-hand scale representing the relative frequencies.

c) Draw a frequency polygon.

d) Draw an ogive.

5-4 measures of central tendency

Graphic representation gives us a general visual description of collected data; there are, however, limitations to its use. For one thing, it is not always possible to present data visually; for another, graphic description does not generally lend itself to the mathematical treatment that is necessary for statistical inference. For these reasons, the statistician generally prefers to use numerical *measures of central tendency*. He calculates some central value from the sample data, which gives him a mental picture of the data, and at the same time, helps him make inferences about the true nature of the population. Among the central values to be discussed in this section are the *arithmetic mean*, the *median*, and the *mode*. Of the three, the arithmetic mean is the most important for inferential purposes.

arithmetic mean

The arithmetic mean, sometimes referred to simply as "the mean," is the sum of the values of all observations divided by the number of observations made. Let n represent the size of a sample containing observations X_1, X_2, . . . , X_n and let $\overline{X}$ (read "X bar") represent the sample mean. The arithmetic mean is symbolically expressed as follows:

$$\overline{X} = \frac{\sum\limits_{i=1}^{n} X_i}{n} \tag{5-2}$$

EXAMPLE 5-2   Find the mean for the following sample of 11 observations.

X_1	X_2	X_3	X_4	X_5	X_6	X_7	X_8	X_9	X_{10}	X_{11}
1	1	2	2	2	3	4	5	5	9	10

Using equation 5-2, we can compute the sample mean.

$$\overline{X} = \frac{\Sigma X}{11} = \frac{1+1+2+2+2+3+4+5+5+9+10}{11}$$
$$= \frac{44}{11} = 4$$

Although there are 11 values for X, there are only seven distinct values. The value 1 appears twice, 2 three times, 5 twice, and the others only once. The same result can be achieved by multiplying each distinct value by the number of times it has occurred and then summing the products thus obtained. That is,

$$\overline{X} = \frac{1(2) + 2(3) + 3(1) + 4(1) + 5(2) + 9(1) + 10(1)}{11} = \frac{44}{11} = 4$$

The values that are not in parentheses are the distinct values; the values inside parentheses represent the number of times that the respective distinct values have occurred. Thus, if the lower-case letter x is used to designate the distinct values and $FR(x)$ to designate the number of times each x has occurred, or the frequency of x, the sample mean $(\overline{X})$ can be expressed as

$$\overline{X} = \frac{\sum\limits_{j=1}^{m} x_j FR(x_j)}{n} \tag{5-3}$$

In the above formula, the subscript j, like i, designates a counting number for identifying each distinct value or class midpoint, and m designates the total number of distinct values or classes.

Equation 5-3 is important because collected data are usually organized into a frequency distribution. Equation 5-2 cannot be used to calculate the mean for data organized in this manner. When organized data are used to compute the mean, the midpoints of the various classes are considered to be the distinct values designated by x. This is done on the assumption that the observations in each class take on the value of the midpoint or are evenly distributed throughout the class. This assumption usually results in some discrepancies in the calculation. The discrepancies are generally negligible, however, since the errors tend to offset each other. Especially for large samples, the advantages gained through the use of a frequency distribution may outweigh the disadvantage of minor inaccuracies. Occasionally, equations 5-2 and 5-3 might give the same result, as the following example shows.

EXAMPLE 5-3   Find the sample mean for the data in tables 5-1 and 5-2 by applying equations 5-2 and 5-3.

If we use equation 5-2, the sample mean for the 20 observations presented in table 5-1 can be computed as follows:

$$\overline{X} = \frac{1}{20}(8+11+13+15+17+18+21+21+23+25+25+25$$
$$+26+29+30+30+30+35+36+42) = \frac{480}{20} = 24$$

We may also use equation 5-3 to compute the sample mean. (See table 5-3.) Although the results will be the same for these particular data, in most instances the two methods will result in slightly different values for $\overline{X}$.

Table 5-3   Computation of sample mean for the data in table 5-2

x	$FR(x)$	$xFR(x)$
10	2	20
15	3	45
20	3	60
25	5	125
30	4	120
35	2	70
40	1	40
Total	20	480

$$\overline{X} = \frac{\Sigma\, xFR(x)}{20} = \frac{480}{20} = 24$$

One property of the mean is that the sum of the differences between the sample mean and the value of each observation in the sample is zero.

Since $\overline{X} = \dfrac{1}{n}\,\Sigma X$, $n\overline{X} = \Sigma X$. Because $\overline{X}$ is a constant, and $\Sigma\overline{X} = n\overline{X}$, by equation 2-4, we obtain

$$\Sigma\,(X - \overline{X}) = \Sigma X - \Sigma\overline{X} = n\overline{X} - n\,\overline{X} = 0 \qquad (5\text{-}4)$$

the median

Another measure of central tendency is the median. The median of a set of observations is usually defined as follows:

> The median is the value that falls in the middle of a group of observations arranged in order of magnitude.

The method of determining the median depends on whether or not the collected data are organized into a frequency distribution.

For data that are unorganized but arranged in an array, if the number of values is *odd*, the middle value will be the median. For instance, in example 5-2, the median occurs at the sixth observation (3); it is the middle value, and there are five values below it and five values above it. If the number of values is *even*, then the average of the two middle values is usually considered the median, even though any value falling between the two middle observations can be taken as the median. For the 10 values 1, 2, 2, 2, 3, 4, 5, 5, 9, and 10, for example, the median falls halfway between the two middle values 3 and 4. The median is 3.5; but it could also be 3.1, 3.9, or any other value between 3 and 4, since each of these values divides the 10 values into two equal groups.

For data that have been organized into a frequency distribution, the median, or *MD*, of a set of n observations can be determined by using the following equation.

$$MD = L_m + \frac{n/2 - CF(m-1)}{FR(m)}(w) \qquad (5\text{-}5)$$

$CF(m-1) = $ the cumulative frequency below the lower limit of the median class. (The median class is the class that contains the median.) $FR(m) = $ the frequency of the median class, $w = $ class interval, and $L_m = $ lower limit of the median class.

EXAMPLE 5-4 Find the median for the data in table 5-2 (page 91). From table 5-2, we can see that the median class is the one with 25 as its midpoint, since the $n/2$th, that is, the 10th, observation falls in this class. Thus we have

Therefore,
$$CF(m-1) = 8 \qquad FR(m) = 5$$
$$w = 5 \qquad L_m = 22.5$$

$$MD = 22.5 + \frac{20/2 - 8}{5}(5) = 22.5 + \left(\tfrac{2}{5}\right)(5) = 24.5$$

the mode

The mode is occasionally used as a measure of central tendency. *The mode is the value that occurs most frequently in a set of observations.* For the 11 values in example 5-2, the mode is 2, which occurs more frequently than any other values involved. When the data are organized into a frequency distribution, it is easy to identify the *modal class,* because it is the class with the highest frequency. Graphically, the modal class is one with the highest column in the histogram. For the data in table 5-2, for instance, the modal class is 22.5 to 27.5.

The mode is very useful in expressing the central tendency for observations on qualitative characteristics such as color, marital status, occupations, and state of birth. In such cases, it is ludicrous to ask for the mean or median of the observations. It does make sense, however, to speak of the modal color of eyes of the people in a community or the modal occupation of the household heads of a neighborhood.

5-4 exercises

1. The employees of a company are classified according to religion, and the company is asked to name the religion preferred by the greatest number of its employees. Which measure of central tendency would be most appropriate in this case?

2. Which formula, equation 5-2 or 5-3, would you use to calculate the arithmetic mean for the data in table 5-1? Why?

3. Compute the arithmetic mean for the data in problem 11 of exercises 5-3 by (a) equation 5-2 and (b) equation 5-3.

4. Compute the arithmetic mean for the data in problem 12 of exercises 5-3 by (a) equation 5-2 and (b) equation 5-3.

5. Verify that $\Sigma (X - \overline{X}) = 0$ by using the following data.

X_1	X_2	X_3	X_4	X_5	X_6	X_7	X_8	X_9	X_{10}
2	2	3	3	3	4	4	5	5	9

6. Compute the arithmetic mean for the data in problem 8 of exercises 5-3 by (a) equation 5-2 and (b) equation 5-3.

7. Find the median for the following data: 1, 3, 5, 7, 9, 11, 13, 15, 17, 19, 20, 22, 24, 26, and 28.

8. Find the median for the following data: 2, 4, 6, 8, 10, and 12.

9. Find the median for the data in problem 11 of exercises 5-3 by equation 5-5.

10. Find the median for the data in problem 12 of exercises 5-3 by equation 5-5.

11. What is the modal class for the data given in problem 11 of exercises 5-3.

12. Identify the modal class for the data in problem 12 of exercises 5-3.

13. What is the mode for the data on racial background as represented in figure 5-4?

14. What is the mode for the data in problem 10 of exercises 5-3?

15. Ten animals in a laboratory are weighed, and their weights are as follows:

    12   20   28   14   26   20   19   21   23   17

    a) Find the arithmetic mean of the 10 weights.
    b) Find the median of the 10 weights.
    c) Find the mode of the 10 weights.

16. An elevator is designed to carry a maximum load of 2000 pounds. If the mean weight of passengers using the elevator is 165 pounds, what is the maximum number of passengers that can be allowed to ride the elevator at the same time? Might it be overloaded if the maximum number of passengers is allowed?

17. *Index numbers,* which are frequently used in business and economics, are measures of average changes in the values (or quantities) of goods and services. The *base* of an index is the period, usually a month or year, *from* which changes are measured, whereas the period *for* which an index number is constructed is referred to as the *given* period. The value for the base period is generally taken as 100, and an increase (or decrease) in value for the given period is shown as a percentage above (or below) 100.

    Suppose that we want an index number for the 1974 prices of three commodities; we will use 1972 as the base year. The prices of these commodities for the two years are given below.

Commodity	1972 price	1974 price
A	$20	$30
B	30	24
C	50	65

For each commodity we calculate a *price relative,* which is the ratio of 1974 price to 1972 price. The price relatives for the three commodities are shown below.

Commodity	Price relative
A	$30/20 = 1.5$
B	$24/30 = 0.8$
C	$65/50 = 1.3$

The mean of these price relatives multiplied by 100 to express it as a percentage gives us the index number desired. Letting I designate such an index number, we have

$$I = \frac{1.5 + 0.8 + 1.3}{3} (100) = \frac{3.6}{3} (100) = 120$$

In general, if P_b denotes the base-period price and P_g denotes the given-period price of any single commodity, and there are n commodities under consideration, then such a price index is computed by

$$I = \frac{\Sigma(P_g/P_b)}{n}$$

a) Construct such an index for comparing the 1974 prices with the 1972 (base-year) prices for the three commodities listed below.

Commodity	1972 price	1974 price
Beef	$1.50	$1.80
Pork	1.20	1.20
Lamb	1.00	1.10

b) Construct such an index for the given year 1974 with 1972 as the base year for the following consumer goods and their prices.

Commodity	1972 price	1974 price
Milk (per quart)	$0.30	$0.33
Soft drinks (per dozen)	1.50	1.65
Vinegar (per quart)	0.30	0.24

18. In (b) of the preceding problem, vinegar is obviously assumed to be as important as the other commodities. It exerts a substantial influence on the price index, even though the amount of vinegar used is generally far less than the amount of either milk or soft drinks. It is the usual practice, therefore, that unless the quantities consumed are about the same for all the commodities concerned, a price index is constructed by treating the quantities as *weights*.

Let us use the base-period quantities (Q_b) as weights to compute the price index for the data of part b of the preceding problem. If the quantities for milk, soft drinks, and vinegar purchased in 1972 are 100 quarts, 20 dozen, and 3 quarts, respectively, then we have

Commodity	1972 quantity (Q_b)	1972 price (P_b)	1974 price (P_g)	Value of 1972 quantity	
				1972 price	1974 price
Milk	100 quarts	$0.30	$0.33	$30.00	$33.00
Soft drinks	20 dozen	1.50	1.65	30.00	33.00
Vinegar	3 quarts	0.30	0.24	0.90	0.72
Total				$60.90	$66.72

The price index for the given year 1974, with 1972 quantities consumed as weights, is the ratio of the total value of 1972 quantities at 1974 prices to the total value at 1972 prices. Thus the *weighted price index* is

$$I = \frac{\$66.72}{\$60.90} = 109.6$$

In general, the weighted price index with the base-period quantities as weights is given by

$$I = \frac{\Sigma P_g Q_b}{\Sigma P_b Q_b}$$

    a) Given the data below, construct a weighted price index for 1974, with 1973 as the base period and 1973 quantities as weights.

Commodity	1973 quantity Q_b	1973 price P_b	1974 price P_g
Corn	300 bushels	$1.5	$1.2
Wheat	500 bushels	2.0	2.5
Oats	200 bushels	1.0	1.2

    b) Given the prices in (a) above, and assuming that the 1974 quantities (Q_g) for corn, wheat, and oats are 200, 400, and 400 bushels, respectively, construct a weighted price index for 1974 with 1974 quantities as weights.

5-5 measures of variability

Having considered the measures of central tendency for collected data, we shall now consider measures of *dispersion,* or *variability.* Two measures of variation will be discussed in this section: the range and the standard deviation.

    The *range* is determined by the two extreme values of the sampled data; it is simply the *difference between the lowest and the highest of the observations.* In example 5-2, the lowest value is 1 and the highest is 10; thus the range is 10−1, or 9. The range for the data in table 5-1 (page 90) is $42−$8, or $34. The range is particularly useful when one wants to know the extent of extreme variations such as the high and low prices of stocks or the high and low temperatures on a given day.

The range is used only in limited situations. Because of its dependence on only two values, the lowest and highest, it tends to increase as the sample size is enlarged. Furthermore, it does not provide a measurement of the variability of the observations relative to the center of the distribution. The disadvantages of the range as a measure of dispersion can be corrected by the use of the standard deviation.

The variability of any distribution is generally viewed in terms of the *deviation* of each observed value (X) from the sample mean ($\overline{X}$), or ($X - \overline{X}$). If the deviations are small, the data are obviously less variable, or less dispersed, than if the deviations are large. Thus, the deviation ($X - \overline{X}$) provides information about the degree of dispersion in a sample. Our task is to derive a formula based upon such deviations so that variability can be measured. We might expect that the average of the deviations would serve this purpose. Unfortunately, this is impossible because some of the deviations are negative, some are positive; and the sum of all the deviations, as indicated in equation 5-4, is equal to zero.

One possible solution to this difficulty is to compute *the average of the absolute values* of the deviations. In fact, the average of absolute deviations has been used as a measure of dispersion. It is, however, unsatisfactory for the purpose of statistical inference because it is not readily amenable to algebraic manipulations. To overcome the difficulty caused by the sign of deviations, statisticians choose to work with the square of the deviations. *The average of the square of the deviations is known as the variance.* If the variance of a set of observations is large, we can say that the data have greater variability than a set of data which have a small variance.

The use of variance as a measure of variability, however, also has its drawbacks. Small variances imply a small variation, but this knowledge is helpful only when several sets of observations are being compared. If we attempt to describe the dispersion of a single set of observations, the variance does not provide immediate help. For instance, what can we say about the variability of a sample with a variance of 25? This latter difficulty lies in the fact that because of the squaring operation, the variance is expressed not in the original unit, but in a *square unit*. It is therefore necessary to restore the original unit by extracting the square root. The measure thus obtained is known as the standard deviation. That is,

> The standard deviation of a set of observations is the square root of the average of the squared deviations from the mean.

The formula used to compute the variance, and subsequently the standard deviation, is determined by whether or not the collected data are organized. Let us first present a basic mathematical definition of the concept for unorganized data and then a more convenient way of calculating the variance and standard deviation for such data.

unorganized data

Let X be a random variable. A sample of n observations $(X_1, X_2, \ldots, X_n)$ is selected. Then the variance of X, designated by $\hat{s}^2$, or $\hat{s}_x^2$, is expressed as follows:

$$\hat{s}^2 = \frac{\sum_{i=1}^{n} (X_i - \overline{X})^2}{n} \tag{5-6}$$

and the standard deviation of X, designated by $\hat{s}$, or $\hat{s}_x$ is

$$\hat{s} = \sqrt{\frac{\sum_{i=1}^{n} (X_i - \overline{X})^2}{n}} \tag{5-7}$$

Since the standard deviation is used as a measurement, only the *positive root* of the variance is taken as the measure of variability.

EXAMPLE 5-5   Find the variance and the standard deviation for the 11 observations in example 5-2, namely, 1, 1, 2, 2, 2, 3, 4, 5, 5, 9, and 10.

From example 5-2, we know that the sample mean is 4. By applying equation 5-6 the variance of X is calculated as

$$\hat{s}^2 = \frac{1}{11}[(1-4)^2+(1-4)^2+(2-4)^2+(2-4)^2+(2-4)^2+(3-4)^2$$

$$+(4-4)^2+(5-4)^2+(5-4)^2+(9-4)^2+(10-4)^2]$$

$$= \frac{9+9+4+4+4+1+0+1+1+25+36}{11} = \frac{94}{11} = 8.55$$

From appendix D (pages 314–317), we find that the square root of 8.55 is 2.924. Thus, the standard deviation ($\hat{s}$) is

$$\hat{s} = \sqrt{8.55} = 2.924$$

The above procedure is quite laborious because the mean must be subtracted from every value of the sample. It is especially cumbersome if the sample is large. For more efficient computation, the following formula may be used.

$$\hat{s}^2 = \frac{\Sigma X^2}{n} - \overline{X}^2 \tag{5-8}$$

Equation 5-8 is derived in the following way. Using equation 5-6 we obtain

$$\hat{s}^2 = \frac{1}{n}\sum(X - \overline{X})^2 = \frac{1}{n}\sum(X^2 - 2\overline{X}X + \overline{X}^2)$$

Because $2\overline{X}$ is a constant, and $\Sigma 2\overline{X}X = 2\overline{X}\Sigma X$, after distributing the summation sign we have

$$\hat{s}^2 = \frac{1}{n}\left(\sum X^2 - 2\overline{X}\sum X + \sum \overline{X}^2\right)$$

Since by equation 5-2 $\Sigma X = n\overline{X}$, and by equation 2-2, $\Sigma \overline{X}^2 = n\overline{X}^2$, we have

$$\hat{s}^2 = \frac{1}{n}\sum X^2 - \frac{2\overline{X}n\overline{X}}{n} + \frac{n\overline{X}}{n} = \frac{1}{n}\sum X^2 - 2\overline{X}^2 + \overline{X}^2 = \frac{1}{n}\sum X^2 - \overline{X}^2$$

Equation 5-8 makes it unnecessary to subtract the arithmetic mean from each observed value of the sample. The variance is simply the mean of the squared values minus the square of the mean.

EXAMPLE 5-6   Using equation 5-8, find the variance and the standard deviation for the following 11 values: 1, 1, 2, 2, 2, 3, 4, 5, 5, 9, and 10.

$$\hat{s}^2 = \frac{1}{11}(1^2 + 1^2 + 2^2 + 2^2 + 2^2 + 3^2 + 4^2 + 5^2 + 5^2 + 9^2 + 10^2) - 4^2$$

$$= \frac{1}{11}(1 + 1 + 4 + 4 + 4 + 9 + 16 + 25 + 25 + 81 + 100) - 16$$

$$= \frac{270}{11} - 16 = 24.55 - 16 = 8.55$$

and the standard deviation is

$$\hat{s} = \sqrt{8.55} = 2.924$$

organized data

Since large samples are usually organized into frequency distributions, some different formula must be developed for computing the standard deviation for organized data. In computing the variance for the 11 values in example 5-5, the squared deviation $(1 - 4)^2$ appears twice, $(2 - 4)^2$ three times, $(5 - 4)^2$ twice, and all other squared deviations appear only once. Thus, the computation procedure for $\hat{s}^2$ may be slightly changed as follows:

$$s^2 = \frac{1}{11} \left[(1-4)^2(2) + (2-4)^2(3) + (3-4)^2(1) + (4-4)^2(1) \right.$$
$$\left. + (5-4)^2(2) + (9-4)^2(1) + (10-4)^2(1) \right]$$
$$= \frac{18 + 12 + 1 + 0 + 2 + 25 + 36}{11}$$
$$= \frac{94}{11} = 8.55$$

Observe that the numbers 1, 2, 3, 4, 5, 9, and 10 in the left-hand parentheses of the numerator are distinct values of the random variable X (designated by the lower-case letter x), the value 4 is the sample mean ($\overline{X}$), the values in the right-hand parentheses are frequencies $[FR(x)]$, and the denominator 11 is the sample size (n). Accordingly, the following formula may be constructed for computing the variance for a frequency distribution.

$$\hat{s}^2 = \frac{\sum_{j=1}^{m} (x_j - \overline{X})^2 FR(x_j)}{n} \tag{5-9}$$

The subscript j again designates a counting number for identifying each distinct value or class midpoint, and m designates the total number of distinct values or classes.

EXAMPLE 5-7 Compute the variance and the standard deviation for the frequency distribution shown in table 5-2 (page 91).

Table 5-4 shows how the variance and standard deviation for the data in table 5-2 are computed.

Table 5-4 Computation of the variance and standard deviation for the data in table 5-2 ($\overline{X} = 24$)

x	$x - \overline{X}$	$(x - \overline{X})^2$	$FR(x)$	$(x - \overline{X})^2 FR(x)$
10	−14	196	2	392
15	− 9	81	3	243
20	− 4	16	3	48
25	1	1	5	5
30	6	36	4	144
35	11	121	2	242
40	16	256	1	256
Total			20	1330

The total in the right-hand column of table 5-4 constitutes the numerator of equation 5-9. Dividing this value by the sample size, we obtain the sample variance as follows:

$$\hat{s}^2 = \frac{1330}{20} = 66.5$$

and the standard deviation is

$$\hat{s} = \sqrt{66.5} = 8.1548$$

sample variance as an estimator

As we mentioned earlier, the sample mean is generally used to make inferences about the population mean. Although we did not explicitly say that the sample mean provides a good estimate of the population mean, we implied that this is the case. Similarly, it would seem reasonable to expect that the sample variance would provide a good estimate of the population variance. Unfortunately, a sample variance computed by the formula given above tends to *underestimate* the population variance, especially when the sample size is small.

There is another formula for sample variance that gives a more accurate estimate of population variance, especially when the sample is small; it is frequently designated simply by s^2.

$$s^2 = \frac{\sum_{i=1}^{n} (X_i - \overline{X})^2}{n-1} \tag{5-10}$$

Observe that equation 5-10 differs from equation 5-6 only in the denominator: n is replaced by $(n-1)$. When n is large, $\hat{s}^2$ and s^2 are approximately equal. The mathematical proof showing s^2 to be a better estimator need not concern us here; it is sufficient to say that statisticians generally prefer s^2 to $\hat{s}^2$, and invariably use s^2 as the estimator of the population variance.

If the sample variance $\hat{s}^2$ is given, it is a simple matter to transform it to s^2. From equations 5-6 and 5-10, we have

$$n\hat{s}^2 = \sum (X - \overline{X})^2 \text{ and } (n-1)s^2 = \sum (X - \overline{X})^2$$

Thus,

$$n\hat{s}^2 = (n-1)s^2$$

Consequently,

$$s^2 = \frac{n}{(n-1)} \hat{s}^2 \tag{5-11}$$

EXAMPLE 5-8   Find the population variance estimator (s^2) for the data in example 5-5. (See page 106.)

In example 5-5, the numerator on the right-hand side of the equation

109

is 94. Thus the estimator s^2 is

$$\frac{94}{10} = 9.4$$

Or, since in example 5-5, $s^2 = 94/11$, by equation 5-11 we obtain the same result as follows:

$$s^2 = \left(\frac{11}{10}\right)\left(\frac{94}{11}\right) = \frac{94}{10} = 9.4$$

Furthermore, the estimator s^2 can be computed from the following short-cut formula, which is derived from equation 5-10.

$$s^2 = \frac{\sum X^2 - \dfrac{(\sum X)^2}{n}}{n - 1} \tag{5-12}$$

EXAMPLE 5-9   Using equation 5-12, compute s^2 for the data in example 5-5.

$$\sum X^2 = 1^2 + 1^2 + 2^2 + 2^2 + 2^2 + 3^2 + 4^2 + 5^2 + 5^2 + 9^2 + 10^2$$
$$= 1 + 1 + 4 + 4 + 4 + 9 + 16 + 25 + 25 + 81 + 100 = 270$$
$$\sum X = 1 + 1 + 2 + 2 + 2 + 3 + 4 + 5 + 5 + 9 + 10 = 44$$

Therefore,

$$s^2 = \frac{270 - (44)^2/11}{11 - 1} = 27 - \frac{1936}{110} = 27 - 17.6 = 9.4$$

The variance computed by equation 5-10 or 5-12 is generally referred to as the *unbiased* estimator of the population variance. Any estimator of a population parameter is called unbiased if the long-run average of the estimator is equal to the parameter. It can be demonstrated that the long-run average of the sample variance s^2 computed by equation 5-10 or 5-12 is equal to the population variance which it is supposed to estimate.

5-5
exercises

1. Which of the following measures does not belong with the group and why?

   a) The range
   b) The average of absolute deviations
   c) The median
   d) The variance
   e) The standard deviation

2. Explain the similarity and the difference between the average of absolute deviations and the standard deviation.

3. Which of the following groups of scores exhibits the least variability and which exhibits the most?

    a) 1, 3, 5, 7, 9, and 11
    b) 1, 2, 3, 9, 10, and 11
    c) 1, 1, 2, 10, 11, and 11

4. Let us measure the variability of the test scores of men and women cited in the first section of this chapter. The mean of the scores for men is the same as that for women, namely, 60. The variance of the test scores for men is

$$\hat{s}^2 = \frac{\Sigma(x - \overline{X})^2 FR(x)}{n} = \frac{2200}{20} = 110$$

Compute the variance ($\hat{s}^2$) of the following test scores for women.

Test scores x	20	30	40	50	60	70	80	90	100
$FR(x)$		1	1	2	3	6	3	2	1	1

5. If the distributions of X and Y have the same mean and range, and the standard deviation of X is 10 and that of Y is 6, what can you conclude about each distribution relative to its mean?

6. The variance of the scores 1, 3, 5, 7, 9, and 11 is $\hat{s}^2 = \frac{70}{6}$.

    a) If the number 2 is added to each of these six scores, what is the variance?
    b) If the number 1 is subtracted from each of these six scores, what is the variance?
    c) What conclusion can you derive from the results of (a) and (b)?

7. Compute the variance and the standard deviation for the data in problem 11, exercises 5-3, by equation 5-9. (For $\overline{X} = 3.0$)

8. Compute the variance and the standard deviation for the data in problem 12, exercises 5-3, by equation 5-9.

9. Compute the variance and the standard deviation for the data in problem 5, exercises 5-4, by (a) equation 5-10 and (b) equation 5-12.

10. Compute the variance and the standard deviation for the data in problem 8, exercises 5-3, by (a) equation 5-6 and (b) equation 5-8.

11. Compute the variance and the standard deviation for the data in problem 9, exercises 5-3, by (a) equation 5-6 and (b) equation 5-8.

12. Compute the unbiased estimator of population variance for the following sample of 11 observations.

X_1	X_2	X_3	X_4	X_5	X_6	X_7	X_8	X_9	X_{10}	X_{11}
1	1	2	2	2	3	3	3	4	4	8

111

13. Convert the sample variance obtained in problem 7 above into an unbiased estimator of population variance.

14. Convert the sample variance obtained in problem 8 above into an unbiased estimator of population variance.

15. Eleven hogs are randomly selected, and their birth weights are obtained as follows:

    14   26   25   15   18   18   23   21   20   24   16

    Using equation 5-10, estimate the variance and the standard deviation of the birth weights of all baby hogs.

16. Using the shortcut method of equation 5-12, compute the variance and the standard deviation for the following data.

    2   5   3   7   6   4   1   9   8

probability distribution $\textstyle{6}$
and population parameters

6-1
introduction

As we have already pointed out, the sample mean $(\overline{X})$ and variance (s^2) are good estimators of the population mean and variance. The population mean and variance are usually represented by the Greek letters μ (mu) and σ^2 (sigma squared). The letter σ (sigma) is used to denote the population standard deviation.

In sections 6-2 and 6-3, attention will be focused on the methods of determining the mean and variance of the population, respectively. As we shall see, the methods for determining the mean and variance for a finite population are quite similar to the methods for computing the mean and variance of a sample. Computing the mean and variance for an infinite population, however, is based on a different concept. The concept is not a new one: it is the notion of a probability distribution.

Any *probability distribution* is a *population distribution;* it is the theoretical distribution of a random variable, which takes on an unspecified value for each single trial. In section 6-4, we shall introduce another kind of theoretical distribution called *sampling distribution*. It is the distribution of a sample statistic such as the sample mean. When a sample is viewed alone, the sample mean is considered a *constant*, because there is only one mean that can be computed from a given sample. From any given population, however, it is possible to draw many, and sometimes an infinite number of, samples of equal size, each of which yields an unspecified value as its mean. Thus, a new kind of random variable is generated; it is the sample mean which may assume an unspecified value for each different sample. If we use X to denote the original random variable and $\overline{X}$ to denote the sample mean, the new random variable, we shall see that the mean and variance of $\overline{X}$ are closely related to the mean and variance of X.

In the last section of this chapter, we shall try to answer two questions that are frequently raised by students of statistics. One concerns the application of the standard deviation; the other relates to the appropriate size of a sample. Answers to these questions are provided in *Tche-*

bycheff's theorem, which can be applied to the population as well as to a set of empirical observations.

<table>
<tr>
<td>

6-1
exercises

</td>
<td>

1. Name the two population parameters that are the counterparts of the sample statistics discussed in the preceding chapter, namely, $\overline{X}$ and s^2.

2. Is the probability function shown in table 4-4 (page 70) a relative frequency distribution or a population distribution? Explain.

3. Is the probability distribution of a random variable such as X designating the IQ of any elementary school pupil a

   a) sampling distribution?
   b) population distribution?
   c) frequency distribution?

4. Is the probability distribution of a sample statistic such as $\overline{X}$ designating the mean IQ of any random sample of n elementary school pupils a

   a) sampling distribution?
   b) population distribution?
   c) relative frequency distribution?

5. Under what condition is a sample mean a constant and under what condition is it a random variable?

6. Give an example to show the difference between an empirical relative frequency distribution and a theoretical probability distribution.

</td>
</tr>
</table>

<table>
<tr>
<td>

6-2
population
mean—
expected
value

</td>
<td>

Knowing how to calculate the mean of the sample, we might imagine that the mean of the population (μ) can be calculated in the same way. That is, we might look at equation 5-2, and replacing sample size (n) by population size (N), write the following new equation.

</td>
</tr>
</table>

$$\mu = \frac{\sum_{i=1}^{N} X_i}{N} \tag{6-1}$$

This formula for finding the population mean is correct provided that the population is finite; that is, that N designates a fixed number of units or observations. When the population is infinite, or when N approaches infinity, the above expression obviously cannot be used for determining the mean of the population.

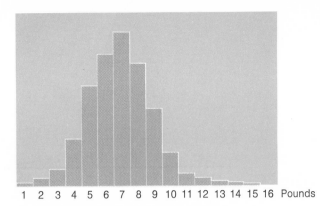

1 2 3 4 5 6 7 8 9 10 11 12 13 14 15 16 Pounds

FIGURE 6-1
Birth weights of babies born alive

Some students may find infinite populations rather difficult to visualize. Perhaps the following example will clarify the concept. Suppose, for instance, that we were interested in the weights of babies born alive in the past, present, and future. It is almost impossible to picture the vast sea of babies that were born, are now being born, or will be born in the future. And if we try to consider the weights of these babies, we would be confronted with billions upon billions of figures, most of which cannot even be determined because countless babies have yet to be born and many of those born in the past may not have been weighed.

Fortunately there is a way to handle such infinite populations. Note how, in figure 6-1, countless numbers of birth weights are grouped into a *distribution*. Such a distribution tells us the probability that any baby will weigh 2 pounds, 3 pounds, 4 pounds, . . . , 15 pounds, and so on. (The weights are rounded to the nearest pound.) We can then picture an infinite population as a *discrete probability distribution*. If the weights are not rounded to the nearest pound, or any subdivision of the pound, we can then consider the weights as values of a continuous random variable, and picture the infinite population as a *continuous probability distribution*, such as the one shown in figure 6-2. Any probability distribution, discrete or continuous, is referred to as a *population distribution*. In contrast, a *sample distribution*, like the one shown in table 5-2 or figure 5-1, is a frequency distribution.

the mean of a probability distribution

Since an infinite population is often viewed as a probability distribution, we can determine the mean of an infinite population by computing the mean of its probability distribution.

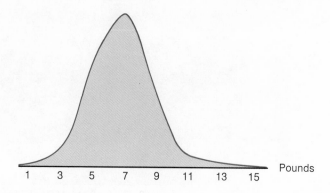

FIGURE 6-2
Birth weights of babies born alive

An infinite population can be generated from a finite set of values or units if sampling is taken *with replacement*. For illustration, let us consider the following example.

EXAMPLE 6-1   Suppose, for instance, that 10 balls numbered 0, 1, 2, ..., 8, and 9 are placed in a bowl. After they are thoroughly mixed, a ball is drawn and its number is noted and designated by X. The ball is then returned to the bowl before the next ball is drawn, and so on. How do we find the value of μ, the population mean of X?

We may consider the 10 digits a finite population, and then compute its mean. We may also consider the 10 digits an infinite population, since the drawing is with replacement and there is no limit, theoretically speaking, to the number of digits that can be drawn. For computing the mean of such an infinite population, it is necessary to first establish a probability distribution for X. The probability distribution of the 10 digits with each digit equally probable is shown below.

X	0	1	2	3	4	5	6	7	8	9
$P(X)$	0.1	0.1	0.1	0.1	0.1	0.1	0.1	0.1	0.1	0.1

First, we use equation 6-1 to compute the mean of the finite population of 10 digits.

$$\mu = \frac{0 + 1 + 2 + 3 + 4 + 5 + 6 + 7 + 8 + 9}{10} = \frac{45}{10} = 4.5$$

This expression for μ can be rearranged to show more clearly that a probability distribution is applied in computing the population mean with the same result.

$$\mu = 0(\tfrac{1}{10}) + 1(\tfrac{1}{10}) + 2(\tfrac{1}{10}) + 3(\tfrac{1}{10}) + 4(\tfrac{1}{10}) + 5(\tfrac{1}{10}) +$$
$$6(\tfrac{1}{10}) + 7(\tfrac{1}{10}) + 8(\tfrac{1}{10}) + 9(\tfrac{1}{10}) = \tfrac{45}{10} = 4.5$$

Observe that the latter expression for μ is simply the sum of the products obtained by multiplying each value of X by the corresponding probability. Symbolically, if m denotes the *number* of distinct values that the discrete random variable X can take on and $P(X_j)$ denotes the probability of each distinct value of X, then the population mean of X is

$$\mu = \sum_{j=1}^{m} X_j P(X_j) = X_1 P(X_1) + X_2 P(X_2) + \ldots + X_m P(X_m) \qquad (6\text{-}2)$$

EXAMPLE 6-2   Find the mean of X which designates the number of heads obtained in the toss of four coins.

The toss of four coins is identical with the toss of one coin four times. As shown in table 4-6 (page 74) there are five possible values for X: 0, 1, 2, 3, and 4, and the corresponding probabilities are $\frac{1}{16}$, $\frac{4}{16}$, $\frac{6}{16}$, $\frac{4}{16}$, and $\frac{1}{16}$. Thus the mean μ for X is

$$\mu = 0(\tfrac{1}{16}) + 1(\tfrac{4}{16}) + 2(\tfrac{6}{16}) + 3(\tfrac{4}{16}) + 4(\tfrac{1}{16}) = 2$$

The mean 2 is the *long-run average* of heads obtained in the toss of four coins. If you toss four coins a few times, or hundreds or even thousands of times, there is no guarantee that the average for all the tosses will be exactly 2. As the number of tosses approaches infinity, however, the average of the tosses comes closer to 2.

the concept of expected value

The long-run average mentioned above is often referred to as the *expected value,* or *mathematical expectation*. The expected value of a digit drawn at random is 4.5, and the expected number of heads in any toss of four coins is 2, though the actual results of a limited number of trials can be quite different from the expectation. In short, the expected value of any discrete random variable X, designated by $E(X)$, is the mean of the probability distribution of X; that is,

$$E(X) = \mu = X_1 P(X_1) + X_2 P(X_2) + \ldots + X_m P(X_m) \qquad (6\text{-}3)$$

The concept of expected value arose originally in games of chance. In the simplest form, the expected value of a game is the sum of the products of the amounts a player stands to win or lose and the corresponding probabilities.

EXAMPLE 6-3   Consider a game involving the toss of a coin. If the coin turns up heads, you win \$2; if the coin turns up tails, you lose \$2. What is the expected value of the game?

It is important to remember that in equation 6-3, the X's are positive if they represent winnings or profits, and negative if they represent losses or deficits. In this example $m = 2$ because there are only two values for X, namely, +\$2 and −\$2. The probability for each X is $\frac{1}{2}$. Consequently, the expected value of the game is

$$E(X) = \$2(\tfrac{1}{2}) - \$2(\tfrac{1}{2}) = 0$$

The game is a fair one, because no player is favored. That is, a game is called *fair*, or *equitable*, if the mathematical expectation is zero; no player will gain or lose anything in the long run. Many games, however, are not equitable; the following example is a case in point.

EXAMPLE 6-4   Suppose that in a lottery 2000 tickets are sold at \$1.00 each, and three prizes are to be awarded. The first prize is a stereo worth \$500, the second is a portable television set worth \$300, and the third is a record player worth \$200. If you have purchased one ticket, what is your expected gain?

If X is used to designate the gain for each ticket, then there are four possible values for X. Three of these four values are positive, \$499, \$299, and \$199 with a probability of 1/2000 for each (\$1 less than the value of each prize because you have to pay \$1 for the ticket); and one is negative, −\$1, with a probability of 1997/2000.
Your expected gain is therefore

$$E(X) = \$499(1/2000) + \$299(1/2000) + \$199(1/2000) - \$1(1997/2000)$$

$$= \$997/2000 - \$1997/2000 = -\$0.5$$

This is obviously not a fair game because you stand to lose, on the average, 50 cents every time you purchase a lottery ticket.

The method for computing the expected value of a continuous random variable involves the use of calculus, and is therefore beyond the scope of this text.

6-2
exercises

1. Let X denote the number of heads obtained in the toss of three coins. What is the expected value of X?

2. Let X denote the number of heads obtained in the toss of two coins. What is the expected value of X?

3. Let X designate the number of spots on the face of a rolled die. What is the expected value of X?

4. Let X designate the sum of spots on the faces of two rolled dice. What is the expected value of X?

5. One thousand lottery tickets are to be sold at \$2.00 each. If the

prize is a TV set with a retail value of $400, and you purchase two tickets, what is your expected gain?

6. Suppose that there is a function $f(X) = X^2$ where X is the number of heads obtained in the toss of two coins and $f(X)$ is the number of dollars you could win if heads came up. What is the amount of money you could expect to win when the two coins are tossed?

7. Mr. Anderson has two million dollars, which he is considering investing in common stocks. X designates the amount of money he will end up with; it has the following probability distribution.

X	$P(X)$
$1 million	0.2
2 million	0.3
3 million	0.2
4 million	0.2
5 million	0.1
Total	1.0

What can Mr. Anderson expect to make if he decides to invest his two million dollars?

8. Suppose that a TV set contains 10 tubes, two of which are defective. Two tubes are selected at random and removed from the set for inspection. Let X be the number of defective tubes found in the sample of two tubes. What is the expected value of X?

9. Assume that the probability of a car to be in an accident in any given year is 0.02. Assume further that the probability distribution of all automobile accidents relative to property damage or financial loss is as follows.

Damage or Loss	Probability
$ 100	0.40
500	0.20
800	0.15
1500	0.10
3000	0.09
10,000	0.06
Total	1.00

What is the expected value of loss or damage in one year due to a car accident?

10. A random variable Y has the following probability distribution. What is the population mean of Y?

Y:	0	1	2	3	4	Total
$P(Y)$:	0.20	0.15	0.25	0.05	0.35	1.00

11. If the probability distribution of the IQ's of all high school graduates is as follows, what is the mean of their IQ's?

IQ's	Probability
42.5–57.5	0.01
57.5–72.5	0.02
72.5–87.5	0.05
87.5–102.5	0.40
102.5–117.5	0.30
117.5–132.5	0.10
132.5–147.5	0.08
147.5–162.5	0.04
Total	1.00

12. Let $Y = X^2$ where X is a random digit. Find the expected value of Y.

6-3 population standard deviation

As we learned in section 5-5, the sample standard deviation is a measure of the variability of collected data from the mean. Likewise, the population standard deviation is used to measure the variation of the values of a random variable (X) from the population mean (μ). Having learned how to obtain the sample variance in equation 5-6, we might suspect that the population variance, or σ^2, can be similarly defined as follows:

$$\sigma^2 = \frac{\sum\limits_{i=1}^{N} (X_i - \mu)^2}{N} \qquad (6\text{-}4)$$

This definition is acceptable, however, only on the condition that the population size (N) is finite. When N is not finite, this expression obviously cannot be used for determining the population variance.

It is important to keep in mind that the *variance is the average of the squared deviations from the mean*. For a sample, the deviations are from the sample mean; for a population, the deviations are from the population mean. In other words, the population variance (σ^2) is the *average* of $(X - \mu)^2$. In equation 6-4, the notation that signifies "the average of" includes the following part of the expression.

$$\frac{\sum\limits_{i=1}^{N}}{N}$$

which can be used only in connection with a finite population. If the population is infinite, we have to use the probability distribution for computing the variance. We may then replace the "average of" notation by the "expected value" sign (E) to express the population variance, since the expected value of any variable is the long-run *average* for the variable. Thus, the expression $E(X - \mu)^2$ is generally used to denote the population variance; that is,

$$\sigma^2 = E(X - \mu)^2$$

We know that

$$(X - \mu)^2 = X^2 - 2\mu X + \mu^2$$

Therefore,

$$E(X - \mu)^2 = E(X^2 - 2\mu X + \mu^2) = E(X^2) - E(2\mu X) + E(\mu)^2$$

Because of the fact that 2μ and μ^2 are constants, and the expected value of the product of a constant and a variable equals the constant multiplied by the expected value of the variable, or $E(2\mu X) = 2\mu E(X)$, and the expected value of a constant is the same constant, or $E(\mu)^2 = \mu^2$, we have

$$\begin{aligned} E(X - \mu)^2 &= E(X^2) - 2\mu E(X) + \mu^2 \\ &= E(X^2) - 2\mu\mu + \mu^2 \\ &= E(X^2) - 2\mu^2 + \mu^2 \end{aligned}$$

Consequently,

$$\sigma^2 = E(X^2) - \mu^2 \tag{6-5}$$

The way this equation can be used to calculate the variance of a population distribution is illustrated in the following two examples.

EXAMPLE 6-5 Let the 10 digits 0, 1, 2, 3, 4, 5, 6, 7, 8, and 9 be designated by X. In a drawing with replacements, the probability for each to occur is $\frac{1}{10}$. What is the variance of X?

Table 6-1 Computation of the variance of X

X	X^2	$P(X)$	$X^2 P(X)$
0	0	0.1	0.0
1	1	0.1	0.1
2	4	0.1	0.4
3	9	0.1	0.9
4	16	0.1	1.6
5	25	0.1	2.5
6	36	0.1	3.6
7	49	0.1	4.9
8	64	0.1	6.4
9	81	0.1	8.1

$$E(X^2) = \Sigma X^2 P(X) = 28.5$$

$$\begin{aligned} \sigma^2 = E(X^2) - \mu^2 &= \Sigma X^2 P(X) - \mu^2 \\ &= 28.5 - (4.5)^2 = 28.5 - 20.25 = 8.25 \end{aligned}$$

In table 6-1, we see that the expected value of X^2 is equal to the sum of the products of X^2 multiplied by the corresponding probabilities of X^2, which are the same as the probabilities of X because of the fact that X^2 happens only if X occurs.

The standard deviation of X can be obtained from appendix D, which shows the square root of 8.25 to be 2.8723.

EXAMPLE 6-6   Find the variance of the random variable X, which designates the number of heads that will occur in the toss of four coins.

The probability distribution of X is shown in table 6-2. The table also shows the process of computing the variance of X by the use of equation 6-5. The mean (μ) of X is equal to 2 (see example 6-2).

Table 6-2   Computation of the variance of X, the number of heads in the toss of four balanced coins

X	X^2	$P(X)$	$X^2P(X)$
0	0	$\frac{1}{16}$	0
1	1	$\frac{4}{16}$	$\frac{4}{16}$
2	4	$\frac{6}{16}$	$\frac{24}{16}$
3	9	$\frac{4}{16}$	$\frac{36}{16}$
4	16	$\frac{1}{16}$	$\frac{16}{16}$

$$E(X^2) = \Sigma X^2 P(X) = \frac{80}{16} = 5$$

$$\sigma^2 = \Sigma X^2 P(X) - \mu^2 = 5 - 2^2 = 5 - 4 = 1 \text{ and } \sigma = \sqrt{1} = 1$$

6-3 exercises

1. Let X designate the number of heads obtained in the toss of three coins. What is the variance of X?

2. Let X designate the sum of spots on the faces of two rolled dice. What is the variance of X?

3. Refer to problem 7, exercises 6-2. Find the variance and the standard deviation of X.

4. Refer to problem 10, exercises 6-2. Find the variance and the standard deviation of Y.

5. Refer to problem 8, exercises 4-5 (page 71). Find the variance and the standard deviation of X.

6. Refer to problem 12, exercises 6-2. Find the variance and the standard deviation of Y.

7. The probabilities for an automobile dealer to sell 0, 1, 2, 3, 4, and 5

cars on a single day are, respectively, 0.10, 0.20, 0.30, 0.25, 0.10, and 0.05. If X is the number of cars sold each day, find the variance and the standard deviation of X.

8. A bag contains four balls marked 0, 2, 4, and 6. A ball is drawn, and after its number is noted it is returned to the bag, and so on.

   a) Do the four numbers constitute a sample or a population?
   b) Find the variance and the standard deviation of X (which designates each number being drawn).

9. Let X be a binomial random variable with $n = 5$ and $p = 0.5$. Find the variance of X.

10. Let X be a binomial random variable with $n = 5$ and $p = 0.1$. What is the variance of X? What is the variance of X if $n = 5$ and $p = 0.9$? What is your generalization with respect to the variance of a binomial random variable?

6-4 sampling distribution of the mean

In a single sample, the sample mean is a *constant* because there is only one value for the mean. Clearly, however, numerous samples of equal size can be taken from the same population, and the values we obtain for $\overline{X}$ may very likely differ from sample to sample. Suppose, for instance, that 10 samples of 30 pupils each are randomly taken from the elementary schools of a certain state; IQ tests are given, and mean IQ's computed for each sample. The following figures are obtained.

$$93 \quad 102 \quad 98 \quad 100 \quad 98 \quad 104 \quad 110 \quad 96 \quad 103 \quad 106$$

These figures reveal a very interesting fact: the sample mean generally varies from sample to sample, and with one exception they all differ from the population mean (μ), which is 100. Furthermore, the overall mean of the 10 samples is

$$\frac{93 + 102 + 98 + 100 + 98 + 104 + 110 + 96 + 103 + 106}{10} = \frac{1010}{10} = 101$$

which is also different from the true mean 100.

The deviations between *what we actually get,* the sample means, and *what we expect,* the population mean, are commonly referred to as *sampling errors* or *chance variations*. Chance variations reflect the uncertainties of life. We cannot be certain, for instance, how many fatal automobile accidents will occur in Los Angeles during the next month. We cannot be certain how many eggs a chicken will lay during the next week, or how many passengers will board a certain commuter train next Monday morning. Fortunately, statistical theory has made it possible to

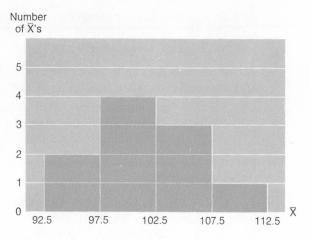

FIGURE 6-3
A histogram for 10 sample means

measure, and to some extent control, such uncertainties or chance fluctuations.

The means of the 10 samples mentioned above range from 93 to 110 with an overall average of 101. If we distribute the 10 means among four classes with interval lengths of five units, namely, 92.5 to 97.5, 97.5 to 102.5, 102.5 to 107.5, and 107.5 to 112.5, we could draw a histogram like the one shown in figure 6-3. This graph is not quite symmetrical, and the center is not equal to the mean 100. Remember that we have only 10 sample means. If there were 1000 samples, the distribution of sample means would undoubtedly become more symmetrical, and the center would be closer to 100. In fact, if a sufficiently large number of samples are taken, and the mean IQ's are not rounded to the nearest unit or any subdivisions thereof, then the sample means would form a smooth bell-shaped symmetrical curve similar to the normal probability distribution. Such a distribution is called the *sampling distribution* of the sample mean.

In actual practice, however, we do not take many samples from the same population. In fact, we usually obtain only a single sample mean for the purpose of analysis. We must therefore resort to other techniques for obtaining information about the sampling distribution of $\overline{X}$. These techniques involve the use of the mean and the standard deviation of the sampling distribution.

The mean and the standard deviation of $\overline{X}$, designated as $\mu_{\overline{X}}$ and $\sigma_{\overline{X}}$, are closely related to the mean and the standard deviation of X, designated as μ_X and σ_X. Let us first use a simple hypothetical example to show the sampling distribution of this new random variable $\overline{X}$ as compared with the population distribution of X, before describing the relationships between $\mu_{\overline{X}}$ and μ_X and between $\sigma_{\overline{X}}$ and σ_X.

Consider a random variable that has the set of values $\{1, 3, 5, 7\}$.

Suppose that a number is randomly selected *with replacement* from these four numbers, and the number thus selected is designated by X. Thus, these four numbers are used to generate an infinite population for X. The infinite population thus generated is, as indicated in the last section, a *population*, or *probability*, *distribution*, which is shown in table 6-3.

Table 6-3 Population distribution of X with the set of values {1, 3, 5, 7}

X	$P(X)$
1	$\frac{1}{4}$
3	$\frac{1}{4}$
5	$\frac{1}{4}$
7	$\frac{1}{4}$
Total	$\frac{4}{4} = 1$

Suppose further that random samples of two numbers each are taken with replacement from the set of values {1, 3, 5, 7}, and a mean is obtained for each sample thus selected. Obviously, an infinite population for $\overline{X}$ can also be generated in this case. Countless samples will be identical, and so will be the sample mean. For obtaining the probability distribution of $\overline{X}$, we need, first of all, to know all the *distinct possible samples* that can be drawn in this experiment. The distinct possible samples and their means are shown in table 6-4.

Table 6-4 Distinct possible two-number samples and their means for X with the set of values {1, 3, 5, 7}

Sample	X_1	X_2	Sample total	Sample mean
1	1	1	2	1
2	1	3	4	2
3	1	5	6	3
4	1	7	8	4
5	3	1	4	2
6	3	3	6	3
7	3	5	8	4
8	3	7	10	5
9	5	1	6	3
10	5	3	8	4
11	5	5	10	5
12	5	7	12	6
13	7	1	8	4
14	7	3	10	5
15	7	5	12	6
16	7	7	14	7

As shown in table 6-4, 16 different possible samples can be drawn with X_1 designating the number of the first selection and X_2 the number of the second. (There are four ways to select X_1 and four ways to select X_2; therefore the total number of sequences to select X_1 and X_2 is $4 \times 4 = 16$.) Although there are 16 different possible samples, there are only *seven distinct sample means*. Thus, the random variable $\overline{X}$ has the set of values $\{1, 2, 3, 4, 5, 6, 7\}$. Of the 16 possible samples, only one has the mean 1, two have the mean 2, three have the mean 3, four have the mean 4, three have the mean 5, two have the mean 6, and only one has the mean 7. Each of the four numbers has the same probability, $\frac{1}{4}$, to be drawn; each of the 16 samples has the same probability, $(\frac{1}{4})(\frac{1}{4}) = \frac{1}{16}$, of being selected. Consequently, the probability of each distinct sample mean $\overline{X}$ is obtained by dividing the number of samples yielding the same mean by 16. The probability for each $\overline{X}$ is shown in the last column of table 6-5.

Table 6-5 Sampling distribution of the mean $\overline{X}$ with $n = 2$ for X having the set of values $\{1, 3, 5, 7\}$

Sample mean $\overline{X}$	Number of samples	Probability $P(\overline{X})$
1	1	$\frac{1}{16}$
2	2	$\frac{2}{16}$
3	3	$\frac{3}{16}$
4	4	$\frac{4}{16}$
5	3	$\frac{3}{16}$
6	2	$\frac{2}{16}$
7	1	$\frac{1}{16}$
Total	16	$\frac{16}{16} = 1$

The probability distribution shown in table 6-5 is called a sampling distribution of the mean $\overline{X}$. Figure 6-4 shows its graphic representation. In general,

> The probability distribution of all distinct values of a given sample statistic from all different possible samples of equal size taken from the same population is referred to as a *sampling distribution*.

the mean of $\overline{X}$

The foregoing discussion should have left no doubt in the mind of the reader that the sample mean $\overline{X}$ is a random variable. As a random variable, $\overline{X}$ has a mean and a standard deviation. We shall now demonstrate

P($\overline{X}$)

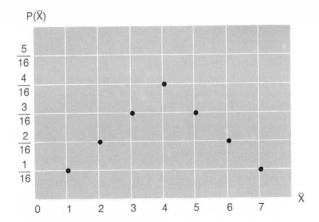

FIGURE 6-4
Sampling distribution of $\overline{X}$ with
$n = 2$ for X having the set of
possible values $\{1, 3, 5, 7\}$

that the mean of the sampling distribution of $\overline{X}$ is identical with the mean of the population distribution of X.

Since the sampling distribution of $\overline{X}$ is a probability distribution, we apply equation 6-2 to calculate the mean of $\overline{X}$, or $\mu_{\overline{X}}$, as follows:

$$\mu_{\overline{X}} = \overline{X}_1 P(\overline{X}_1) + \overline{X}_2 P(\overline{X}_2) + \ldots + \overline{X}_m P(\overline{X}_m) \tag{6-6}$$

EXAMPLE 6-7   Compute the mean of the sampling distribution of $\overline{X}$ as shown in table 6-5.

By applying equation 6-6, the mean of $\overline{X}$ is obtained below.

$$\mu_{\overline{X}} = 1(\tfrac{1}{16}) + 2(\tfrac{2}{16}) + 3(\tfrac{3}{16}) + 4(\tfrac{4}{16}) + 5(\tfrac{3}{16}) + 6(\tfrac{2}{16}) + 7(\tfrac{1}{16})$$

$$= \frac{(1 + 4 + 9 + 16 + 15 + 12 + 7)}{16} = \frac{64}{16} = 4$$

The mean of the population distribution of X, or μ_X, can be similarly computed by applying equation 6-2 to table 6-3.

$$\mu_{\overline{X}} = 1(\tfrac{1}{4}) + 3(\tfrac{1}{4}) + 5(\tfrac{1}{4}) + 7(\tfrac{1}{4})$$

$$= \frac{(1 + 3 + 5 + 7)}{4} = \frac{16}{4} = 4$$

Thus, we have $\mu_{\overline{X}} = \mu_X$, or

$$E(\overline{X}) = E(X) \tag{6-7}$$

The fact that the mean of $\overline{X}$ is equal to the mean of X can be further demonstrated as follows:

$$E(\overline{X}) = E\left(\frac{X_1 + X_2 + \ldots + X_n}{n}\right)$$

$$= \frac{E(X_1) + E(X_2) + \ldots + E(X_n)}{n}$$

$$= \frac{nE(X)}{n}$$

$$= E(X)$$

the standard error

Like the standard deviation of any random variable, the standard deviation of the random variable $\overline{X}$ measures the variability, or dispersion, of the sampling distribution of $\overline{X}$. The standard deviation of the sample mean, or $\sigma_{\overline{X}}$, is commonly called the *standard error* of $\overline{X}$.

EXAMPLE 6-8 Compute the standard error of the mean for the sampling distribution shown in table 6-5.

The standard error of the sampling distribution of $\overline{X}$ shown in table 6-5 is derived from the calculations in table 6-6.

Table 6-6 Computation of the standard error of the mean for the sampling distribution in table 6-5

Sample mean $\overline{X}$	Probability $P(\overline{X})$	Mean squared $\overline{X}^2$	Product $\overline{X}^2 P(\overline{X})$
1	$\frac{1}{16}$	1	$\frac{1}{16}$
2	$\frac{2}{16}$	4	$\frac{8}{16}$
3	$\frac{3}{16}$	9	$\frac{27}{16}$
4	$\frac{4}{16}$	16	$\frac{64}{16}$
5	$\frac{3}{16}$	25	$\frac{75}{16}$
6	$\frac{2}{16}$	36	$\frac{72}{16}$
7	$\frac{1}{16}$	49	$\frac{49}{16}$
Total	$\frac{16}{16}$	$\Sigma \overline{X}^2 P(\overline{X}) =$	$\frac{296}{16} = 18.5$

$$\sigma_{\overline{X}}^2 = E[(\overline{X} - \mu)^2] = E(\overline{X}^2) - \mu^2$$

$$= \Sigma \overline{X}^2 P(\overline{X}) - \mu^2 = 18.5 - 4^2 = 2.5$$

and

$$\sigma_{\overline{X}} = \sqrt{2.5} = 1.5811$$

Table 6-6 shows that the variance of $\overline{X}$ is 2.5. The variance of X can be similarly calculated by applying equation 6-5 to the population distribution of X shown in table 6-3:

$$\sigma_X^2 = E(X^2) - \mu^2 = \Sigma X^2 P(X) - \mu^2$$
$$= 1^2(\tfrac{1}{4}) + 3^2(\tfrac{1}{4}) + 5^2(\tfrac{1}{4}) + 7^2(\tfrac{1}{4}) - 4^2$$
$$= \frac{(1 + 9 + 25 + 49)}{4} - 16 = \frac{84}{4} - 16 = 21 - 16 = 5$$

Thus,

$$\sigma_X = \sqrt{5}$$

Observe that the variance of X is 5 and the variance of $\overline{X}$ is 2.5. The variance of $\overline{X}$ is equal to the variance of X divided by the sample size (in this case, $n = 2$). That is,

$$\sigma_{\overline{X}}^2 = \frac{\sigma_X^2}{n} \tag{6-8}$$

and the standard error of $\overline{X}$ is

$$\sigma_{\overline{X}} = \frac{\sigma_X}{\sqrt{n}} \tag{6-9}$$

The proof of equations 6-8 and 6-9 need not concern us here. It is sufficient to say now that the standard error of the mean plays an important and fundamental role in statistics, for it measures the variability of the sampling distribution of $\overline{X}$, that is, the chance variations of the sample mean from the true mean μ. The ratio $\sigma/\sqrt{n}$ shows that the standard error is affected by two factors: If σ_X, or simply σ, is large, it means great variability in the population from which samples are taken, and the sampling distribution of the sample mean can be expected to have a proportionally large variation, which is reflected by a large standard error. On the other hand, the larger the sample size, the smaller the standard error; then the chance variations of the sample mean will be smaller, and the sample mean can be expected to be closer to the population mean. Chance variations, therefore, can be controlled by varying the size of the sample.

6-4 exercises

1. Suppose that a bag contains three numbers: 2, 4, and 6. A sample of two numbers is drawn with replacement, and the mean of the sample is designated by $\overline{X}$.
   a) Find the sampling distribution of $\overline{X}$, and present it graphically.
   b) Find the standard error of $\overline{X}$ on the basis of the sampling distribution.
   c) Find the variance of X and then compute the standard error of $\overline{X}$ by equation 6-9.

2. Suppose that a random variable X has the following probability distribution.

$$X = 1 \quad 2 \quad 3$$
$$P(X) = \tfrac{1}{3} \quad \tfrac{1}{3} \quad \tfrac{1}{3}$$

a) Find the population variance of X.

b) Let $\overline{X}$ be the mean of a random sample of two observations taken with replacement from this population. Find the sampling distribution of $\overline{X}$ and present it graphically.

c) Find the standard error of $\overline{X}$ on the basis of the sampling distribution and then verify it by using equation 6-9.

3. Refer to problem 11, exercises 6-2.

a) What is the variance of X, where X designates the IQ of any high school graduate?

b) If a sample of 25 high school graduates is randomly selected, and $\overline{X}$ is used to designate the mean of the sample, find the standard error of $\overline{X}$.

c) Find the mean of $\overline{X}$.

4. Refer to problem 8, exercises 6-3. If a sample of two balls is drawn with replacement, find (a) the sampling distribution of the sample mean $\overline{X}$; (b) $E(\overline{X})$; and (c) the variance and the standard error of $\overline{X}$.

5. The variance of a random variable Y is known to be 225. If $\overline{Y}$ is the mean of a random sample of 36 observations for Y, find the standard error of $\overline{Y}$.

6. The breaking strength of a certain rope is known to have a mean of 2000 pounds and a variance of 2500 pounds. Let $\overline{X}$ be the mean of any sample of 100 ropes selected at random. Find the expected value and the standard error of $\overline{X}$.

7. Suppose that the typing speed of college freshmen has a mean of 40 words per minute and a variance of 100 words. Let $\overline{X}$ be the mean speed of any random sample of 25 freshmen. Find the expected value and the standard error of $\overline{X}$.

8. Let X be the lifetime mileage of a certain brand of automobile tires. Assume that the mean and variance of X are, respectively, 30,000 and 40,000 miles. If a random sample of 16 tires is selected, what will be the expected value and the standard error of the sample mean?

6-5
tchebycheff's theorem

As we have learned, the standard deviation of a random variable is used to measure the variations of the values of the random variable from its mean. The smaller the standard deviation, the more likely we are to get a value close to the mean; the larger the standard deviation, the more likely we are to get a value far away from the mean. This important idea is expressed in a useful and interesting theorem developed by the Russian mathematician, Tchebycheff. Briefly, Tchebycheff's theorem can be stated as follows:

> The probability that the value of a random variable will lie within k standard deviations from the mean is at least $1 - 1/k^2$, where k is equal to or greater than 1. Symbolically, this theorem can be expressed
>
> $$P(|X - \mu| \leq k\sigma) \geq 1 - \frac{1}{k^2} \qquad (6\text{-}10)$$
>
> where μ and σ are, respectively, the mean and the standard deviation of the random variable X.

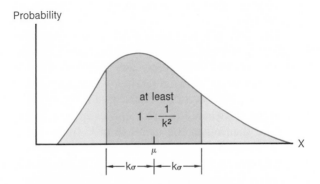

FIGURE 6-5
An illustration of Tchebycheff's theorem

Tchebycheff's theorem is illustrated in figure 6-5. Note that an interval is constructed by marking off a distance of $k\sigma$ on either side of the mean μ. The probability that any potential observation for X will fall within the constructed interval is at least $1 - 1/k^2$. Accordingly, the probability that the values of X will lie within two standard deviations from μ is at least $1 - \frac{1}{2^2} = 1 - \frac{1}{4} = 0.75$; the probability that the values of X will lie within 10 standard deviations from μ is at least $1 - \frac{1}{10^2} = 1 - \frac{1}{100} = 0.99$.

When $k = 1$, $1 - 1/k^2 = 0$. That is, the probability that the values of X will fall within one standard deviation from the mean is *at least* zero, a most uninformative result. The theorem is very conservative because it puts emphasis upon the idea of "at least." It applies to all distributions, skewed or symmetrical. Although the above discussion and notation involve population distributions only, the theorem can be applied to sample distributions as well.

Furthermore, Tchebycheff's theorem is often applied to sampling distributions. For a sampling distribution of $\overline{X}$, the theorem is expressed as

$$P(|\overline{X} - \mu| \leq k\frac{\sigma_X}{\sqrt{n}}) \geq 1 - \frac{1}{k^2} \tag{6-11}$$

Observe that in equation 6-11 the notations $\overline{X}$ and $\sigma_X/\sqrt{n}$ replace X and σ_X, respectively, in equation 6-10; otherwise, the two expressions are identical. Equation 6-11 states that the *probability that the mean of any sample of* n *observations selected at random will lie within* k *standard errors from the true mean* μ *is at least* $1 - 1/k^2$.

Two quantities involved in equation 6-11 are of particular interest for analysis. One is $k\sigma_X/\sqrt{n}$, which represents the *sampling error*. It is the difference between the observed sample statistic $\overline{X}$ and its expected value, namely, the population mean μ. Given the population standard deviation σ_X, the sampling error is directly related to k and inversely related to n. The other is $1 - 1/k^2$, which measures the *level of confidence* that $\overline{X}$ will not differ from μ by more than the stated error. The larger the value of k, the greater will be the confidence, assuming a given magnitude of error. However, when k increases, the quantity $k\sigma_X/\sqrt{n}$ will also increase unless n is increased sufficiently so that a given error is maintained. Further, the error can even be reduced if the increase in n can more than offset the increase in k. In fact, We can make n so large that the level of confidence, $1 - 1/k^2$, is made close to 100 percent, while the error is reduced to an arbitrarily small number. This relationship is generally known as the *law of large numbers*.

EXAMPLE 6-9   Suppose that a sample is to be taken from a population, whose distribution pattern is not known. We are told that the population variance is 400. If it is specified that the sampling error shall not be greater than 4, and the level of confidence not lower than 99 percent, how large should the sample be in order to meet these specifications?

To answer this question, we first solve the following equation to obtain the value for k.

$$1 - \frac{1}{k^2} = 0.99$$

Multiplying each term on both sides of the equation by k^2, we get

$$k^2 - 1 = 0.99k^2 \qquad \text{so} \qquad k^2 - 0.99k^2 = 1$$

that is,

$$0.01k^2 = 1 \qquad \text{or} \qquad k^2 = \frac{1}{0.01} = 100$$

Therefore,

$$k = 10$$

Since the population variance σ^2 is given as 400, its standard deviation (σ) is 20. Thus,

$$\frac{10(20)}{\sqrt{n}} = 4 \qquad \text{or} \qquad 10(20) = 4\sqrt{n}$$

$$\frac{200}{4} = \sqrt{n} \qquad \text{or} \qquad 50 = \sqrt{n}$$

Consequently,

$$n = 2500$$

As mentioned previously, Tchebycheff's theorem is very conservative. Generally speaking, a sample size computed on the basis of this theorem is the maximum required to meet the specifications. As will be seen in the next chapter, when the population distribution is symmetrical and bell-shaped, the probability that an observed value will lie within only 3 standard deviations from the mean is even greater than 99 percent. Under such circumstances, therefore, a much smaller sample is sufficient to meet the same set of specifications.

6-5 exercises

1. According to the Tchebycheff theorem, what is the probability that the value of a random variable will lie no more than three standard deviations from its mean?

2. According to the Tchebycheff theorem, what is the probability that the value of a random variable will lie

   a) more than 10 standard deviations below or above its mean?
   b) more than two standard deviations below or above its mean?

3. On the basis of the Tchebycheff theorem, find the probability that the value of a sample mean will lie more than three standard errors from the expected value of the sample mean.

4. Suppose that X is a random variable with a mean of 100 and a variance of 225. That is, $\mu = 100$ and $\sigma^2 = 225$. Answer the following questions according to the Tchebycheff theorem.

   a) At least how many percent of all potential observations of X will fall between 85 and 115?
   b) At least how many percent of all potential observations of X will fall between 70 and 130?

5. As indicated in the text, the Tchebycheff theorem applies to the distribution of sample observations as well. Let $RF(X)$ designate the relative frequency, or the proportion of times, that X occurs in

the sample. We have

$$RF(|X - \overline{X}| \leq ks) \geq 1 - \frac{1}{k^2} \qquad (6\text{-}12)$$

where $\overline{X}$ and s are, respectively, the sample mean and sample standard deviation. Then, if a sample of 100 observations yields a mean of 50 and a standard deviation of 5, how many observations will have values

   a) less than 40 or more than 60?
   b) between 35 and 65?

6. Refer to problem 5 above. Suppose that the test scores of a randomly selected sample of 50 students have a mean of 60 points and a standard deviation of 10 points. At least how many students scored between 10 and 110 points?

7. The personnel director of a large corporation wishes to know the average starting salary of last year's college graduates. It is known from past experience that the variance of starting salaries is $900. Since he does not have the time to interview all graduates, he plans to interview only a sample of them. Meanwhile, he has two specifications: first, he wants the sample mean to differ from the true mean within $10.; second, he wants the probability that the sample mean will lie within $10 of the true mean to be at least 99 percent. How large a sample should he have?

8. Suppose that the distribution of the weights of turkeys is known to have a standard deviation of two pounds. The purchasing agent of a large group of chain food stores wishes to estimate the population mean weight by randomly selecting a sample of turkeys supplied by the vendor. The agent does not want the sample mean (which he will use as an estimate) to differ from the true mean by more than one pound, and he does want the probability that the sample mean will not deviate by more than one pound to be at least 96 percent. How large should the sample size be in order to meet the agent's specifications?

7-1
introduction

This chapter deals with the *normal distribution,* which is by far the most important probability distribution in statistics. Normal distributions involve random variables that are considered to be continuous in nature, and for this reason a normal distribution is referred to as a *continuous probability distribution.*

A continuous random variable has a sample space with an infinitely large number of real values that can be arranged on a continuous scale. The dial shown in figure 7-1 provides a good example of such a scale. The circumference of the circle is marked with 0, 0.25, 0.50, 0.75, and 1.0. When the spinner at the center is spun, it circles around, and eventually stops at some position. X, which designates such a position, is a continuous random variable with the sample space of all the real numbers between 0 and 1.

It is interesting to point out that the probability of any continuous random variable to take on a particular value is virtually zero. Suppose, for instance, that the spinner in figure 7-1 stops at the position that reads 0.12. What was the probability that X would stop at this position? To answer this question, we need to know the "width" represented by 0.12; for the width, which is expressed as a fraction of 1 (that is, interval from 0 to 1) would be the probability. However, we have

$$P(0.11999 < X < 0.12001) = 0.00002$$

and

$$P(0.119999999999 < X < 0.120000000001) = 0.000000000002$$

Both these values could be considered expressions of the probability that $X = 0.12$. Further, we could continue and make the width infinitesimally small. In the final analysis, therefore, we have to conclude that $P(X = 0.12) = 0$. Accordingly, $P(X \leq 0.12) = P(X < 0.12) = 0.12$. In general, $P(X \leq a) = P(X < a)$, and $P(X \geq a) = P(X > a)$, where a is any value in the sample space of the continuous random variable X.

Because of the fact that the normal bell-shaped curve depicts the

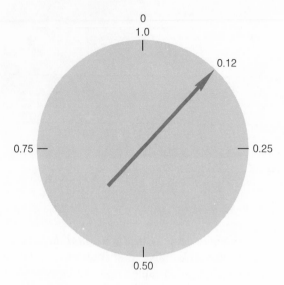

FIGURE 7-1
A dial

probability function of a continuous random variable, we cannot refer to any particular point on the curve as the probability of X. To determine probabilities, we have to refer to intervals, such as the interval between a and b in figure 7-2. The shaded area under the curve gives the probability that the random variable will assume any value between a and b. (A continuous probability function is often referred to as a *probability density function*. The term "density" is borrowed from physics, where the word is used in much the same way statisticians use the word "probability.")

The normal distribution curve is symmetrical around its mean (μ). (See figure 7-3. If we folded the paper along the dotted line, the two halves of the curve would coincide.)

Wherever we turn, we are confronted with continuous variables.

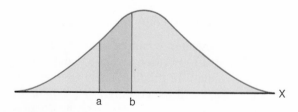

FIGURE 7-2
A probability density

136

Heights and weights, laboratory measurements of error, and automobile speeds, are typical examples of continuous variables. Successive measurements for each of the above examples could conceivably show a continuum of possibilities. The weight measurement of a laboratory animal, for instance, may be 5 pounds, 5.5 pounds, 5.25 pounds, 5.255 pounds, or 5.25525 pounds, depending on the degree of accuracy desired or the measuring device used. Generally, quantities of this nature are rounded to the nearest whole unit or to a few decimal points. Nevertheless, the variables involved are still continuous in nature.

The normal distribution is of great importance because it can be used as an approximation to a number of other distributions; for example, the binomial distribution approaches the normal pattern as *n* becomes large. When *n* is large, it is quite troublesome to work out probabilities for a binomial random variable. Fortunately, probabilities for intervals of values based on the normal function can be used as approximations to binomial probabilities, which saves a considerable amount of work. The normal distribution is an excellent approximation not only to the binomial but also to several other distributions.

Another reason for the importance of the normal distribution lies in the fact that numerous random variables appear to follow a pattern of distribution that is similar to the normal curve. Errors of measurements, the life spans of light bulbs, and intelligence quotients are not only continuous but also normal random variables. Furthermore, the distribution of the sample mean is approximately normal regardless of the shape of the population distribution as long as the sample size is large. This interesting idea is known as the *central limit theorem*. The theorem is of paramount importance in inferential statistics because, as will be seen shortly, the sample size that is considered large enough for applying the normality to the sampling distribution of the mean is rather easy to obtain in most practical situations.

Thus far we have been concerned with only the nature and importance of the normal distribution. In the next section of this chapter, attention will be directed to the methods of determining probabilities for normal

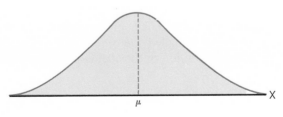

FIGURE 7-3
A normal curve

random variables, and the subsequent section will deal with the applications of these methods. The central limit theorem and the technique of normal approximation to the binomial will be considered in the remainder of the chapter.

7-1
exercises

1. Look at the following random variables, and identify those that are continuous.

    a) X designating the weight of any calf at birth

    b) Y designating the number of telephone calls received at a switchboard during a given period of the day

    c) Z designating the height of any tree

2. What are some of the characteristics of the normal distribution?

3. Why is the normal distribution so important in statistical analysis?

4. Use the dial in figure 7-1 as an example to explain how a continuous random variable can be made a discrete one.

7-2
the standard
normal
distribution

The normal distribution has two parameters, the mean (μ) and the standard deviation (σ). These two parameters determine the particular shape and location of the normal curve. When distributions have the same mean but different standard deviations (see figure 7-4a), the curves will have an identical center; if the standard deviation is small, the apex of the curve will be higher; if the standard deviation is large, the curve will be flatter. On the other hand, if distributions have identical standard deviations but different means (see figure 7-4b), the shapes of the curves remain the same while the curve itself shifts along the X axis. Thus, the normal distribution curve is completely defined by its mean and standard deviation.

To obtain probabilities for particular intervals of values, it is necessary to know the probability distribution, especially the cumulative probability distribution, of the random variable. There are so many normal variables, however, that it is quite impractical to develop a different probability distribution for each one. Fortunately, a probability distribution exists that can be applied to each and every possible normal random variable: it is called the *standard normal distribution*. It is the probability distribution of the standard normal variable Z, which is defined as

$$Z = \frac{X - \mu}{\sigma} \tag{7-1}$$

where X is the ordinary normal random variable, which has the mean μ and the standard deviation σ. Observe that *the Z ratio is the difference*

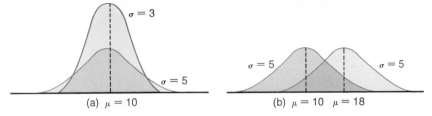

FIGURE 7-4
Normal distributions showing (a) identical
means but different standard deviations,
(b) different means but identical standard deviations

between the observed value of X *and its mean in terms of its standard
deviation.* Thus the value of Z is the same as the number of standard
deviations. Let us use a simple example to illustrate the transformation
of an ordinary normal variable to the standard normal variable for the
purpose of determining probabilities.

EXAMPLE 7-1   It is believed that the scores from a certain achievement
test administered to all high school graduates are normally distributed,
with a mean of 500 points and a standard deviation of 100 points. What
is the probability that a score selected at random will be higher than
700 points?

Let X designate the test score achieved by any student. X has a mean
of 500 and a standard deviation of 100. Applying equation 7-1, we transform X to Z as follows:

$$Z = \frac{700 - 500}{100} = \frac{200}{100} = 2$$

The above question is now changed to: What is the probability that any
score selected at random is higher than two standard deviations *above*
the mean? For the time being, we answer this question by referring to
figure 7-5. The figure shows that the area under the curve within two
standard deviations from the mean is 95.45 percent, and thus the two
tails beyond two standard deviations include only $1 - 0.9545$, or 0.0455,
of the area under the curve. Because of symmetry, one-half of 0.0455, or
0.02275 rounded to 0.0228, which is the area under the curve to the right
of $+2$ standard deviations, is the probability that any score selected at
random will be higher than 700 points.

The Z ratio has made the task of computing probabilities for any normal random variable easy indeed. The expression $Z < 2$ means that the
observations are less than 2σ; and the expression $-1 < Z < 3$ means
that the observation is between -1σ and $+3\sigma$. Since the test scores in

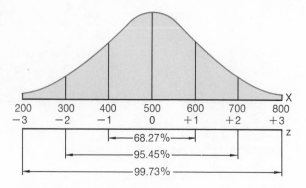

FIGURE 7-5
Transformation of test scores (X)
into Z values with percentages of
the distribution within 1, 2, and 3
standard deviations from the mean
($\mu_X = 500$ and $\sigma_X = 100$)

the above example are normally distributed with the mean and standard deviation known, any test score can be easily changed to a Z value and the desired probabilities determined accordingly. As shown in figure 7-5, 68.27 percent of all potential observations fall within one standard deviation from the mean, 95.45 percent within two standard deviations, and 99.73 percent within three standard deviations.

Since the Z score is expressed as the deviation of the observed value of X from its mean, the center of the Z distribution represents no deviation; that is, the mean of Z is 0. Furthermore, because of the fact that the Z score is in standard deviation units, the standard deviation of Z is 1. For instance, "two standard deviations above 0" is the same as "Z is 2." There is one and only one probability distribution for the standard normal variable Z, and the distribution is completely defined by the mean 0 and the standard deviation 1.

Figure 7-5 cannot be used to determine all probabilities because it provides only a few whole numbers for Z. For decimal values of Z, which are the most frequently encountered, the standard normal distribution table is indispensable.

The cumulative standard normal distribution table is presented in appendix E. The first column and the top row present the z values with two decimal places. (Observe that we have used the capital letter Z to designate the ratio computed from sample data; and the lower-case letter z to denote the value in the table of appendix E.) For instance, the first number in the first column is 0.0 and the first number in the top row is 0.00, making a z value of 0.00. The cumulative probability for $z = 0.00$ is 0.500, the corresponding table entry. This is obviously the case since the mean of z is 0 and the cumulative probability to the mean is one-half of

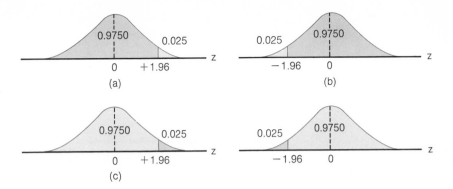

FIGURE 7-6
Standard normal distributions
showing (a) $P(Z \leq 1.96)$, (b)
$P(Z \geq -1.96)$, (c) $P(Z \geq 1.96)$,
and (d) $P(Z \leq -1.96)$

the total area under the normal curve. Appendix E is constructed so that
the table entries are cumulative probabilities for z *beyond* its mean, and
therefore are always larger than 50 percent except for $z = 0$, when the
cumulative probability is exactly 50 percent. To illustrate further the use
of the standard normal table, let us consider another example.

EXAMPLE 7-2 Find the following probabilities: (1) $P(Z \leq 1.96)$;
(2) $P(Z \geq -1.96)$; (3) $P(Z \geq 1.96)$; and (4) $P(Z \leq -1.96)$.

1) $P(Z \leq 1.96)$ In the first column of appendix E, we locate the value 1.9.
   Move horizontally to the right until reaching the column headed 0.06.
   The table entry 0.9750 is the cumulative probability $P(Z \leq 1.96)$.
   Graphically, the probability is represented by the shaded area in (a)
   of figure 7-6. It is the cumulative probability from $-\infty$ to $+1.96$
   for Z.

2) $P(Z \geq -1.96)$ Because of the symmetrical nature of the standard
   normal distribution, the probability cumulated from $-\infty$ to $+1.96$
   must be equal to the probability cumulated from $+\infty$ to -1.96. The
   shaded area in (b) must be identical with the shaded area in (a). (Look
   again at figure 7-6.) That is, $P(Z \geq -1.96) = P(Z \leq 1.96) = 0.9750$.

3) $P(Z \geq 1.96)$ This probability is represented by the shaded area in
   (c) of figure 7-6. Since the total area under the curve is 1, the shaded
   area must be $1 - 0.9750 = 0.0250$. That is, $P(Z \geq 1.96) = 1 - P$
   $(Z \leq 1.96) = 1 - 0.9750 = 0.0250$.

4) $P(Z \leq -1.96)$ Note the shaded area in (d) of figure 7-6. The probability is also 0.0250. It is obvious by now that $P(Z \leq -1.96) =$
   $1 - P(Z \geq -1.96) = 1 - 0.9750 = 0.0250$.

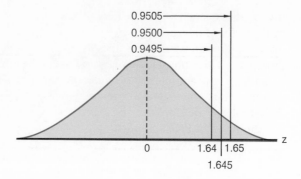

FIGURE 7-7
Determination of $P(Z \le 1.645)$
by interpolation

It is very important to remember that each table entry in appendix E is the probability cumulated from the left to the corresponding *positive* z value, or cumulated from the right to the corresponding *negative z* value. The operation of subtracting the table entry from 1 is not required if we want to determine a probability for Z less than a positive value or greater than a negative value. It is necessary to subtract the table entry from 1 if the desired probability is for Z greater than a positive value or less than a negative value.

Observe that the table entries for z in appendix E contain no more than two decimal places. For values of Z that have more than two digits after the decimal, interpolation will be required. If interpolation is needed, it is usually assumed that the distribution is *uniform* within the interval involved in the interpolation. The following example shows this procedure.

EXAMPLE 7-3 Determine the probability for Z to be less than 1.645, or $P(Z \le 1.645)$. Appendix E provides probabilities for Z less than 1.64 and 1.65, but not 1.645. Interpolation is therefore necessary to solve this problem. From the table we obtain

$$P(Z \le 1.64) = 0.9495 \text{ and } P(Z \le 1.65) = 0.9505$$

the difference of which is 0.0010. Since we want a cumulative probability for a value that is halfway between 1.64 and 1.65 (1.645), we must look for a probability that is halfway between 0.9495 and 0.9505. That is, we must divide 0.0010, the difference between 0.9505 and 0.9495, by 2; and add the result, or 0.0005, to 0.9495 to obtain the desired probability, which is 0.9500. (Figure 7-7 shows the interpolation.)

Inversely, we may obtain the z value by interpolation when a cumulative probability is given. To do this, we first go to the table and find the two entries slightly below and above the probability that is given. A ratio of the difference between this probability and the lower table entry to the difference between the lower and higher table entries is thus obtained. This ratio is then applied to the difference between the corresponding two z values. The desired z value is obtained by adding the fraction thus calculated to the z value corresponding to the lower of the two table entries previously identified. This procedure is illustrated in example 7-5 in the next section.

**7-2
exercises**

1. Find the area under the standard normal curve that lies
   a) to the left of $z = 0$ e) to the right of $z = -1.25$
   b) to the left of $z = 1.35$ f) between $z = 0$ and $z = 1.2$
   c) to the left of $z = -1.56$ g) between $z = 1.0$ and $z = 2.0$
   d) to the right of $z = 1.0$ h) between $z = -1.1$ and $z = 1.6$

2. Find the following probabilities.
   a) $P(Z \leq 2.0)$ e) $P(1.0 \leq Z \leq 1.89)$
   b) $P(Z \leq 1.45)$ f) $P(-1.4 \leq Z \leq 1.75)$
   c) $P(Z \geq -1.76)$ g) $P(-2.15 \leq Z \leq 0.55)$
   d) $P(Z \geq -1.65)$

3. Given that Z is the standard normal score, find
   a) $P(Z \leq 2.326)$ d) $P(-2.326 \leq Z \leq 2.326)$
   b) $P(Z \leq 2.575)$ e) $P(-2.575 \leq Z \leq 2.575)$
   c) $P(Z \leq -2.326)$

4. Given that Z is the standard normal score, find
   a) $P(Z \geq 2.326)$ c) $P(-1.96 \leq Z \leq 1.96)$
   b) $P(Z \geq 2.575)$ d) $P(-1.645 \leq Z \leq 1.645)$

5. Find the value of z for each of the following areas under the standard normal curve.
   a) To the left of z is 0.9949.   e) To the left of z is 0.9412.
   b) To the left of z is 0.9951.   f) To the left of z is 0.0582.
   c) To the right of z is 0.01.   g) To the right of z is 0.2810.
   d) To the right of z is 0.005.   h) To the right of z is 0.0228.

6. Find the value of z for each of the following areas under the standard normal curve.
   a) Between $-z$ and z is 0.9000.
   b) Between $-z$ and z is 0.9800.
   c) Between $-z$ and z is 0.9950.

7. A random variable (X) is normally distributed, with a mean of 100 and a standard deviation of 15. Find the probability that

    a) X is less than 80.5   d) X is between 91 and 109
    b) X is greater than 116.5   e) X is between 85 and 97
    c) X is less than 112

8. A random variable (X) is normally distributed, with a mean of 70 and a standard deviation of 5. Find the probability that

    a) X is greater than 66   c) X is greater than 71 *and* less than 75
    b) X is greater than 63   d) X is greater than 79 *or* less than 61

9. A normal random variable (X) has a mean of 500. If the probability that X is greater than 600 is 0.0228, what is the value of the standard deviation?

10. A normal random variable (X) has a standard deviation of 10. If the probability that X is less than 75 is 0.9938, what is the value of the mean?

7-3
some
applications
of the
Z ratio

The cumulative distribution table is all we need to determine probabilities for any random variable that is normally or approximately normally distributed. Let us now consider a few concrete examples to illustrate the application of the standard normal Z value.

EXAMPLE 7-4   The IQ's of elementary school children are believed to be normally distributed, with a mean of 100 and a variance of 225. If a pupil is selected at random, what is the probability that he will have an IQ (a) greater than 136; (b) less than 73; or (c) between 85 and 136?

a) Let the IQ of any pupil be designated by X. The probability that any pupil has an IQ greater than 136, or $P(X > 136)$, is obtained by first changing the variable X to the standard normal variable Z. Since

$$\frac{X - \mu}{\sigma} = \frac{136 - 100}{\sqrt{225}} = \frac{36}{15} = 2.4,$$

we have

$$P(X > 136) = P(Z > 2.4)$$

From appendix E, we find that $P(Z \leq 2.4) = 0.9918$. Therefore,

$P(Z > 2.4) = 1 - P(Z \leq 2.4) = 1 - 0.9918 = 0.0082$ (See figure 7-8a.)

b) The probability that any pupil has an IQ less than 73 is

$$P(X < 73) = P\left(\frac{X - \mu}{\sigma} < \frac{73 - 100}{15}\right) = P(Z < -1.8)$$
$$= 1 - P(Z \leq 1.8) = 1 - 0.9641 = 0.0359 \text{ (See figure 7-8b.)}$$

c) The probability that any pupil has an IQ between 85 and 136 is expressed as follows:

$$P(85 \leq X \leq 136) = P\left(\frac{85 - 100}{15} \leq \frac{X - \mu}{\sigma} \leq \frac{136 - 100}{15}\right)$$
$$= P(-1 \leq Z \leq 2.4)$$

which, as shown in figure 7-8c, is equal to

$$P(Z \leq 2.4) - P(Z \leq -1) = P(Z \leq 2.4) - [1 - P(Z \leq 1)]$$
$$= 0.9918 - (1 - 0.8413) = 0.9918 - 0.1587 = 0.8331$$

EXAMPLE 7-5   Let X be the service life of a certain brand of automobile battery; it is known that X is normally distributed with a mean of 1000 days and a standard deviation of 100 days. If the manufacturer does not want to replace more than 10 percent of the batteries sold, how long should the length of the guarantee be?

To answer this question, we must keep in mind that the manufacturer is not concerned about the service lives of the batteries that fall in the

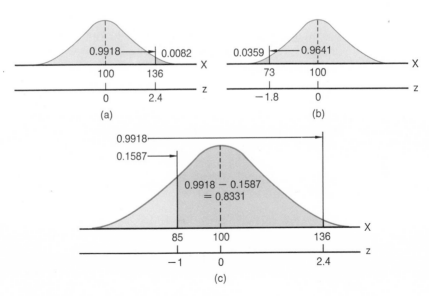

FIGURE 7-8
Determination of (a) $P(X > 136)$,
(b) $P(X < 73)$, and
(c) $P(85 \leq X \leq 136)$

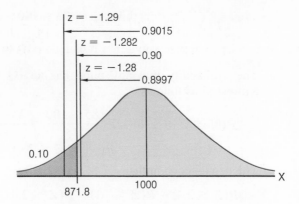

FIGURE 7-9
Distribution of service lives of
automobile batteries ($\mu = 1000$
and $\sigma = 100$)

upper end of the distribution, but only about the 10 percent in the lower end. In order to determine the X value that cuts off the lower 10 percent of the distribution, we need, first of all, to find the value of the standard normal variable that cuts off the lower 10 percent of the standard normal distribution.

From appendix E, we find that corresponding to the cumulative probabilities 0.8997 and 0.9015, within which the number 0.9000 falls, are the z values 1.28 and 1.29. The ratio of the difference between 0.9000 and 0.8997 (or 0.0003) to the difference between 0.9015 and 0.8997 (or 0.0018) is $0.0003/0.0018 = \frac{3}{18} \cong 0.2$. Applying this ratio to the difference between the two z values, or 0.01, we obtain the fraction 0.002, which is added to 1.28 to arrive at a z value of 1.282. Since the percentage 0.9000 refers to the upper part of the distribution, the z values should be negative values as shown in figure 7-9. Thus, by interpolation, we find that the z value that cuts off the lower 10 percent (or the upper 90 percent) of the standard normal distribution is -1.282. Thus,

$$\frac{X - 1000}{100} = -1.282$$

Solving for X, we get
$$X - 1000 = -1.282(100) = -128.2$$
Hence,
$$X = 871.8$$

Accordingly, the manufacturer should guarantee his batteries for 872 days. In so doing, he can expect to replace no more than 10 percent of the batteries.

EXAMPLE 7-6   The lengths of a certain variety of cockroaches at maturity are believed to have a mean of 1.5 inches and a standard deviation of 0.2 inches. On the assumption that the distribution of the lengths of these cockroaches is normal, find the probability that any cockroach found at random is (1) longer than 2 inches; (2) shorter than 1.25 inches; and (3) between 1.4 and 1.6 inches long.

1) Longer than 2 inches:
$$P\left(Z > \frac{2.0 - 1.5}{0.2}\right) = P(Z > 2.5) = 1 - P(Z \le 2.5)$$
$$= 1 - 0.9938 = 0.0062$$

2) Shorter than 1.25 inches:
$$P\left(Z < \frac{1.25 - 1.5}{0.2}\right) = P(Z < -1.25) = 1 - P(Z \le 1.25)$$
$$= 1 - 0.8944 = 0.1056$$

3) Between 1.4 and 1.6 inches long:
$$P\left(\frac{1.4 - 1.5}{0.2} \le Z \le \frac{1.6 - 1.5}{0.2}\right) = P(-0.5 \le Z \le 0.5)$$
$$= P(Z \le 0.5) - P(Z \le -0.5) = P(Z \le 0.5) - [1 - P(Z \le 0.5)]$$
$$= 0.6915 - (1 - 0.6915) = 0.6915 - 0.3085$$
$$= 0.3830$$

EXAMPLE 7-7   Suppose that radar is used to check the speed of automobiles passing a particular checkpoint on the San Diego Freeway at certain periods of the day. It is assumed that the hourly speed of automobiles, or X, passing this checkpoint at the specific period of the day is normally distributed with a mean of 60 miles and a standard deviation of 5 miles. Determine the following: (1) the percentage of cars traveling below 50 miles per hour; (2) the percentage of cars exceeding the legal limit of 65 miles per hour; and (3) the percentage of cars traveling between 60 and 77.45 miles per hour.

1) The percentage of cars with an hourly speed less than 50 miles:
$$P(X < 50) = P\left(\frac{X - \mu}{\sigma} < \frac{50 - 60}{5}\right) = P(Z < -2)$$
$$= 1 - 0.9772 = 0.0228$$

2) The percentage of cars with an hourly speed more than 65 miles:
$$P(X > 65) = P\left(\frac{X - \mu}{\sigma} > \frac{65 - 60}{5}\right) = P(Z > 1)$$
$$= 1 - P(Z \le 1) = 1 - 0.8413 = 0.1587$$

3) The percentage of cars with an hourly speed between 60 and 77.45 miles:

$$P(60 \leq X \leq 77.45) = P\left(\frac{60-60}{5} \leq Z \leq \frac{77.45-60}{5}\right)$$
$$= P(0 \leq Z \leq 3.49) = P(Z \leq 3.49) - P(Z \leq 0) = 0.9998 - 0.5000$$
$$= 0.4998$$

7-3 exercises

1. Let X be the IQ of any college student. Assume that X is normally distributed with a mean of 107 and a variance of 225. If a college student is selected at random, what is the probability that the student has an IQ
   a) greater than 125?   e) between 104 and 140?
   b) greater than 131?   f) between 77 and 92?
   c) less than 98?   g) between 110 and 131?
   d) less than 110?

2. A nationally administered achievement test is known to have a mean score of 500 points and a standard deviation of 100 points. If we assume that the test scores are normally distributed, what is the probability that a randomly selected student will score
   a) over 750 points?   c) between 400 and 600 points?
   b) below 300 points?   d) between 550 and 750 points?

3. An English instructor has found that the length of time required by students to finish a final examination is normally distributed with a mean of 110 minutes and a standard deviation of 10 minutes.
   a) What is the probability that a randomly selected English student will complete the examination in less than two hours?
   b) What is the probability that a randomly selected English student will complete the examination in 125 minutes or longer?
   c) If there are 50 students in the class, how many of them will complete the examination within one hour and 50 minutes?

4. Suppose that the weight gain for chickens fed on a certain ration over a period of one month has a mean of 80 grams and a standard deviation of eight grams. If 10,000 chickens are fed the ration for a month, how many of these chickens will gain 76 grams or more? Assume that the weight gain is normally distributed.

5. The hourly wages of workers in a particular trade are believed to be normally distributed with a mean of $5.50 and a standard deviation of $0.50. Suppose that a worker in that trade is randomly selected. Find the probability that he earns
   a) more than $7.00 per hour
   b) less than $4.75 per hour
   c) between $4.90 and $6.45 per hour

6. In a large corporation, the weekly sales volume per salesman has a mean of $10,000 and a variance of $250,000. Assume that the sales volume has a normal distribution. Find the following.
   a) If a salesman is selected at random, what is the probability that his sales volume is between $9500 and $11,000 per week?
   b) If a sales commission of 5 percent is paid to each salesman, what percentage of salesmen will earn more than $550 per week?

7. The hourly output of production workers in a large factory is assumed to be normally distributed with a mean of 240 units and a standard deviation of 20 units. If 10,000 production workers are employed in the factory, find the following.
   a) How many workers have an output of more than 250 units per hour?
   b) If any worker who turns out less than 200 units per hour must receive further training, how many will receive training?

8. Tomato output per plant has a mean of 12 pounds and a standard deviation of two pounds. It is believed that the tomato output per plant has a normal distribution. Find the following.
   a) If a tomato plant is selected at random, what is the probability that it will yield an output of 15 pounds or more?
   b) If there are 10,000 tomato plants on a farm, how many plants will yield more than 11 pounds each?

9. Suppose that the average length of patient confinement in a hospital is 10 days, and that the standard deviation of the length of patient confinement is two days. On the assumption that such lengths are normally distributed, find the following.
   a) What is the probability that the next patient to be admitted will stay longer than nine days?
   b) If 100 patients are admitted today, how many will still be in the hospital two weeks later?

10. Assume that the sugar content per orange is normally distributed, with a mean of 0.5 ounces and a standard deviation of 0.05 ounces. What is the probability that an orange selected at random will have a sugar content between 0.54 and 0.615 ounces?

7-4 the central limit theorem

As we learned in section 6-4, the population distribution of a random variable is closely related to the sampling distribution of the sample mean. A legitimate question may now be raised: if X is normally distributed, what is the distribution pattern for $\overline{X}$? In this section we shall try to answer both this question and one that is even more important:

what is the distribution pattern of $\overline{X}$ when X is not normally distributed?

Briefly, if X is a normal random variable with a mean μ and a standard deviation σ, then the mean of n observations taken from the population of X will also be a normal random variable with the mean μ and the standard deviation $\sigma/\sqrt{n}$. This is the case regardless of the sample size. Accordingly, whenever X is distributed normally, the sample mean will have a normal distribution. In determining probabilities for $\overline{X}$, the standard normal Z value is computed as follows:

$$Z = \frac{\overline{X} - \mu}{\sigma/\sqrt{n}} \qquad (7\text{-}2)$$

where μ is the population mean and σ is the population standard deviation. Recall that in Section 6-4 the random variable $\overline{X}$ has a mean μ and a standard deviation $\sigma/\sqrt{n}$, called standard error. Thus the Z score here is expressed as the difference between the sample mean and the population mean in terms of the standard error.

EXAMPLE 7-8 Suppose that the test scores on a certain standard examination are normally distributed with a mean of 60 and a variance of 256. A random sample of 16 scores is selected, and the sample mean computed. What is the probability that $\overline{X}$ will be greater than 70?

The desired probability here is $P(\overline{X} > 70)$. With a standard deviation $\sigma = \sqrt{256} = 16$, and sample size $n = 16$, the probability is:

$$P(\overline{X} > 70) = P\left(\frac{\overline{X} - \mu}{\dfrac{\sigma}{\sqrt{n}}} > \frac{70 - 60}{\dfrac{16}{\sqrt{16}}}\right) = P\left(Z > \frac{10}{4}\right)$$

$$= 1 - P(Z \leq 2.5) = 1 - 0.9938 = 0.0062$$

The distributions of many random variables, however, do not even resemble the normal pattern. How is the sample mean distributed if the random variable is not normal? A very important theorem has been developed just to deal with this problem. It is called the *central limit theorem,* which is stated briefly below.

> If X is any random variable with a mean μ and standard deviation σ, the sampling distribution of a sample statistic such as the mean $\overline{X}$ will be approximately normal with a mean μ and standard deviation $\sigma/\sqrt{n}$ regardless of the shape of the distribution of X as long as the sample size is sufficiently large.

According to this theorem, the ratio computed from a sufficiently large sample by equation 7-2 will also be approximately normal with a mean 0 and a standard deviation 1 even if the random variable X is not normally distributed.

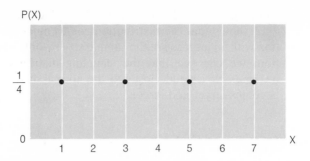

FIGURE 7-10
Population distribution of X with
the set of possible values $\{1, 3, 5, 7\}$ (source: table 6-3)

How large must the sample size be in order to be considered "sufficiently" large? It must be pointed out that any value we may use is arbitrary, and that the larger the value of n, the better the approximation will be. In general, however, the tendency for the distribution of $\overline{X}$ to become normal is quite strong even though the samples are of only moderate size. In most instances, a sample of 30 observations is considered large enough to permit the application of the normal probability distribution to $\overline{X}$. Since 30 is a number that is not difficult to obtain in most sampling situations, this theorem is of paramount practical significance. Thus, whenever n is equal to 30 or more, we can use equation 7-2 to compute the Z value for determining the desired probability for $\overline{X}$ regardless of the shape of the population distribution of X.

We shall now use a simple example to demonstrate the tendency for the distribution of $\overline{X}$ to become closer to the normal pattern as the sample size becomes larger.

In table 6-3 (see page 125) the random variable X with a set of values $\{1, 3, 5, 7\}$ is uniformly distributed. In figure 7-10, we can see that the graphic representation of this distribution does not resemble the bell-shaped normal curve. Suppose that a random sample of two observations is taken *with replacement* from this population, and the two observations yield a sample mean $\overline{X}$. The probability distribution of $\overline{X}$ is presented in table 6-5. Its graphic representation (see figure 7-11) is much closer to the normal distribution curve even though the sample size is only two.

Let us now increase the sample size from two to three, and see how this increase will affect the shape of the distribution. When $n = 3$, there will be 4^3, or 64, different possible sequences (or samples). (See table 7-1.) Out of these 64 different samples, there are 10 *distinct* possible sample means. The probability distribution of the 10 distinct possible means is shown in table 7-2 and in figure 7-12. A comparison of figures 7-11 and 7-12 shows that the distribution comes much closer to the normal function with an increase of the sample size by only one. From this demonstration, we can see that the sampling distribution of the sample mean becomes closer and closer to the normal distribution as the sample size becomes larger and larger.

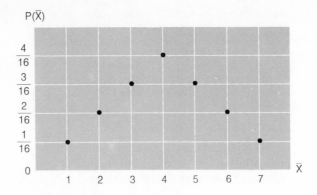

FIGURE 7-11
Sampling distribution of $\overline{X}$ when X
has the set of possible values
$\{1, 3, 5, 7\}$ and $n = 2$ (source:
table 6-5)

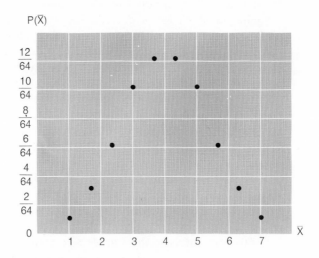

FIGURE 7-12
Sampling distribution of $\overline{X}$ when X
has the set of possible values
$\{1, 3, 5, 7\}$ and $n = 3$

Table 7-1   Possible samples and sample means for X with the set of values $\{1, 3, 5, 7\}$ and $n = 3$

Possible samples $X_1\ X_2\ X_3$	Sample total	Sample mean $(\overline{X})$	Number of samples	Possible samples $X_1\ X_2\ X_3$	Sample total	Sample mean $(\overline{X})$	Number of samples
1 1 1	3	1	1	1 5 7			
				1 7 5			
1 1 3				5 1 7			
1 3 1	5	$1\frac{2}{3}$	3	5 7 1			
3 1 1				7 1 5			
				7 5 1	13	$4\frac{1}{3}$	12
1 3 3				3 5 5			
3 1 3				5 3 5			
3 3 1	7	$2\frac{1}{3}$	6	5 5 3			
1 1 5				3 3 7			
1 5 1				3 7 3			
5 1 1				7 3 3			
3 3 3				1 7 7			
1 3 5				7 1 7			
1 5 3				7 7 1			
3 1 5				3 5 7			
3 5 1	9	3	10	3 7 5	15	5	10
5 1 3				5 3 7			
5 3 1				5 7 3			
1 1 7				7 3 5			
1 7 1				7 5 3			
7 1 1				5 5 5			
1 5 5				3 7 7			
5 1 5				7 3 7			
5 5 1				7 7 3	17	$5\frac{2}{3}$	6
1 7 3				5 7 5			
1 3 7				7 5 5			
3 1 7	11	$3\frac{2}{3}$	12	5 5 7			
3 7 1							
7 1 3				5 7 7			
7 3 1				7 5 7	19	$6\frac{1}{3}$	3
3 3 5				7 7 5			
3 5 3							
5 3 3				7 7 7	21	7	1
			32				32

The total number of samples is $32 + 32 = 64$

Table 7-2 Sampling distribution of the mean for X with the set of values $\{1, 3, 5, 7\}$ and $n = 3$

Sample mean $(\overline{X})$	Probability $P(\overline{X})$
1	$\frac{1}{64}$
$1\frac{2}{3}$	$\frac{3}{64}$
$2\frac{1}{3}$	$\frac{6}{64}$
3	$\frac{10}{64}$
$3\frac{2}{3}$	$\frac{12}{64}$
$4\frac{1}{3}$	$\frac{12}{64}$
5	$\frac{10}{64}$
$5\frac{2}{3}$	$\frac{6}{64}$
$6\frac{1}{3}$	$\frac{3}{64}$
7	$\frac{1}{64}$
Total	$\frac{64}{64} = 1$

7-4 exercises

1. A standard achievement test has a mean of 500 points and a standard deviation of 100 points. The test scores are believed to have a normal distribution. What is the probability that any sample mean computed from 25 test scores will have a value greater than 555?

2. Let X be the IQ of any college student. It is believed that X has a normal distribution with a mean of 107 and a standard deviation of 15. If a sample of 25 college students is selected at random, find the probability that the sample mean is

   a) greater than 110
   b) below 102.5
   c) between 98 and 111.5

3. Suppose that the age distribution for the employees of a large corporation has a mean of 35 years and a standard deviation of six years. It is believed that the distribution is not normal. If a sample of 36 employees is selected at random, and their average age is computed, what is the probability that the average age of the sample is

   a) over 37.5 years?　　c) between 34.25 and 34.75 years?
   b) below 33 years?　　d) between 36 and 37.75 years?

4. Suppose that annual family income in the United States has a mean of $10,000 and a standard deviation of $3000. It is known that the annual income per family is not normally distributed. A sample of 100 families

is selected at random, and the average income of these 100 families, or $\overline{X}$, is computed. Find the following probabilities.

a) $P(\overline{X} \leq \$11,200)$ c) $P(\$10,150 \leq \overline{X} \leq \$13,000)$
b) $P(\overline{X} \geq \$10,450)$ d) $P(\$7000 \leq \overline{X} \leq \$9,400)$

5. The distribution of the 10 random digits 0, 1, 2, . . . , and 9 is referred to as *uniform*, since the probability for each digit to appear is 0.1. Suppose that a random sample of 100 digits is selected either by the use of the random digit table or by the goldfish-bowl method with replacement, and a sample mean is computed. Find the following probabilities.

a) $P(\overline{X} \leq 4.84)$ c) $P(4.18 \leq \overline{X} \leq 4.87)$
b) $P(\overline{X} \geq 4.79)$ d) $P(4.00 \leq \overline{X} \leq 4.90)$

6. Refer to problem 8, exercises 6-3. If a sample of 400 observations is taken, and a sample mean obtained, find the following probabilities.

a) $P(\overline{X} \leq 3.25)$ c) $P(2.75 \leq \overline{X} \leq 3.375)$
b) $P(\overline{X} \geq 2.875)$ d) $P(2.625 \leq \overline{X} \leq 3.375)$

7. Suppose that a random variable has the following uniform probability distributions.

$$X = 3 \quad 6 \quad 9$$
$$P(X) = \tfrac{1}{3} \quad \tfrac{1}{3} \quad \tfrac{1}{3}$$

a) If a random sample of two observations for X is taken with replacement, and a sample mean is obtained, find the probability distribution of $\overline{X}$. Represent the distribution graphically.
b) Increase the sample size to three and find the probability distribution of $\overline{X}$.
c) After comparing (a) and (b) above, what conclusion can you reach with regard to the probability distribution of the sample mean?

8. Refer to problem 7 above. A sample of 36 observations for X is taken and the sample mean $\overline{X}$ computed. Find the following probabilities.

a) $P(\overline{X} \geq 7.5)$ c) $P(5.6 \leq \overline{X} \leq 6.45)$
b) $P(\overline{X} \leq 4.75)$ d) $P(6.75 \leq \overline{X} \leq 7.75)$

7-5 normal approximation to the binomial distribution

The distribution of a continuous random variable can be represented by a smooth curve. This is not true of the distribution of a binomial random variable, which takes on the value of a whole number. The binomial distribution, however, approaches the shape of continuous normal probability distribution as the size of the sample increases.

Figure 7-13 shows three histograms: $n = 4$, $n = 10$, and $n = 30$. All three graphs are based on $p = 0.5$. The rectangles represent probabilities

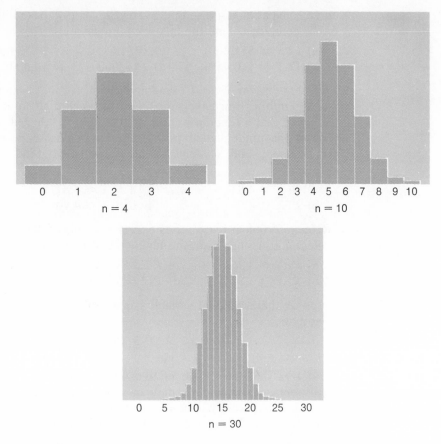

FIGURE 7-13
Binomial probability distributions
for $p = 0.5$ and $n = 4$, 10, and 30

for the various values of X in each of the three distributions. As the value of n increases, the rectangles become vertical lines with no space in between, and the distribution becomes a bell-shaped normal curve like the one shown in figure 7-3. Accordingly, the normal distribution is frequently used as an approximation to the binomial as long as the sample size is sufficiently large. Normal approximation to the binomial is, in fact, an application of the central limit theorem discussed in the previous section.

In chapter 3, we defined the binomial random variable X as *the number of successes in* n *trials;* as such it is a sample characteristic, or statistic, and its probability distribution is a sampling distribution. As we have seen, when n increases, the probability distribution of X becomes closer and closer to the normal pattern. The tendency for the binomial to approach the normal distribution is more rapid and pronounced when the probability of success, or p, is close to 0.5, as is the case in figure 7-13. Nevertheless, this tendency still exists when p takes on other values. In fact, whenever n is sufficiently large, we can apply the normal distribution as an approximation to the binomial regardless of the value of p.

For satisfactory approximations, however, the value of n should be greater when p is farther away from 0.5 than when $p = 0.5$. A good rule of thumb for using the normal approximation is that both np and $n(1 - p)$ are greater than 5.

Since the normal distribution has two parameters, the mean (μ) and the variance (σ^2), it is necessary to identify a mean and variance for the binomial random variable X whenever normal approximation is intended. The formulas for μ and σ^2 of X can be derived on the basis of the mean and variance of the Bernoulli random variable (see section 4-6). The required quantities for the mean and variance of the Bernoulli random variable W are expressed in table 7-3.

Table 7-3   Quantities for the mean and variance of the Bernoulli random variable W

W	$P(W)$	W^2	$W \cdot P(W)$	$W^2 \cdot P(W)$
0	$1 - p$	0	0	0
1	p	1	p	p
Total	1		$\Sigma W \cdot P(W) = p$	$\Sigma W^2 \cdot P(W) = p$

Using equations 6-3 and 6-5, we obtain the mean and variance of W as follows:

$$E(W) = \Sigma W \cdot P(W) = 0(1 - p) + 1(p) = 0 + p = p \qquad (7\text{-}3)$$

and

$$\begin{aligned} \sigma_W^2 &= E(W^2) - [E(W)]^2 = \Sigma W^2 \cdot P(W) - p^2 \\ &= 0^2(1 - p) + 1^2(p) - p^2 \\ &= p - p^2 = p(1 - p) \end{aligned} \qquad (7\text{-}4)$$

That is, the mean of the Bernoulli random variable W is equal to p, the probability of success, and the variance of W is the product of p and $(1 - p)$, or pq, where q is the probability of failure.

The mean and variance of the binomial random variable X can be expressed in terms of the mean and variance of W, respectively. Since $X = W_1 + W_2 + \ldots + W_n$, the mean of X is

$$\begin{aligned} E(X) &= E(W_1 + W_2 + \ldots + W_n) \\ &= E(W_1) + E(W_2) + \ldots + E(W_n) \\ &= p + p + \ldots + p = np \end{aligned} \qquad (7\text{-}5)$$

For deriving the variance of X on the basis of W, we need an important theorem which can be stated as follows:

The variance of the sum of a set of *independent* variables is equal to the sum of the variances of the variables.

The student should keep this theorem in mind, for it will be referred to later in the text. According to this theorem, the variance of the sum of the W's is equal to the sum of the variances of the W's. Accordingly,

$$\sigma_X^2 = \sigma_{(W_1 + W_2 + \ldots + W_n)}^2 = \sigma_{W_1}^2 + \sigma_{W_2}^2 + \ldots + \sigma_{W_n}^2$$
$$= pq + pq + \ldots + pq \text{ (by equation 7-4)} \qquad (7\text{-}6)$$
$$= npq$$

EXAMPLE 7-9   Find the mean and variance of X, the number of heads obtained in the toss of four balanced coins by the use of equations 7-5 and 7-6.

In this problem we have $n = 4$, $p = 0.5$, and $q = 0.5$. Accordingly, the mean of X is

$$E(X) = np = 4(0.5) = 2$$

Check the above with the answer in example 6-2. The variance of X also checks with the answer in example 6-6:

$$\sigma_X^2 = npq = 4(0.5)(0.5) = 1$$

For a binomial random variable, it is correct to use the expression $P(X = c)$ when c is a whole number. When we want to approximate the binomial distribution with a normal curve, it is necessary to use an interval of values for X because the probability of a continuous random variable equal to any single value is zero. This implies that a *correction for continuity* is required in a normal approximation to the binomial. It is done by representing any whole number c by the interval from $(c - \frac{1}{2})$ to $(c + \frac{1}{2})$. For instance, 7 is represented by the interval from 6.5 to 7.5; and the three numbers 6, 7, and 8 are represented by the interval from 5.5 to 8.5. The following example illustrates the procedure to obtain a normal approximation with the required correction for continuity.

EXAMPLE 7-10   It has been found that a serum is 50 percent effective in curing a certain disease; that is, 50 percent of those who contract the disease are cured by injections of the serum. Using the normal approximation, find the probability that in a random sample of 15 patients who have the disease, 4 to 6 patients will be cured by the serum.

Figure 7-14 presents both the binomial probabilities represented by the rectangles for $n = 15$ and $p = 0.5$ and the normal curve used as an approximation to the binomial distribution. The probability of X equal to 4, 5, or 6 is represented by the three rectangles lying over $X = 4, 5,$

158

Probability

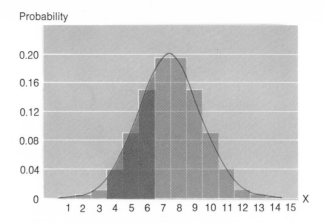

FIGURE 7-14
Normal approximation to the binomial distribution for $n = 15$,
$p = 0.5$

and 6, which, from appendix A, is found to be $0.0417 + 0.0916 + 0.1527 = 0.2860$. The shaded area represents the probability obtained by a normal approximation.

In this example, the mean of X is $np = 15(0.5) = 7.5$ and the standard deviation is $\sqrt{npq} = \sqrt{15(0.5)(0.5)} = \sqrt{3.75} = 1.9365$. Thus,

$$
\begin{aligned}
P(3.5 \leq X \leq 6.5) &= P\left(\frac{3.5 - 7.5}{1.9365} \leq Z \leq \frac{6.5 - 7.5}{1.9365}\right) \\
&= P(-2.066 \leq Z \leq -0.516) \\
&= P(Z \leq -0.516) - P(Z \leq -2.066) \\
&= 1 - P(Z \leq 0.516) - [1 - P(Z \leq 2.066)] \\
&= -P(Z \leq 0.516) + P(Z \leq 2.066) \\
&= -0.6972 + 0.9806 = 0.2834
\end{aligned}
$$

which is quite close to the exact probability 0.2860 obtained above by using the binomial probability distribution.

7-5
exercises

(Except for problems 7 and 8, correction for continuity is not required in this exercise.)

1. Suppose that a certain medicine is 80 percent effective in curing a particular kind of disease. That is, out of every 100 patients who have the disease and receive the medicine, 80 are expected to recover. Let X be the number of patients in a random sample of 100 who will recover after treatment. Find the following probabilities by normal approximation.

a) More than 80 will recover, or $P(X > 80)$
b) $P(80 < X < 90)$
c) $P(70 < X < 75)$

2. It has been found that 10 percent of all high school graduates enroll in a four-year college or university. If we consider a certain high school graduating class of 400 students as a random sample, then

    a) what is the probability that 50 or more of the graduates will enroll in a four-year college or university?
    b) what is the probability that 35 or fewer graduates will enroll in a four-year college or university?

3. Suppose that 50 percent of all high school graduates will attend community colleges. If the graduating class of 100 students of a certain high school is taken as a random sample, what is the probability that

    a) at least 45 graduates will attend community college?
    b) no more than 54 graduates will attend community colleges?

4. It is assumed that all chickens are equally divided between hens and roosters. If 10,000 eggs are hatched, what is the probability that

    a) at least 5100 chickens will be hens?
    b) no more than 4975 chickens will be hens?

5. Ten percent of all the units produced by a certain production process are defective. What is the probability that in a sample of 2500 units produced by the process

    a) at least 262 units will be defective?
    b) at most 241 units will be defective?
    c) between 244 and 268 will be defective?

6. Sixty percent of all registered voters are Democrats and 40 percent are Republicans. In a sample of 150 randomly selected voters, what is the probability that at least one-half of them will be Republicans?

7. A balanced coin is tossed 10 times. Find the probability that either six, seven, or eight heads will occur by the

    a) binomial distribution
    b) normal approximation method with correction for continuity

8. In a large corporation 20 percent of the employees are members of minority races. In a random sample of 30 employees, what is the probability that either three, four, or five of them will belong to minority groups?

    a) Compute the probability by the method of normal approximation with correction for continuity.
    b) Compute the probability by the binomial distribution.

statistical inference

**8-1
introduction**

Now that we have acquired the basic concepts of probability and sampling distribution, we are ready to discuss the methods and principles of *statistical inference*. Statistical inference is usually discussed under two headings: hypothesis testing and estimation. *Hypothesis testing* refers to the process involved in accepting or rejecting statements about population parameters, whereas *estimation* deals with estimating the values of population parameters.

A *statistical hypothesis* is a tentative statement about the value of a population parameter, or parameters. We call such a statement tentative because we are not certain of the true values of the parameters in question. Business executives, scientists, military strategists, politicians, and people in every walk of life make decisions based on statistical hypotheses. Some simply guess the values of parameters, while others make inferences about them on the basis of incomplete sample data. Hypothesis testing can tell us whether a tentative statement is supported or repudiated by evidence from the sample.

In the next four sections we shall present the basic principles and methods of hypothesis testing concerning the population mean, the difference between two means, and the population proportion. Statistical estimation will be examined in the final section of the chapter.

**8-2
essential
elements of
hypothesis
testing**

formulation of hypotheses

There are two types of hypotheses. One, called the *null hypothesis*, is formed primarily to determine whether it can be rejected. Such a hypothesis is conventionally designated by the symbol H_o, with the letter "H" suggesting hypothesis and the subscript $_0$ signifying "null" or "nought." For instance, we may wish to determine whether the mean IQ of the pupils of a certain school district is different from 100. Because we must presume that it is 100, we have a null hypothesis $H_0: \mu = 100$.

The reasoning involved in determining whether a null hypothesis can be rejected bears a remarkable resemblance to that involved in a criminal court proceeding. The court presumes that the accused is innocent, or *not guilty,* until his guilt is established. The prosecution collects and presents evidence in an attempt to repudiate the not-guilty presumption, the "null hypothesis." If the "null hypothesis" is not rejected, there will be no conviction, and the accused will be free. This does not necessarily mean that the accused is really not guilty. The truth may be that he is guilty, but the available evidence is not sufficient to warrant a conviction.

In the example mentioned above, the mean IQ is "the accused." The null hypothesis to be tested is that the mean IQ is *not different* from 100. The researcher plays the role of the prosecutor; he collects evidence by drawing a sample from the relevant population. An attempt is made to use the evidence contained in the sample to reject the null hypothesis and to support his belief that the mean IQ is different from 100.

The other type of hypothesis is called the *alternative hypothesis,* commonly designated by the symbol H_1. The alternative hypothesis is accepted when the null hypothesis is rejected. In the test to determine whether the mean IQ of the pupils is different from 100, the alternative hypothesis is $H_1: \mu \neq 100$. That is, the mean IQ is not equal to 100. A test of an alternative hypothesis that does not specify the direction of the difference (the mean IQ could be either larger or smaller than 100) is called a *two-tailed test.* Obviously, an alternative hypothesis may also involve a one-sided alternative; a test for such a hypothesis is called a *one-tailed test.* If the alternative hypothesis is $H_1: \mu > 100$, the null hypothesis is rejected only if the collected evidence indicates a sufficiently large value for μ.

level of significance

We learned earlier that samples drawn from the same population are seldom identical. Differences between samples are due to chance, to the accidental factors that determine the selection of observations in any sample. Suppose that we set up two hypotheses, say, $H_0: \mu = 100$ and $H_1: \mu \neq 100$. To test our hypotheses, we draw a sample and calculate its mean. Because of the random nature of sampling, our sample mean may (though it is not very likely) deviate so far from the expected mean (100) that we will reject the null hypothesis even though it is in fact true. How much, then, does a sample mean have to deviate from the expected mean before we can feel justified in rejecting the null hypothesis? In other words, when can a sample mean be considered "significantly deviant"?

The answer to this question depends on the level of error we are willing to tolerate—that is, the probability that our sample has yielded such a significantly different value from the hypothesized value because of chance factors. If the null hypothesis is rejected only when a sample mean as extreme as the one we have obtained would occur no more than five in 100 times, then the level of error will be 0.05. (Other values such as

0.01 and 0.02 are also used.) Accordingly, 0.05 is called the *level of significance*.

The level of significance, then, is the probability of rejecting a true null hypothesis, or making what is called a *type I error*; this probability is commonly designated by the Greek letter α (alpha). The value of α affects the decision whether any difference between the observed sample value and the hypothesized population value will be considered significant, or too extreme to be attributable to chance. The term refers to the state of being *statistically significant*, and is not used in the sense of "importance" or "meaningfulness."

The selection of the value of α is an arbitrary matter; it depends on how much risk we are willing to take in erroneously rejecting a true null hypothesis. The greater the risk we are willing to take, the larger the value of α will be. Once the value of α is selected, a *decision rule* can be set up to specify the conditions under which the null hypothesis will be rejected. As will be seen in the next section, the setting up of the decision rule is an essential step in hypothesis testing.

Since α is the probability of making a type I error, why shouldn't we choose a value that is as small as possible? In answering this question, we have to realize that as α decreases, the probability that we will accept a false null hypothesis increases. The error of not rejecting the null hypothesis when it is false is called a *type II error*; the probability of making a type II error is commonly designated by β (the Greek letter beta). The two types of error are inversely related; as α decreases β increases, and vice versa.

test statistic

With α and a decision rule specified, we can calculate the test statistic. A test statistic is a value based on the sample data that is computed for the purpose of determining whether the null hypothesis will be rejected. In the test to see whether the mean IQ is different from μ_0, the parameter value under the null hypothesis, we first compute the sample mean $\overline{X}$ and then convert the sample mean into the standard normal Z value by equation 7-2.

$$Z = \frac{\overline{X} - \mu_0}{\sigma/\sqrt{n}}$$

Here the capital letter Z designates a test statistic.

Test statistic values are divided into two categories: one is called the *rejection region*, and the other the *acceptance* or *nonrejection region*. The rejection region has a total probability equal to α if the null hypothesis is true, and a probability equal to $1 - \beta$ if it is false. Once a sample is taken, the value of the test statistic is then obtained. If the test statistic assumes a value in the rejection region, the null hypothesis is rejected;

otherwise, it is accepted. When the null hypothesis is true, the probability of making the erroneous decision by rejecting it is α, and the probability of making the correct decision by accepting it is $1 - \alpha$. When the null hypothesis is false, the probability of making the erroneous decision by accepting it is β, and the probability of making the correct decision by rejecting it is $1 - \beta$. These two states for the null hypothesis and the two possible decisions for the researcher form a two-way table as shown in table 8-1.

Table 8-1   A decision table

Decision	H_0 true	H_0 false
Reject H_0	Type I error (α)	Correct decision ($1 - \beta$)
Accept H_0	Correct decision ($1 - \alpha$)	Type II error (β)

Observe that α is involved only under a true null hypothesis; β is involved only under a true alternative hypothesis.

The value that separates the rejection and nonrejection regions is called the *critical value*. The critical value is in the rejection region, which is also called the *critical region*. Thus, if the test statistic takes on the critical value, the null hypothesis is to be rejected. The question of how the critical value is determined will be answered in the next section.

8-2
exercises

1. Explain the similarity and difference between the testing of a hypothesis and the estimation of a population parameter.

2. What is meant by a statistical hypothesis? Why is a hypothesis referred to as "null"?

3. Define type I and type II errors.

4. Define α and β. Is α plus β equal to 1?

5. Distinguish between the rejection and nonrejection region in the test of statistical hypotheses.

6. Explain the following terms.
   - a) critical value
   - b) critical region
   - c) level of significance
   - d) one-tailed test
   - e) two-tailed test
   - f) test statistic

7. In the test of the null hypothesis $\mu = 100$, the alternative hypothesis can be one of the following.

  a) $\mu = 110$ d) $\mu < 100$

  b) $\mu = 90$ e) $\mu \neq 100$

  c) $\mu > 100$

Which of these five tests are one-tailed? Which are two-tailed?

8. Suppose that the mean score of a standard test was known to be 500 in the past. Formulate a null and alternative hypothesis for each of the following situations.

  a) An educator claims that the mean score is no longer equal to 500.

  b) Another educator claims that the mean score is now higher than 500.

  c) Still another educator claims that the mean score is now lower than 500.

8-3 tests concerning population means

Hypotheses may be classified as exact and inexact. An *exact hypothesis* is one that specifies a unique value for a population parameter, whereas an *inexact hypothesis* states that the parameter may assume any one of a set of values. For instance, $H_0: \mu = 100$ and $H_1: \mu = 120$ are two exact hypotheses, and $H_0: \mu \leq 100$ and $H_1: \mu > 100$ are inexact hypotheses. Similarly, $H_1: \mu \neq 100$ is also an inexact hypothesis. In this section we shall use a simple example to illustrate the procedure of testing some exact hypotheses; then the testing of inexact hypotheses will be considered.

  In making any statistical inference about a population mean on the basis of a sample mean it is important to remember the central limit theorem discussed in the previous chapter. If X is normally distributed, then $\overline{X}$ will also be normally distributed regardless of the sample size. If X is not normally distributed, $\overline{X}$ will still have a nearly normal distribution if the sample size is large, that is, with n equal to 30 or more. Thus, whenever n is large, $\overline{X}$ is treated as normal regardless of the distribution pattern of the population.

  Inferences about the population mean are made under two different circumstances: (1) The population standard deviation (σ) is *known* to the researcher when he conducts his tests, and (2) the population standard deviation is *unknown* and has to be estimated from a sample standard deviation (s) that is computed from a large sample.

population standard deviation known

The following three examples are based on the assumption that the population standard deviation (σ) is known to the researcher.

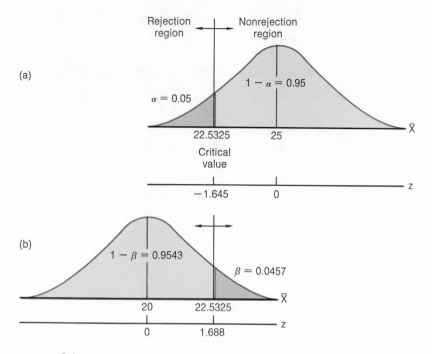

FIGURE 8-1
Sampling distributions of $\overline{X}$ for
$\sigma_{\overline{X}} = 6/4 = 1.5$ showing the rejection region for (a) $\mu = 25$ and
(b) $\mu = 20$

EXAMPLE 8-1   Suppose that the nicotine content of cigarettes is normally distributed with a mean of 25 and a standard deviation of 6 milligrams. A new manufacturing process is developed for decreasing the
nicotine content to 20 milligrams with the standard deviation unchanged.
A sample of 16 cigarettes produced by the new process yields a mean of
22 milligrams, or $\overline{X} = 22$. Will you reject the null hypothesis in favor of the
alternative at $\alpha = 0.05$? What is the probability of making a type II error?
In this test, the two competing hypotheses are

$$H_0: \mu = 25 \quad \text{and} \quad H_1: \mu = 20$$

With the level of significance specified at 0.05, the critical z value for the
test statistic is -1.645, since -1.645 is the standard normal value that
cuts off the lower 5 percent of the z distribution as shown in figure 8-1a.
(Observe that we use the lower-case letter z to denote the critical standard
normal score.) The null hypothesis will be rejected only if the test statistic is a sufficiently low value, in view of the fact that the alternative
hypothesis specifies a lower value than the null hypothesis for the

population parameter μ. We then set up the following *decision rule:*

$$\text{Reject } H_0 \text{ if } Z \leq -1.645$$

The test statistic Z, which is based on the null hypothesis, or $\mu = 25$, is computed as follows:

$$Z = \frac{\overline{X} - \mu}{\sigma/\sqrt{n}} = \frac{22 - 25}{6/\sqrt{16}} = \frac{-3}{1.5} = -2.0$$

Since Z turns out to be less then -1.645 (that is, it falls in the rejection, or critical, region) we reject H_0.

Corresponding to -1.645, the critical value for the test statistic Z, the critical value for $\overline{X}$ is 22.5325. (This value is computed on the basis of the equality $(\overline{X} - 25)/1.5 = -1.645$. Solving this for $\overline{X}$, we have $\overline{X} - 25 = -1.645(1.5) = -2.4675$. Thus, $\overline{X} = 25 - 2.4675 = 22.5325$.) It follows that the null hypothesis is rejected if $X \leq 22.5325$. The value of β is the probability of $\overline{X}$ being greater than 22.5325 when in fact $\mu = 20$, or H_1 is true. That is,

$$\beta = P(\overline{X} > 22.5325 \,|\, \mu = 20) = P\left(Z > \frac{22.5325 - 20}{6/4}\right) = P(Z > 1.688)$$

$$= 1 - P(Z \leq 1.688) = 1 - 0.9543 = 0.0457$$

which is represented by the shaded area in figure 8-1b. It is the probability of accepting H_0 when the alternative hypothesis is true, or $\mu = 20$.

Since a level of significance α is associated only with an exact null hypothesis, any hypothesis testing must be based on an exact null hypothesis in order to determine the critical value. In actual practice, however, the null hypothesis is often stated as an inexact one, such as $\mu \geq \mu_0$ or $\mu \leq \mu_0$. Fortunately, in such cases it is entirely legitimate to ignore the inequality sign and simply treat the null hypothesis as $\mu = \mu_0$. Such a simplification will by no means alter the outcome of the test. The shaded area in figure 8-2 under curve a is the level of significance associated with $\mu = \mu_0$. The shaded area becomes smaller and smaller as the mean becomes less and less than μ_0, which is shown under curves b and c. With curve d, which lies farther to the left, the upper tail extending beyond the critical value C is not even discernible. This implies that if we can reject $H_0: \mu = \mu_0$ with a probability of α, then we can reject any hypothesis $\mu < \mu_0$ with an error probability less than α. That is, if we can be $(1 - \alpha)$ confident that we are not making any error in rejecting $H_0: \mu = \mu_0$, we can be more confident in rejecting any hypothesis $\mu < \mu_0$. For this reason we shall always treat $H_0: \mu \leq \mu_0$ or $\mu \geq \mu_0$ as if it is $\mu = \mu_0$.

EXAMPLE 8-2 Let X be the starting monthly salary for anyone who has just graduated from a four-year college. It is suspected that the mean monthly salary is more than \$800, and not \$800 or less as some have

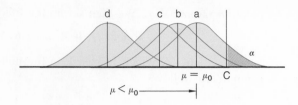

FIGURE 8-2
Distributions under $H_0: \mu \leq \mu_0$
showing that α is not increased by
treating H_0 like $\mu = \mu_0$

predicted. Assume that the variance of X is known to be $2500. A sample of 100 graduates is randomly taken and the sample mean found to be $810. Shall we reject the null hypothesis at $\alpha = 0.01$?

The two competing hypotheses to be tested in this example are

$$H_0: \mu \leq \$800 \quad \text{and} \quad H_1: \mu > \$800$$

As indicated above, the null hypothesis here can be treated as $\mu = \$800$ without increasing the value of α, which is specified at 0.01. Appendix E shows that the critical value for Z which cuts off the upper 1 percent (or lower 99 percent) of the standard normal distribution is 2.326. That is, the decision rule for this test is

$$\text{Reject } H_0 \text{ if } \frac{X - \mu}{\sigma / \sqrt{n}} \geq 2.326$$

Since the observed sample mean $\overline{X}$ is 810, the mean μ under H_0 is 800, the standard deviation is $\sqrt{2500} = 50$, and the sample size is 100, the value of Z is

$$Z = \frac{\overline{X} - \mu}{\sigma / \sqrt{n}} = \frac{810 - 800}{50 / \sqrt{100}} = \frac{10}{5} = 2$$

which is less than the critical value 2.326. Accordingly, we do not reject H_0. Figure 8-3 shows the critical value and the rejection region.

Examples 8-1 and 8-2 involve the so-called one-tailed test. In the following example the two-tailed test will be illustrated.

EXAMPLE 8-3   Suppose that the IQ's of all pupils in the elementary schools of the country are known to be normally distributed with a mean of 100 and a variance of 225. The superintendent of a certain district wishes to find out whether the pupils in his district score higher or lower on the test. For this purpose, a random sample of 25 pupils is selected. (1) Design a decision rule to test the hypothesis that the mean IQ of the

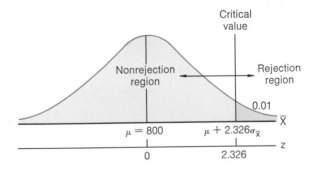

FIGURE 8-3
Sampling distribution of $\overline{X}$ showing the rejection region for a one-
tailed test with $\alpha = 0.01$

pupils in this district is 100 against the alternative that it is *not* 100 at
$\alpha = 0.05$. (2) On the basis of the decision rule, do you reject the null
hypothesis if the sample yields a mean of 105?

1) The null hypothesis to be tested in this case is $H_0: \mu = 100$. Since the
purpose of the test is to see whether the mean IQ is different from 100,
the alternative hypothesis is $H_1: \mu \neq 100$. That is, the null hypothesis
will be rejected if the test statistic is either a significantly high or a
significantly low value. The level of significance (α) is divided into
two equal parts, one under the upper tail and the other under the lower
tail, as shown in figure 8-4. The critical values for Z that cut off the
upper 2.5 percent and lower 2.5 percent (or the lower 97.5 percent
and the upper 97.5 percent) of the standard normal distribution are
1.96 and -1.96, respectively. Accordingly, the required decision
rule for this test is

$$\text{Reject } H_0 \text{ if } \frac{\overline{X} - 100}{\sqrt{225/25}} \geq 1.96$$

$$\text{or } \frac{\overline{X} - 100}{\sqrt{225/25}} \leq -1.96$$

2) Since $\overline{X} = 105$, the test statistic Z is found to be

$$\frac{105 - 100}{\sqrt{225/25}} = \frac{5}{3} = 1.667$$

which, according to the decision rule set up in (1) above, falls in the
nonrejection region. Accordingly, we do not reject H_0.

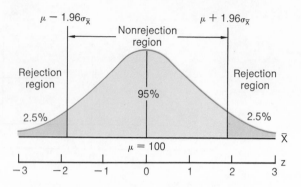

FIGURE 8-4
Sampling distribution of $\overline{X}$ showing the rejection and nonrejection regions for a two-tailed test with $\alpha = 0.05$

population standard deviation unknown

The population standard deviation, however, is usually not known to the researcher. Fortunately, when the population standard deviation is unknown, the methods of testing hypotheses are identical with those discussed above as long as the *sample size is large*. This is the case because the larger n becomes, the closer the sample standard deviation will come to σ. Thus, for a large n (not less than 30), the test statistic approximately becomes

$$Z = \frac{\overline{X} - \mu}{s/\sqrt{n}}$$

Accordingly, we may compare this Z ratio with the critical z value for the purpose of testing hypotheses about μ. (The methods of hypothesis testing for small samples with the population standard deviation unknown are discussed in the next chapter.)

EXAMPLE 8-4 The average typing speed of secretaries employed by a large company is believed to be 60 words per minute. The personnel director of the company claims that the new training program for secretaries has increased their average typing speed to more than 60 words per minute. A random sample of 36 secretaries shows that the average speed $(\overline{X})$ is 62 words per minute and the standard deviation (s) is 9 words. If the level of significance is specified at 0.05, shall we accept the claim of the personnel director?

The null and alternative hypotheses involved in this test are

$$H_0: \mu = 60 \quad \text{and} \quad H_1: \mu > 60$$

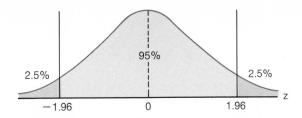

FIGURE 8-5
The standard normal distribution

With a large sample size ($n = 36$) and $\alpha = 0.05$, we can use the standard normal score $z_{0.05} = 1.645$ as the critical value for the test. Thus, the decision rule is

$$\text{Reject } H_0 \text{ if } Z \geq 1.645$$

Since the test statistic Z is

$$Z = \frac{\overline{X} - \mu}{s/\sqrt{n}} = \frac{62 - 60}{9/\sqrt{36}} = \frac{2}{1.5} = 1.33$$

we do not reject H_0. The evidence provided in the sample does not support the claim of the personnel director. In other words, the increase in speed to 62 words per minute can be attributed to chance and is not statistically significant.

EXAMPLE 8-5   A sugar producer packs sugar into paper bags, each of which is supposed to hold 5 pounds, or 80 ounces, of sugar. A random sample of 100 bags is taken periodically to see if the bags contain the proper amount. If the sample mean $\overline{X}$ differs significantly from 80 ounces, the packing process is considered to be functioning improperly. Suppose that a sample of 100 bags yields a mean of 79 ounces and a standard deviation of 5 ounces. Is the process functioning properly for $\alpha = 0.05$?

Here we have a two-tailed test, since the average weight can be either above or below 80 ounces per bag when the process is functioning improperly. The two hypotheses to be tested are

$$H_0 : \mu = 80 \quad \text{and} \quad H_1 : \mu \neq 80$$

For a two-tailed test with $\alpha = 0.05$ and a large sample, the critical values are 1.96 and -1.96. Thus the decision rule is

$$\text{Reject } H_0 \text{ if } Z \geq 1.96 \text{ or } Z \leq -1.96$$

The test statistic Z is computed as

$$Z = \frac{79 - 80}{5/\sqrt{100}} = \frac{-1}{0.5} = -2$$

Since Z is less than -1.96, we reject H_0. The sample evidence shows that the packing process is not functioning properly. Figure 8-5 shows the rejection region for the test statistic Z for a two-tailed test with $\alpha = 0.05$.

8-3 exercises

1. Use your own words to answer the following questions.
    a) What is the difference between exact and inexact hypotheses?
    b) Under what condition is the Z score used in tests concerning the mean of a single population when the population variance is unknown?

2. In a sample of 30 responses, X is the number of Yes answers to the question "Are you in favor of capital punishment?" Let p be the true proportion of those who favor capital punishment. Find the values of α and β for the following tests.

H_0	H_1	Decision rule
a) $p = 0.7$	$p = 0.5$	Reject H_0 if $X \leq 10$
b) $p = 0.4$	$p = 0.3$	Reject H_0 if $X \leq 7$
c) $p = 0.5$	$p = 0.7$	Reject H_0 if $X \geq 18$
d) $p = 0.3$	$p = 0.5$	Reject H_0 if $X \geq 13$

3. It is suspected that the IQ's of pupils of a certain ethnic group are, on the average, eight points higher than the average for all pupils throughout the nation. We know that, for all pupils, the mean is 100 and the standard deviation is 15. Tests given to a sample of 25 pupils selected at random from the ethnic group yield a mean IQ of 104. Assuming that the IQ's have a normal distribution, test the hypothesis $H_0 : \mu = 100$ against the alternative $H_1 : \mu = 108$ at $\alpha = 0.05$. Also, find the value of β.

4. A company that manufactures natural fibers claims that their fibers have a mean breaking strength of 40 pounds and a standard deviation of eight pounds. A buyer suspects that the mean breaking strength is only 37 pounds. A random sample of 64 fibers yields a mean of 38 pounds. Shall the buyer reject $H_0 : \mu = 40$ in favor of $H_1 : \mu = 37$, if the level of significance is 0.01?

5. Look again at problem 4. What is the value of β? What is the value of β if $\alpha = 0.10$?

6. A flour manufacturer packs flour into paper bags, each of which is supposed to hold 10 pounds, or 160 ounces. Some customers have complained that the bags hold only 9.5 pounds, or 152 ounces. A

test is conducted to determine whether the complaint is warranted. The weight of each bag is normally distributed, with a standard deviation of five ounces. A sample of 16 bags yields a mean weight of 156 ounces. Shall we reject the null hypothesis of $\mu = 160$ ounces at $\alpha = 0.10$?

7. In problem 6 above, find the value of β. What will be the value of β if (a) the sample size (n) is 4 and (b) $n = 64$?

8. A standard college entrance examination is administered nationally every year. In the past the scores in this examination were normally distributed with a mean of 500 or fewer points and a standard deviation of 100 points. It is believed that because of improved secondary school education the mean score has increased to more than 500 points. To test the null hypothesis $\mu \le 500$ against the alternative $\mu > 500$, a sample of 100 scores is randomly selected. The sample has a mean score of 520 points. Shall we reject the null hypothesis at $\alpha = 0.02$?

9. A hosiery manufacturer is considering replacing his old sewing machine with a new one. The old machine produces, at most, an average of 300 pairs of stockings per hour, with a standard deviation of 30 pairs. The hourly output of such sewing machines is believed to have a normal distribution. The supplier of the new machine claims that its average output per hour is more than 300 pairs. The new machine is tested during a 25-hour period, and its average hourly output measured at 310 pairs. If the level of significance is 0.05, shall we reject the null hypothesis $\mu \le 300$?

10. Employees who contract a certain disease, and receive conventional medical treatment for it, are on sick leave for an average of 15 days. A medical research team claims that a new treatment has been developed that will reduce the average period of sick leave. Assume that the sick leave period has a normal distribution and a standard deviation of three days. Shall we reject the null hypothesis $\mu = 15$ at $\alpha = 0.01$ if a sample of 16 patients who have received the new treatment shows an average sick leave of exactly 13 days?

11. The norms established for a reading comprehension test show that high school freshmen should average a score of 85 with a variance of 100. A school superintendent suspects that the reading comprehension of freshmen has decreased. A sample of 64 randomly selected scores yields a mean of 83. If the level of significance is specified at 0.05, shall we conclude that the reading comprehension of the new freshmen has significantly decreased?

12. A public utility company wishes to determine whether its new work schedule has significantly reduced the waiting time for customer

service. The waiting time was at least 30 minutes in the past. A random sample of 144 observations yields a mean of 28 minutes and a standard deviation of 12 minutes ($s = 12$). Shall we reject the null hypothesis $\mu \geq 30$ in favor of the alternative $\mu < 30$ at $\alpha = 0.01$?

13. A vendor sells chickens in large quantities to a restaurant chain. He claims that the mean weight of the chickens in a large shipment is at least 30 ounces. However, the purchase agent for the restaurant chain suspects that it may be less. A random sample of 100 chickens is selected; the observations yield a mean of 27 ounces and a variance of 144 ounces ($s^2 = 144$). Given a significance level of 0.05, shall we reject the vendor's claim?

14. Suppose that an airline wishes to determine whether the average weight of suitcases carried by passengers between New York and London is more than 30 pounds. A sample of 49 passengers is randomly selected and their suitcases weighed. The mean is 32 pounds with a standard deviation of 5.6 pounds ($s = 5.6$). If α is specified at 0.01, shall we conclude that the average weight of suitcases is more than 30 pounds?

15. A psychologist wishes to test the hypothesis that a reaction time has an average of 0.30 seconds against the alternative that it is different from 0.30 seconds. A random sample of 64 observations is selected, and it yields a mean of 0.33 seconds and a standard deviation of 0.072 seconds, or $s = 0.072$. At $\alpha = 0.02$, shall we reject the null hypothesis that the average reaction time is 0.30 seconds in favor of the alternative that it is different from 0.30 seconds?

8-4 testing the difference between two means

In statistics, we are frequently asked to decide whether the difference between two sample means is large enough to indicate that the samples were taken from two different populations, or whether the difference is small enough to be attributed to chance. Suppose, for instance, that a company employs both men and women for the same kind of work. If a sample of men yields an average hourly output of 65 units, and a sample of women yields an average hourly output of 70 units, we may want to determine whether the difference between the two averages can be attributed to chance, or whether there is a significant difference in the performance between the two types of workers. Similarly, we may want to determine, on the basis of sample information, whether the average amount of wheat harvested per acre on land that was treated with a new fertilizer is equal to the amount harvested on land treated with the old fertilizer. In like manner, we may want to compare the average IQ's of students in two disciplines or the average service lives of light bulbs produced by two companies.

To test hypotheses about differences between two means, such as those described above, we must introduce a new random variable: $\overline{D}$. $\overline{D}$ is defined as

$$\overline{D} = \overline{X}_1 - \overline{X}_2$$

where $\overline{X}_1$ is the mean of a sample taken from a certain population and $\overline{X}_2$ is the mean of a sample taken from another population. The population means are μ_1 and μ_2, respectively. The sampling distribution of $\overline{D}$ has two parameters, the mean and the standard deviation. The population mean of $\overline{D}$ is $\mu_1 - \mu_2$, since

$$E(\overline{D}) = E(\overline{X}_1 - \overline{X}_2) = E(\overline{X}_1) - E(\overline{X}_2) = \mu_1 - \mu_2 \qquad (8\text{-}1)$$

The standard deviation of $\overline{D}$ is computed on the basis of an important theorem, which reads as follows: If $\overline{X}_1$ and $\overline{X}_2$ are the means of two independent samples of sizes n_1 and n_2 selected at random from two populations with variances σ_1^2 and σ_2^2, respectively, then the standard deviation of $\overline{D}$, or $\sigma_{\overline{D}}$, is

$$\sigma_{\overline{D}} = \sqrt{\frac{\sigma_1^2}{n_1} + \frac{\sigma_2^2}{n_2}} \qquad (8\text{-}2)$$

which is also referred to as the *standard error of the difference* between two sample means.

If the original random variables X_1 and X_2 are normally distributed, then $\overline{X}_1$ and $\overline{X}_2$ are normally distributed; accordingly, $\overline{D}$, or $\overline{X}_1 - \overline{X}_2$, will have a normal distribution. If the original random variables X_1 and X_2 are not normally distributed, $\overline{X}_1$ and $\overline{X}_2$ will still be approximately normally distributed as long as n_1 and n_2 are large. Thus, for large n_1 and n_2 (30 or more) $\overline{D}$ will have an approximately normal distribution with a mean $\mu_1 - \mu_2$ and a variance $\sigma_1^2/n_1 + \sigma_2^2/n_2$ regardless of the distribution pattern of the populations from which the two samples are drawn.

Tests about differences between two means, like tests about a single mean, are made under two different circumstances: (1) The population variances (σ_1^2 and σ_2^2) are *known* to the researcher, and (2) the population variances are *not known* and have to be estimated from sample variances (s_1^2 and s_2^2) which are computed from two large samples.

population variances known

The null hypothesis formulated to test the difference between two population means (usually designated by δ, the Greek letter delta) is $H_0: \mu_1 - \mu_2 = 0$, or $\delta = 0$, and accordingly $E(\overline{D}) = \delta = 0$.

Thus, the test statistic Z, based on known population variances, is expressed in the following form

$$Z = \frac{\overline{X}_1 - \overline{X}_2}{\sqrt{\dfrac{\sigma_1^2}{n_1} + \dfrac{\sigma_2^2}{n_2}}} \tag{8-3}$$

The testing procedure here is exactly the same as in the preceding section with only the test statistic and the standard error defined differently.

EXAMPLE 8-6   Suppose that we want to determine whether the average nicotine content of cigarette brand I is equal to that of brand II. For the purpose of testing this hypothesis, two samples of 100 cigarettes each are randomly selected from the two brands and two sample means obtained: $\overline{X}_1 = 22$ milligrams and $\overline{X}_2 = 24$ milligrams. The variances of the two brands are known to be $\sigma_1^2 = 36$ and $\sigma_2^2 = 64$. Determine whether the two brands of cigarettes have the same mean at $\alpha = 0.05$.

The two hypotheses involved in this test are

$$H_0 : \mu_1 - \mu_2 = 0 \quad \text{and} \quad H_1 : \mu_1 - \mu_2 \neq 0$$

With the level of significance specified at 0.05, the decision rule again is

$$\text{Reject } H_0 \text{ if } Z \geq 1.96 \text{ or } Z \leq -1.96$$

The test statistic Z is computed as follows:

$$Z = \frac{\overline{X}_1 - \overline{X}_2}{\sqrt{\dfrac{\sigma_1^2}{n_1} + \dfrac{\sigma_2^2}{n_2}}} = \frac{22 - 24}{\sqrt{\dfrac{36}{100} + \dfrac{64}{100}}} = \frac{-2}{\sqrt{\dfrac{100}{100}}} = \frac{-2}{1} = -2$$

which is less than -1.96. Accordingly, we reject H_0 and conclude that the nicotine content of brand I is not the same as that of brand II.

EXAMPLE 8-7   Two models of automobiles are compared for braking ability. Two random samples of 81 cars each are selected, and the distance required to stop when the brake is applied at 60 miles per hour measured and recorded. Suppose that the variance for model I is known to be $\sigma_1^2 = 150$ feet and that for model II is $\sigma_2^2 = 174$ feet. The two sample means are $\overline{X}_1 = 85$ feet and $\overline{X}_2 = 90$ feet. Can we conclude that $\overline{X}_1$ is significantly less than $\overline{X}_2$ at $\alpha = 0.01$?

The null and alternative hypotheses involved in this test are

$$H_0 : \mu_1 = \mu_2 \quad \text{and} \quad H_1 : \mu_1 < \mu_2$$

At $\alpha = 0.01$, the critical value for the test statistic Z is -2.326. Thus the decision rule is

$$\text{Reject } H_0 \text{ if } \frac{\overline{X}_1 - \overline{X}_2}{\sqrt{\dfrac{\sigma_1^2}{n_1} + \dfrac{\sigma_2^2}{n_2}}} \leq -2.326$$

The value of the test statistic is computed as follows:

$$Z = \frac{85 - 90}{\sqrt{\dfrac{150}{81} + \dfrac{174}{81}}} = \frac{-5}{\sqrt{\dfrac{324}{81}}} = \frac{-5}{\sqrt{4}} = -2.5$$

which is less than the critical value -2.326. Accordingly, we reject H_0.

population variances unknown

The procedures involved in testing hypotheses about the difference between two means when the population variances are unknown are essentially the same as those discussed above, as long as the samples are large (neither n_1 nor n_2 can be smaller than 30). When σ_1^2 and σ_2^2 are not known, they are estimated with s_1^2 and s_2^2, respectively. As a result, the *estimated standard error* of $\overline{D}$, or $s_{\overline{D}}$, becomes

$$s_{\overline{D}} = \sqrt{\frac{s_1^2}{n_1} + \frac{s_2^2}{n_2}} \tag{8-4}$$

Under the null hypothesis that μ_1 does not differ from μ_2 (that is, $\mu_1 - \mu_2 = 0$, or $\delta = 0$), the test statistic Z used to test the hypothesis is

$$Z = \frac{\overline{X}_1 - \overline{X}_2}{\sqrt{\dfrac{s_1^2}{n_1} + \dfrac{s_2^2}{n_2}}} \tag{8-5}$$

This quantity is compared with the critical z score to determine whether μ_1 is equal to μ_2.

EXAMPLE 8-8 Suppose that we want to compare the service lives (in miles) of two brands of automobile tires. Two independent samples are selected at random. A sample of 100 tires from brand I yields a mean $(\overline{X}_1)$ of 36,000 miles and a variance (s_1^2) of 20,000,000 miles; a sample of 64 tires from brand II yields a mean $(\overline{X}_2)$ of 35,000 miles and a variance (s_2^2) of 10,240,000 miles. At $\alpha = 0.05$, can we conclude that μ_1, the average service life of brand I tires, is longer than μ_2, the average service life of brand II tires?

The null and alternative hypotheses to be tested in this example are

$$H_0: \mu_1 - \mu_2 = 0 \quad \text{and} \quad H_1: \mu_1 - \mu_2 > 0$$

At $\alpha = 0.05$ the critical z score is again 1.645. The decision rule therefore is

$$\text{Reject } H_0 \text{ if } Z \geq 1.645$$

The value of Z is obtained by applying equation 8-5 as follows:

$$Z = \frac{36,000 - 35,000}{\sqrt{\dfrac{20,000,000}{100} + \dfrac{10,240,000}{64}}}$$

$$= \frac{1000}{\sqrt{200,000 + 160,000}}$$

$$= \frac{1000}{600}$$

$$= 1.667$$

Since Z is greater than 1.645, we reject H_0. That is, the average service life of brand I tires is significantly longer than that of brand II, and the observed difference by 1000 miles cannot be attributed to chance.

EXAMPLE 8-9   An experimenter fed two groups of hogs two different diets. The results are as follows:

	Diet I	Diet II
Number of hogs in each group	$n_1 = 100$	$n_2 = 100$
Average weight gained (lbs.)	$X_1 = 185$	$X_2 = 190$
Variance	$s_1^2 = 180$	$s_2^2 = 220$

It is assumed that all factors other than the diets affecting weight are the same for both groups. Test the null hypothesis that the two diets have the same effect on weight against the alternative hypothesis that they have different effects at $\alpha = 0.01$.

Let μ_1 and μ_2 be the true mean weights gained by using diet I and diet II, respectively. The two competing hypotheses involved in this test are

$$H_0 : \mu_1 = \mu_2 \quad \text{and} \quad H_1 : \mu_1 \neq \mu_2$$

This is a two-tailed test because the alternative hypothesis states only that the two diets will have different effects on weights; thus, we can reject H_0 if $\overline{X}_1$ is either significantly smaller or greater than $\overline{X}_2$. At $\alpha = 0.01$, the critical z values are 2.575 and -2.575. The decision rule is

$$\text{Reject } H_0 \text{ if } Z \geq 2.575 \text{ or } Z \leq -2.575$$

Since the Z value is

$$Z = \frac{185 - 190}{\sqrt{\dfrac{180}{100} + \dfrac{220}{100}}} = \frac{-5}{2} = -2.5$$

which is smaller than 2.575 but greater than -2.575, we do not reject H_0. In other words, at $\alpha = 0.01$ the difference of five pounds can be attributable to chance.

8-4
exercises

1. A study was conducted to determine whether pupils of two ethnic groups, I and II, have different average IQ's. It is assumed that the variances of pupils' IQ's in groups I and II are, respectively, 225 and 196. A sample of 25 pupils is taken from group I ($n_1 = 25$), and a sample of 28 pupils is taken from group II ($n_2 = 28$). On the basis of the difference between the two sample means $\overline{X}_1 = 102$ and $\overline{X}_2 = 98$, test the null hypothesis that the pupils of the two ethnic groups have identical average IQ's against the alternative that the two averages are different at $\alpha = 0.05$.

2. A certain large corporation employs both men and women to do the same kind of work. It is hypothesized that the average hourly output of men is less than that of women. Suppose that the following information is provided by a company research team.

	Men	Women
Variance in units	$\sigma_1^2 = 70$	$\sigma_2^2 = 74$
Sample size	$n_1 = 36$	$n_2 = 36$
Sample mean in units	$\overline{X}_1 = 150$	$\overline{X}_2 = 153$

Is the average hourly output of men significantly lower than that of women at $\alpha = 0.05$? (Assume that the two samples are independent.)

3. A business analyst wishes to compare the effectiveness of two methods of training factory employees to perform a certain operation. Two independent and randomly selected groups of workers are separately trained under method I and method II. The average time required for each group to perform the operation after training and other relevant data are given below.

	Method I	Method II
Variance in minutes	$\sigma_1^2 = 200$	$\sigma_2^2 = 276$
Sample size	$n_1 = 24$	$n_2 = 36$
Sample mean in minutes	$\overline{X}_1 = 45$	$\overline{X}_2 = 55$

a) Test the null hypothesis that the two methods are equally effective against the alternative that they have different effectiveness at $\alpha = 0.02$.

b) What assumption do you have to make in order to conduct this test?

4. A county assessor's office desires to determine whether there is any significant difference between the home appraisals made by two of its employees. Two samples of 144 homes each are selected independently, and each appraiser is asked to appraise the homes in the sample assigned to him. The mean value of homes appraised by appraiser I is $\overline{X_1} = \$30,000$ and the mean value of homes appraised by appraiser II is $\overline{X_2} = \$30,900$. It is assumed that the variance of home values for each appraiser is identical; that is, $\sigma_1^2 = \sigma_2^2$, which is known to be $\$18,000,000$.

Is there any significant difference between the appraisals at $\alpha = 0.01$?

5. Suppose a researcher wishes to determine whether a diet supplemented with a growth hormone can significantly increase the weight gain of chickens. For this purpose, two independent groups of chickens are selected at random—one group fed with a usual diet and the other fed a diet with a growth hormone. The weight gains for the two groups are recorded three months after the respective diets have been in use. The relevant data are as follows:

	Group I (usual diet)	Group II (hormone diet)
Variance in ounces	$\sigma_1^2 = 128$	$\sigma_2^2 = 128$
Sample size	$n_1 = 100$	$n_2 = 100$
Sample means in ounces	$\overline{X_1} = 16$	$\overline{X_2} = 19$

With the significance level specified at 0.05, can we conclude that a diet supplemented with a growth hormone will increase the weight gain of chickens?

6. A manufacturer claims that the nylon cord his company produces is stronger than cotton cord. Two independent samples of 36 lengths each are randomly selected, and the results recorded below.

	Nylon cord	Cotton cord
Average breaking strength	$\overline{X_1} = 105$ pounds	$\overline{X_2} = 101$ pounds
Variance	$s_1^2 = 74$ pounds	$s_2^2 = 70$ pounds

Shall we conclude that the nylon cord is indeed stronger than the cotton cord at $\alpha = 0.05$?

7. A nutritionist wishes to compare the effectiveness of two reducing diets. The following data are obtained from two independent samples.

	Diet I	Diet II
Sample size	$n_1 = 40$	$n_2 = 60$
Average weight loss in pounds	$\overline{X}_1 = 9$	$\overline{X}_2 = 11$
Sample variance	$s_1^2 = 50$	$s_2^2 = 60$

At $\alpha = 0.10$, is there sufficient evidence that diet I produces a smaller weight loss than diet II?

8. An experiment is designed to compare the average service lives of two brands of auto batteries. The following data are obtained from two independent samples.

	Brand I	Brand II
Sample size	$n_1 = 36$	$n_2 = 36$
Average service lives in months	$\overline{X}_1 = 38$	$\overline{X}_2 = 35$
Sample variance	$s_1^2 = 41$	$s_2^2 = 40$

At $\alpha = 0.05$, test the hypothesis that the average service lives of the two kinds of batteries are the same against the alternative that they are different.

9. A standard examination is administered to two random samples of college freshmen. One sample contains all male students, and the other, independent of the first, contains female students. The two samples yield the following data.

	Men	Women
Sample size	$n_1 = 72$	$n_2 = 36$
Sample average score	$\overline{X}_1 = 80$	$\overline{X}_2 = 84$
Sample variance	$s_1^2 = 64$	$s_2^2 = 40$

At $\alpha = 0.02$, do the above data provide sufficient evidence to indicate a difference in the test scores of male and female freshmen?

8-5 tests concerning population proportions

There are situations in which we may want to decide whether the population proportion, usually designated by the symbol p, is equal to a certain percentage. In most cases, we will use the *sample proportion,* or the *number of successes in* n *trials,* for purposes of inference. If an event has occurred X times in n trials, the sample proportion (usually designated by the symbol $\hat{p}$) is X/n. This fraction may be used to estimate the population proportion (p), or the probability of a success. If, for example, 70 of 100 T.B. patients show immediate improvement upon receiving a new vaccine, we can use X/n, or 0.70, as an estimate of the true proportion of T.B. patients who will improve with the new vaccine.

For testing hypotheses about a population proportion (p), it is more convenient to use the binomial random variable X than the sample proportion X/n, because X itself is a test statistic. In this section, we shall first illustrate the procedure of testing hypotheses about p for small values of n by using the cumulative binomial probability table in appendix B. For large values of n, the method of normal approximation to the binomial will be illustrated.

EXAMPLE 8-10 An ordinary coin is tossed 20 times to determine whether it is balanced. Let X be the number of heads obtained in the 20 tosses. If $X = 7$, can we reject the null hypothesis that the coin is balanced in favor of the alternative that heads will occur less than 50 percent of the time at $\alpha = 0.0577$?

The two competing hypotheses involved in this example are

$$H_0 : p = 0.5 \quad \text{and} \quad H_1 : p < 0.5$$

At $\alpha = 0.0577$, the critical value for X is 6 because, as appendix B shows, the column headed 0.5 in the table for $n = 20$ gives a cumulated probability of 0.0577 where $X = 6$. That is,

$$P(X \leq 6 \,|\, n = 20, p = 0.5) = 0.0577$$

Accordingly, the decision rule is

$$\text{Reject } H_0 \text{ if } X \leq 6$$

Since the observed value for X is 7, we do not reject H_0.

EXAMPLE 8-11 Suppose that none of the available vaccines for pneumonia is more than 60 percent effective; that is, no one drug cures more than 60 of every 100 patients. A drug manufacturing company claims that a new vaccine has been developed which is more than 60 percent effective in curing pneumonia patients. Of 15 pneumonia patients, 10 are cured after receiving an injection of the new vaccine. Should we accept the company's claim at a significance level of 0.0271?

The null and alternative hypotheses involved in this test are

$$H_0 : p \leq 0.6 \quad \text{and} \quad H_1 : p > 0.6$$

Recall that we can treat an inexact null hypothesis as exact; so we treat H_0 here as $p = 0.6$. For determining the critical value for X, the number of patients cured by the new medicine, we first locate the number 0.9729 (or $1 - 0.0271$), the probability of accepting H_0 when it is true. We find that this value is located in the intersection of row 12 and column 0.6 in the table for $n = 15$ of appendix B. That is,

$$P(X \leq 12 \,|\, n = 15, p = 0.6) = 0.9729$$

That implies that

$$P(X \geq 13 \mid n = 15, p = 0.6) = 1 - 0.9729 = 0.0271$$

which is exactly the specified level of significance. Therefore, the decision rule is

$$\text{Reject } H_0 \text{ if } X \geq 13$$

Since the observed X value is 10, we do not reject H_0. Accordingly, we do not accept the claim of the manufacturer about the effectiveness of his newly developed vaccine.

Observe that in the above examples, the values of α have four decimal places. This is done for the sake of convenience, since the table entries in appendix B have four decimal places.

Appendix B, however, provides only a few values for n. For hypothesis testing involving values of n not provided in the table, the method of normal approximation may be used as long as n is large enough so that both np and $n(1 - p)$ are greater than 5. For sufficiently large values of n, the binomial random variable X is approximately normally distributed with a mean np and a variance $np(1 - p)$. The following example illustrates the procedure of normal approximation for inferences about a population proportion.

EXAMPLE 8-12   Suppose that a certain news program ordinarily attracts 50 percent of all those who watch television during the period while the program is on the air. The regular anchorman has resigned, and a new man has been hired to replace him. The network management wants to determine whether the new anchorman has increased or decreased the percentage of those who watch the program. A telephone survey is conducted. It is found that 45 out of the 100 people who watch T.V. while the news program is on the air, watch this particular program. Test the hypothesis that the percentage of those who watch the program remains unchanged at $\alpha = 0.05$.

The two hypotheses involved in this test are

$$H_0 : p = 0.5 \quad \text{and} \quad H_1 : p \neq 0.5$$

Since $n = 100$, a rather large sample, we can use normal approximation in this test. At $\alpha = 0.05$, the critical values for Z are 1.96 and -1.96. The decision rule is therefore

$$\text{Reject } H_0 \text{ if } Z \geq 1.96 \text{ or } Z \leq -1.96$$

The mean of X is $np = 100(0.5) = 50$, and the standard deviation of X is $\sqrt{np(1 - p)} = \sqrt{100(0.5)(0.5)} = 5$, according to equations 7-5 and 7-6. Thus the test statistic Z is

$$\frac{X - np}{\sqrt{np(1 - p)}} = \frac{45 - 50}{5} = -1$$

which is less than 1.96 and greater than -1.96. Consequently, we do not reject H_0.

1. A drug manufacturer claims that his newly developed medicine is more than 90 percent effective in relieving muscular pain. In a sample of 100 people who suffer muscular pain, the medicine provided relief for 95. Test the null hypothesis that the medicine is 90 percent effective against the alternative that the medicine is more than 90 percent effective at $\alpha = 0.05$.

2. Suppose that we wish to test the hypothesis that the proportion of all stenographers who are productive at a given time is 80 percent ($p = 0.80$) against the alternative that it is less than 80 percent ($p < 0.80$). For this purpose a sample of 64 stenographers is randomly selected, and it is observed that 45 of them are in a productive state. If the level of significance is 0.025, shall we reject the null hypothesis?

3. A market researcher wishes to determine whether housewives prefer cooking oil I or cooking oil II. Thirty housewives were interviewed, and 18 of them indicated their preference for cooking oil I. Can we conclude that housewives in general are in favor of cooking oil I if the level of significance is 0.0495?

4. A foreign automobile manufacturer claims that 20 percent of the potential American buyers of foreign cars prefer his products. A sample of 100 such potential buyers is taken at random, and among them 15 indicate their preference for his products. At $\alpha = 0.05$, test the null hypothesis that $p = 0.20$ against the alternative that $p \neq 0.20$.

5. In the past, 10 percent of the mail-order solicitations for a certain charity resulted in financial contributions. It is claimed that because of the new affluence more than 10 percent of this year's solicitations will result in financial contributions. For testing this claim, 100 people are randomly selected and solicited; 15 responded with donations. Can we conclude that people have become significantly more generous at $\alpha = 0.02$?

6. Suppose that in the past 10 percent of all smokers preferred cigarette brand I. A promotional campaign is conducted to improve its sales. A random sample of 400 smokers is selected and interviewed to determine whether the campaign has been effective. Among the 400 smokers interviewed, 50 indicate their preference for brand I. Test the null hypothesis that the campaign has not been effective at $\alpha = 0.01$.

7. Consider p, the true proportion of registered voters who are against

the death penalty. Suppose that in the past p has been equal to 50 percent or less. Now there is reason to believe that p has increased. A random sample of 20 voters yields a sample proportion of 55 percent. Can we conclude that the true proportion remains unchanged rather than increased at $\alpha = 0.0207$?

8. Past records show that 20 percent of all girls are married within one year after their graduation from high school. It is suspected that the percentage has changed. Two hundred and twenty-five girls were selected and interviewed, and it was found that 54 of them were married within one year after their graduation from high school. Test the hypothesis that the proportion of girls who marry within one year after their high school graduation remains 20 percent ($p = 0.20$) against the alternative that it has changed ($p \neq 0.20$) at $\alpha = 0.05$.

9. It is suggested that professors have become more lenient in grading their students. In the past, 80 percent of all freshmen received C or better grades. A survey of the most recent freshman class shows that 8100 of the 10,000 freshmen in the sample received C or better grades. Is it true that professors have become more lenient if the level of significance is specified at 0.01?

10. Recall that under the normal curve 99.73 percent (almost certainty) of the total area is within *three* standard deviations from the mean. Suppose that a balanced coin is tossed n times. Find the interval of the proportion of heads within three standard deviations from the mean np for the following values of n.

    a) For $n = 100$
    b) For $n = 10,000$
    c) For $n = 1,000,000$
    d) What generalization can you make about the proportion of heads with respect to the value of n?

8-6 statistical estimation

Estimation may be further divided into two categories: point estimation and interval estimation. Example 5-3 (see page 99) shows that the mean of the sample of wages earned by 20 college students on a Saturday is $24. In the absence of other data, this figure may be used as an estimate of μ, the true mean of wages earned by all college students on any Saturday. If a coin is tossed 1000 times and 600 heads appear, then 600/1000, or 0.6, may serve as an estimate of p, the probability that this particular coin will turn up heads. Estimates of this kind are called *point estimates*, since they consist of only a single value, or a single point, on the number scale.

The principal disadvantage in using a point estimate is that it fails to show how close it is to the parameter value to be estimated. In other words, it does not provide any measure of precision in the estimation, or any indication relative to the magnitude of the deviation between the estimate and the parameter value. In order to provide some degree of precision of the estimate, the procedure of interval estimation is frequently used.

The procedure of determining an interval of values, which will include the true value of a population parameter with a probability of $1 - \alpha$, is referred to as *interval estimation*. The symbol α as defined here refers to the probability that the interval will *not* include the true value of the parameter, and is not to be confused with the α that designates a level of significance in hypothesis testing. In estimation, α is the probability of making an *error in the estimation*. Thus $1 - \alpha$ is a measure of confidence that the estimation will be a correct one, and for this reason the interval obtained for estimation purposes is generally known as the $1 - \alpha$ confidence interval for the population parameter in question.

The primary objective of this section is to illustrate the procedure of constructing confidence intervals for the population mean, the difference between two means, and the population proportion.

estimating the population mean

When a sample mean $(\overline{X})$ is computed and used as an estimate of the population mean (μ), there will very likely be an error in the estimation, the magnitude of which is $\overline{X} - \mu$. As shown in figure 8-6, a proportion $1 - \alpha$ of the area under the standard normal curve falls between $-z_{\alpha/2}$ and $z_{\alpha/2}$. Thus, using the fact that the sampling distribution of $\overline{X}$ can be approximated by the normal distribution with the mean μ and the standard error $\sigma_{\overline{X}} = \dfrac{\sigma}{\sqrt{n}}$, we find from figure 8-6 that $\overline{X}$ will differ from μ by less than $z_{\alpha/2} \cdot \dfrac{\sigma}{\sqrt{n}}$ just about $1 - \alpha$ of the time. That is,

> The probability that a sample mean will differ from the population mean by less than $z_{\alpha/2} \cdot \dfrac{\sigma}{\sqrt{n}}$ is $1 - \alpha$.

The fact that the difference between $\overline{X}$ and μ, or $\overline{X} - \mu$, is less than $z_{\alpha/2} \cdot \dfrac{\sigma}{\sqrt{n}}$ can be expressed as follows:

$$z_{\alpha/2} \cdot \frac{\sigma}{\sqrt{n}} > \overline{X} - \mu > - z_{\alpha/2} \cdot \frac{\sigma}{\sqrt{n}}$$

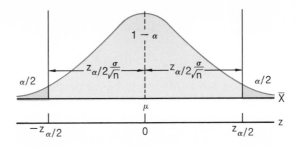

FIGURE 8-6
Sampling distribution of the mean
($\overline{X}$)

If we multiply each of the terms of this expression by -1, we must reverse all the negative, positive, and inequality signs. The expression will then become

$$-z_{\alpha/2} \cdot \frac{\sigma}{\sqrt{n}} < -\overline{X} + \mu < z_{\alpha/2} \cdot \frac{\sigma}{\sqrt{n}}$$

Adding $\overline{X}$ to each of the terms, we obtain

$$\overline{X} - z_{\alpha/2} \cdot \frac{\sigma}{\sqrt{n}} < \mu < \overline{X} + z_{\alpha/2} \cdot \frac{\sigma}{\sqrt{n}} \tag{8-6}$$

That is, when a sample mean is computed from a sample of size n, the probability is $1 - \alpha$ that the interval from $\overline{X} - z_{\alpha/2} \cdot \frac{\sigma}{\sqrt{n}}$ to $\overline{X} + z_{\alpha/2} \cdot \frac{\sigma}{\sqrt{n}}$ will include the population mean μ. Such an interval is called the $1 - \alpha$ *confidence interval* of the mean of the population.

The quantity $\overline{X} - z_{\alpha/2} \cdot \frac{\sigma}{\sqrt{n}}$ is called the *lower limit* and $\overline{X} + z_{\alpha/2} \cdot \frac{\sigma}{\sqrt{n}}$ the *upper limit* of the interval. Observe that the width of the interval varies directly with the population standard deviation σ and inversely with the sample size n, given the mean $\overline{X}$ and the confidence level $1 - \alpha$.

EXAMPLE 8-13   Referring to example 8-3, construct a 95 percent confidence interval for μ, the true mean of the IQ's of all pupils in the elementary schools of the country, if $\overline{X}$ is found to be 108.

Since in this example $\sigma = 15$, $z_{\alpha/2} = z_{0.05/2} = z_{0.025} = 1.96$, and $\sigma_{\overline{X}} = \frac{\sigma}{\sqrt{n}} = \frac{15}{\sqrt{25}} = 3$, by equation 8-6, the 95 percent confidence interval for μ based on $\overline{X} = 108$ is as follows:

$$108 - 1.96(3) < \mu < 108 + 1.96(3)$$
$$108 - 5.88 < \mu < 108 + 5.88$$

or

$$102.12 < \mu < 113.88$$

Equation 8-6 is obviously based on the assumption that the population standard deviation is known. When the population standard deviation is unknown, it must be estimated with the sample standard deviation. Therefore, σ in equation 8-6 is replaced by s. This is permissible as long as the sample size is sufficiently large, that is, 30 or more. Thus, when the sample standard deviation is used in place of the population standard deviation, the $1 - \alpha$ confidence interval for the population mean becomes

$$\overline{X} - z_{\alpha/2} \cdot \frac{s}{\sqrt{n}} < \mu < \overline{X} + z_{\alpha/2} \cdot \frac{s}{\sqrt{n}} \tag{8-7}$$

EXAMPLE 8-14   Refer to example 8-5. Using the sample mean $\overline{X} = 79$ ounces, construct a 98 percent confidence interval for μ, the true mean of the weights of the sugar bags.

In this example we have $n = 100$, $s = 5$; therefore, the standard error is $s = \dfrac{5}{\sqrt{100}} = 0.5$. Since $\alpha = 1 - 0.98 = 0.02$, $z_{\alpha/2} = z_{0.02/2} = z_{0.01} = 2.326$. Accordingly, the 98 percent confidence interval for μ is

$$79 - 2.326(0.5) < \mu < 79 + 2.326(0.5)$$
$$79 - 1.163 < \mu < 79 + 1.163$$

or

$$77.837 < \mu < 80.163$$

estimating the difference between two means

If the sample means $\overline{X}_1$ and $\overline{X}_2$ are independent, the procedure for constructing a confidence interval for δ, the true difference between the two population means μ_1 and μ_2, is the same as that for constructing a confidence interval for the mean of a single population. The only difference is in the random variable: $\overline{X}$ is replaced by $\overline{D}$, the difference between two sample means, or $\overline{X}_1 - \overline{X}_2$. Since the standard error of $\overline{D}$ is $\sqrt{\dfrac{\sigma_1^2}{n_1} + \dfrac{\sigma_2^2}{n_2}}$ (by equation 8-2) where σ_1^2 and σ_2^2 are the two population variances, and n_1 and n_2 are the sizes of the two samples yielding the means $\overline{X}_1$ and $\overline{X}_2$, respectively, the $1 - \alpha$ confidence interval for the true difference δ between the two population means is expressed as follows:

$$\overline{D} - z_{\alpha/2} \sqrt{\frac{\sigma_1^2}{n_1} + \frac{\sigma_2^2}{n_2}} < \delta < \overline{D} + z_{\alpha/2} \sqrt{\frac{\sigma_1^2}{n_1} + \frac{\sigma_2^2}{n_2}} \tag{8-8}$$

EXAMPLE 8-15   In example 8-6, the following quantities are given.

	Brand I	*Brand II*
Sample size	$n_1 = 100$	$n_2 = 100$
Sample mean	$\overline{X}_1 = 22$	$\overline{X}_2 = 24$
Population variance	$\sigma_1^2 = 36$	$\sigma_2^2 = 64$

Find the lower and upper limits of the 95 percent confidence interval for the true difference between the average nicotine content of cigarette brand I and that of brand II.

The confidence interval for the true difference between the two means is computed by applying equation 8-8 as follows:

$$(22 - 24) - 1.96 \sqrt{\frac{36}{100} + \frac{64}{100}} < \delta < (22 - 24) + 1.96 \sqrt{\frac{36}{100} + \frac{64}{100}}$$

$$-2 - 1.96(1) < \delta < -2 + 1.96(1)$$

or

$$-3.96 < \delta < -0.04$$

Therefore the lower and upper limits of the 95 percent confidence interval for the true difference are -3.96 and -0.04, respectively.

It is hardly necessary to point out that when the population variances are unknown, the sample variances s_1^2 and s_2^2 shall be used to substitute for σ_1^2 and σ_2^2, respectively, as long as both n_1 and n_2 are large. Accordingly, the $1 - \alpha$ confidence interval for the true difference between two population means is expressed in the following form.

$$\overline{D} - z_{\alpha/2} \sqrt{\frac{s_1^2}{n_1} + \frac{s_2^2}{n_2}} < \delta < \overline{D} + z_{\alpha/2} \sqrt{\frac{s_1^2}{n_1} + \frac{s_2^2}{n_2}} \qquad (8\text{-}9)$$

EXAMPLE 8-16   Find the 95 percent confidence interval for δ, the true difference between the mean weights gained by using the two diets mentioned in example 8-9.

By applying equation 8-9, we compute confidence interval as follows:

$$(185 - 190) - 1.96 \sqrt{\frac{180}{100} + \frac{220}{100}} < \delta < (185 - 190) + 1.96 \sqrt{\frac{180}{100} + \frac{220}{100}}$$

$$-5 - 1.96(2) < \delta < -5 + 1.96(2)$$

$$-5 - 3.92 < \delta < -5 + 3.92$$

or

$$-8.92 < \delta < -1.08$$

estimation of the population proportion

Just as interval estimation of the population mean is based on the sample mean, interval estimation of the population proportion p is based on the sample proportion $\hat{p}$, or $\dfrac{X}{n}$, where X is a binomial random variable designating the number of successes in n trials. As indicated previously, when n is large, X is approximately normal with a mean $\mu = np$ and a standard deviation $\sqrt{np(1 - p)}$. When the population proportion is unknown, it has to be estimated from the sample proportion $(\hat{p})$. Thus the estimated standard deviation of the binomial random variable X is

$$s = \sqrt{n\hat{p}(1 - \hat{p})}$$

Accordingly, for a large n, the approximated $1 - \alpha$ confidence interval for μ or np, the true mean of X, can be expressed as follows:

$$X - z_{\alpha/2}\sqrt{n\hat{p}(1 - \hat{p})} < np < X + z_{\alpha/2}\sqrt{n\hat{p}(1 - \hat{p})} \qquad (8\text{-}10)$$

Furthermore, if we divide each of the terms of this expression by n, we have

$$\frac{X}{n} - z_{\alpha/2}\sqrt{\frac{\hat{p}(1 - \hat{p})}{n}} < p < \frac{X}{n} + z_{\alpha/2}\sqrt{\frac{\hat{p}(1 - \hat{p})}{n}}$$

or

$$\hat{p} - z_{\alpha/2}\sqrt{\frac{\hat{p}(1 - \hat{p})}{n}} < p < \hat{p} + z_{\alpha/2}\sqrt{\frac{\hat{p}(1 - \hat{p})}{n}} \qquad (8\text{-}11)$$

where the ratio $\hat{p}(1 - \hat{p})/n$ is the estimated variance of the sample proportion $\hat{p}$.

EXAMPLE 8-17   Suppose that a marketing researcher wishes to estimate the true proportion of housewives who prefer a particular brand of detergent. A random sample of 100 housewives is selected, and 64 of the 100 housewives indicate that they prefer that brand of detergent. Find the 99 percent confidence interval for p, the true proportion of all housewives who prefer that brand of detergent.

Since the sample proportion $\hat{p}$ is $\frac{64}{100}$ and $1 - p$ is $\frac{36}{100}$, and $z_{0.01/2} = 2.575$, by equation 8-11, the 99 percent confidence interval for p is

$$0.64 - 2.575\sqrt{\frac{0.64(0.36)}{100}} < p < 0.64 + 2.575\sqrt{\frac{0.64(0.36)}{100}}$$
$$0.64 - 2.575(0.048) < p < 0.64 + 2.575(0.048)$$
$$0.64 - 0.1236 < p < 0.64 + 0.1236$$

or

$$0.5164 < p < 0.7636$$

8-6 exercises

1. What is interval estimation? How is it different from point estimation?

2. The symbol α is usually used in both hypothesis testing and interval estimation. Explain what α designates when used in interval estimation.

3. Suppose that a psychologist wishes to make an interval estimation of the true mean of the IQ's of pupils in a certain ethnic group. It is known that IQ's are normally distributed with a standard deviation of 15. Construct a 95 percent confidence interval for the true mean (μ) on the basis of a sample of 25 pupils with a sample mean of 105.

4. A flour manufacturing company packs flour into paper bags. The company wishes to estimate the true mean weight of the bags. A sample of 36 bags yields a sample mean of 24.5 pounds and a standard deviation of 1.5 pounds. Find the 99 percent confidence interval for μ, the true mean weight of the flour bags.

5. Referring to problem 3, exercises 8-4, determine the 98 percent confidence interval for the true difference in the effectiveness of the two teaching methods.

6. Referring to problem 4, exercises 8-4, determine the 99 percent confidence interval for the true difference between the average appraised values of homes.

7. Using the data given in problem 8, exercises 8-4, determine the 95 percent confidence interval for the true difference between the average service lives of the two brands of automobile batteries.

8. Using the data given in problem 9, exercises 8-4, construct a 98 percent confidence interval for the true difference between the two means.

9. Suppose that a telephone interview is conducted to determine whether or not the respondent is a homeowner. A random sample of 100 people is selected from the telephone directory, and 50 of them are found to be homeowners. Find the 98 percent confidence interval for μ, the true mean of homeowners in a sample of 100.

10. In a randomly selected sample of 64 freshman girls, 32 of them are identified as married. Determine the 95 percent confidence interval for p, the true proportion of all freshman girls who are married.

the t test

9

9-1
introduction

We shall continue to discuss the techniques of hypothesis testing in this chapter. Up to now the discussions have involved situations in which either the researcher has a knowledge of the population standard deviation or the sample used is large. In most cases, however, the researcher possesses only sample data and the sample is sometimes small. Fortunately, there are techniques available to test hypotheses under these conditions.

As indicated in section 8-3, when the population standard deviation is unknown, it has to be estimated by the sample standard deviation. We shall now consider procedures of hypothesis testing and estimation concerning population means and the difference between two means when the standard deviation computed from a small sample is used to replace the standard deviation of the population.

Recall that in testing a hypothesis about a population mean when the population standard deviation is unknown, the ratio

$$\frac{X - \mu}{s / \sqrt{n}}$$

is used as a Z ratio as long as the sample is large. When the sample involved is small, however, s will very likely be quite different from σ; and, consequently, the above ratio can no longer be used as a Z ratio.

The problem of estimating a population parameter from a small sample was tackled with success by W. S. Gosset, who, under the pen name "Student," published a theoretical sampling distribution in 1908. He referred to the ratio mentioned above as t. It is also known as Student's t; and its sampling distribution is often referred to as the Student's t distribution.

Because the mathematical expression of the t function is rather complicated, it is omitted here. Instead, we shall concentrate on some of its characteristics, a few related concepts, and its application to small-sample cases. Section 3 deals with tests concerning the population mean; section 4 is concerned with tests involving the difference between two means.

To avoid confusion, the capital letter T is used to designate the test statistic. For testing about a population mean, we have

$$T = \frac{\overline{X} - \mu}{s/\sqrt{n}} \qquad (9\text{-}1)$$

which is to be computed from sample data. The lower-case letter t is used only to denote the critical value taken from appendix F. The test statistic T used in testing hypotheses about the difference between two means will be defined later.

9-2
the *t* distribution

The t distribution is based on the assumption that the population from which the sample is taken has a normal, or at least approximately normal, distribution. Given this fact, it is permissible for the researcher to use the t distribution for testing hypotheses about the population mean or the difference between two means, even though he does not know the population standard deviation. If there is a drastic departure from normality in the distribution of the population from which the sample is drawn, it is advisable to use some of the *nonparametric methods* that will be discussed later in this text. (See chapter 13.)

Like the standard normal z distribution, the t distribution is also continuous, bell-shaped, and perfectly symmetrical. The only difference between the two distributions is that the t distribution has greater variability; the t curve is more spread out in the tails and is flatter around the center. Figure 9-1 compares the two types of curves. As the sample size increases, the t curve approaches the normal curve; when n approaches infinity, the t curve becomes identical with the normal curve. In other words, as n becomes larger, the estimator s becomes closer to σ; when the sample size approaches the population size, s approaches σ, and there is no longer any difference between the t and z ratios.

The greater variability of the t distribution can be explained by referring to the t ratio. While the z ratio contains only *one* random variable, namely $\overline{X}$ (n and σ are constants), the t ratio contains *two* random variables, namely, $\overline{X}$ and s, which are independent of each other. Thus when $\overline{X}$ is very small, s may be large, and vice versa. Consequently, t is more variable than z.

In short, the t distribution has a mean 0 and a standard deviation that is usually greater than 1; it approaches 1 when the sample size n approaches infinity. Thus, unlike the z score, which has only one distribution with the mean 0 and the standard deviation 1, the t ratio has a *family* of distributions, each of which has the same mean (0) but a different standard deviation depending on the value of n. Here we need to introduce another important concept, namely, the concept of *degrees of freedom*.

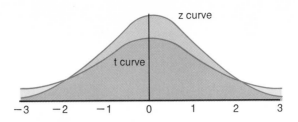

FIGURE 9-1
Comparison of the standard normal
(*z*) and *t* distributions

It is not necessary for us to discuss this concept in great detail here. All we need to know now is that the number of degrees of freedom, commonly designated as ν (the Greek letter nu), is equal to $n - 1$, or one less than the sample size. It is the sole parameter of the *t* distribution. That is to say that the shape of the *t* curve is completely defined when the sample size, and consequently the number of degrees of freedom, is given.

Since for each possible value of ν there is a different distribution of *t*, a *t* table similar to the standard normal *z* table would run as many pages as the values of ν. Fortunately, such a table is unnecessary in most cases. A *t* table need contain only the frequently used probabilities such as those shown in the top two rows of appendix F. Row Q lists the probabilities equal to the upper-tail (or lower-tail) area of the distribution for ν that must be used in a one-tailed test. Each of the probabilities listed in row 2Q is equal to the sum of both the upper- and lower-tail areas of the distribution with ν degrees of freedom. The row 2Q is used for a two-tailed test; the values of ν are listed in the first column of the table. The table entry in a given row under a specified probability is the critical *t* value that cuts off the rejection region of the distribution with ν degrees of freedom. Although only positive *t* values are entered in the table, the negative values are implied because of the fact that the *t* distribution is perfectly symmetrical around $t = 0$. The following example is designed to illustrate the use of the *t* table, appendix F.

EXAMPLE 9-1　Suppose that a certain problem involves a level of significance of 0.05 and a sample of 20 observations. Find the critical *t* values for (a) a one-tailed test with the rejection region in the upper-tail area, (b) a one-tailed test with the rejection region in the lower-tail area, and (c) a two-tailed test.

a) With $\alpha = 0.05$, and $\nu = 20 - 1 = 19$, we find that the table entry in row 19 and column 0.05 for Q in appendix F is 1.729. That is, the critical *t* value with 19 degrees of freedom and a 0.05 level of significance, which is usually designated as $t_{(19, \, 0.05)}$, is 1.729. It is the *t* value that cuts off the upper 5 percent of the *t* distribution.

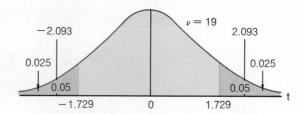

FIGURE 9-2
Critical *t* values for one- and two-
tailed tests involving a distribution
with $\nu = 19$ and $\alpha = 0.05$

b) Since the *t* distribution is symmetrical around $t = 0$, -1.729 is the critical *t* value that cuts off the lower 5 percent of the distribution.

c) For a two-tailed test, the α value is the value for 2Q. We find that $t_{(19,\ 0.05/2)} = \pm 2.093$, which is the table entry in row 19 and column 0.05 for 2Q.

Figure 9-2 shows the critical *t* values for both one-tailed and two-tailed tests at $\alpha = 0.05$ for a *t* distribution with the degrees of freedom $\nu = 19$.

We can see for ourselves that the *t* distribution with a large *n* approaches the *z* distribution by referring to the *t* table, appendix F. We know that the *z* value that cuts off the upper 5 percent of the *z* distribution is 1.645. Looking at the *t* table, we find that the *t* values that cut off the upper 5 percent of the *t* distributions for $\nu = 10, 20, 30$, and 60 are 1.812, 1.725, 1.697, and 1.671, respectively. The *t* value becomes closer and closer to 1.645 as the value of ν becomes larger and larger. When ν approaches infinity, the *t* value that cuts off the upper 5 percent of the *t* distribution becomes 1.645, the same as the *z* value.

9-2
exercises

1. What are some of the characteristics of the *t* curve?

2. List the similarities and differences between the *t* distribution and the standard normal distribution.

3. How many random variables are involved in the *t* ratio? What are they?

4. Suppose that a random variable X has a mean (μ) of 100 and that a sample of 100 observations yields a mean ($\overline{X}$) and a standard deviation (s). If the standard deviation of X is assumed to be 10, then the variability of the T ratio will depend only on $\overline{X}$. Thus we have

$$T = \frac{110 - 100}{10/\sqrt{100}} = \frac{10}{1} = 10 \quad \text{for } \overline{X} = 110$$

and

$$T = \frac{150 - 100}{10/\sqrt{100}} = \frac{50}{1} = 50 \quad \text{for } \overline{X} = 150$$

Since the standard deviation is unknown and its estimate s is a variable which may assume very low or very high values, the value of T will depend on both $\overline{X}$ and s. What will be the value of T if

a) $s = 20$ for $\overline{X} = 110$?
b) $s = 5$ for $\overline{X} = 150$?
c) From the above problems, what conclusions can you make about the variability of T?

5. Recall that the standard normal variable z has only one distribution with a mean of 0 and a standard deviation of 1. How many distributions does the ratio t have?

6. How is the t distribution defined? If a certain test involves a sample of 30 observations, what is the number of degrees of freedom?

7. For a distribution with 10 degrees of freedom, find the critical t value that cuts off each of the following areas under the curve.

a) The upper 2.5 percent   c) The upper 0.005
b) The lower 5 percent   d) The lower 0.01

8. Suppose that a sample of n observations is selected for the normal random variable X, which has a mean (μ) of 10.

a) If $n = 16$, $\overline{X} = 5$, and $s = 4$, what is T?
b) If $n = 25$, $\overline{X} = 8$, and $s = 10$, what is T?
c) If $n = 25$, $\overline{X} = 14$, and $T = 1.25$, what is s?
d) If $n = 9$, $s = 6$, and $T = -2.5$, what is $\overline{X}$?

9. Suppose that a certain test involves a 0.10 level of significance and a sample of 25 observations. Find the critical t value under each of the following conditions and show each answer graphically.

a) A one-tailed test with the rejection region in the upper-tail area
b) A one-tailed test with the rejection region in the lower-tail area
c) A two-tailed test

9-3 tests about the population mean

When a sample is small, the sample variance (s^2) can deviate a great deal from the population variance (σ^2)—so much so that a decision made on the basis of the z score could contain a serious error. Under such circumstances, the t test, the test involving the use of the t distribution, must be used. The test statistic T is compared with the critical t value (taken from appendix F), and inferences can then be made about a population mean.

As pointed out earlier, the only condition that has to be met before applying the *t* test is that the original random variable X be normally distributed.

EXAMPLE 9-2   Because of the generous trade-in allowances offered by a certain automobile dealer, his average profit margin per car sold is suspected to be below the national average of $500. A study is conducted to determine whether this is actually the case. A random sample of 25 sales shows a sample mean of $485 and a standard deviation of $45. Assuming that the profit margin for each car sold by this dealer is normally distributed, shall we conclude that his average profit margin is indeed significantly lower than $500 at $\alpha = 0.05$?

The two hypotheses to be tested are

$$H_0 : \mu = 500 \quad \text{and} \quad H_1 : \mu < 500$$

The *t* table, appendix F, shows that for a one-tailed test at $\alpha = 0.05$ and 24 degrees of freedom, the critical *t* value is -1.711. It is a negative value because the alternative hypothesis states that the mean is less than $500. That is, the critical region is under the lower tail of the *t* distribution. Thus the decision rule is

$$\text{Reject } H_0 \text{ if } T \leq -1.711$$

For $\overline{X} = 485$ the test statistic T is computed as

$$T = \frac{485 - 500}{45/\sqrt{25}} = \frac{-15}{9} = -1.667$$

which is greater than the critical *t* value. Accordingly, we shall not reject H_0. Observe that we would reject H_0 if the z score -1.645 were used as the critical value. Figure 9-3 shows the *t* distribution and the rejection region for this test.

EXAMPLE 9-3   A vending machine is supposed to discharge 8 ounces of coffee if the correct coins are inserted. To test whether the machine is operating properly, 16 cups of coffee are taken from the machine and measured. It is found that the mean and standard deviation of the 16 measurements are 7.5 and 0.8 ounces, respectively. Test the null hypothesis that the machine is operating properly ($\mu = 8$ ounces) against the alternative that it is not at two levels of significance: 0.01 and 0.05.

The hypotheses to be tested in this example are

$$H_0 : \mu = 8 \quad \text{and} \quad H_1 : \mu \neq 8$$

At $\alpha = 0.01$ and $16 - 1$, or 15, degrees of freedom, the critical *t* values for a two-tailed test are 2.947 and -2.947. Thus the decision rule is

$$\text{Reject } H_0 \text{ if } T \geq 2.947 \text{ or } T \leq -2.947$$

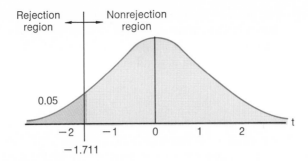

FIGURE 9-3
The t distribution for $\nu = 24$
showing the rejection region for
example 9-2

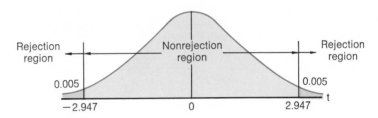

FIGURE 9-4
The t distribution for $\nu = 15$
showing the rejection region with
$\alpha = 0.01$ for example 9-3

Since the T value is

$$T = \frac{7.5 - 8}{0.8/\sqrt{16}} = \frac{-0.5}{0.2} = -2.5$$

which is greater than -2.947, but smaller than 2.947, we do not reject H_0 at the 1 percent level. Figure 9-4 shows the t distribution and rejection regions for this test.

At $\alpha = 0.05$, the critical t values are 2.131 and -2.131. Since $T = -2.5$, we reject H_0 at the 5 percent level.

It is hardly necessary to remind the reader that for small samples the confidence interval is constructed by the use of the t scores consistent with the α value and the degrees of freedom. The method of constructing a confidence interval for the true mean of a population is the same as that shown in equation 8-2 except that *t* values rather than *z* scores are used. Accordingly, the $(1 - \alpha)$ confidence interval for μ on the basis of a mean $\overline{X}$ computed from a small sample is given below.

$$\overline{X} - t_{(v,\ \alpha/2)}\frac{s}{\sqrt{n}} < \mu < \overline{X} + t_{(v,\ \alpha/2)}\frac{s}{\sqrt{n}} \qquad (9\text{-}2)$$

EXAMPLE 9-4 Referring to example 9-3 above, determine the 99 percent confidence interval for μ.

At $\alpha = 0.01$, the *t* values that cut off 0.5 of 1 percent at each end of the *t* distribution with 15 degrees of freedom are 2.947 and -2.947. By equation 9-2 the 99 percent confidence interval of μ is found to be

$$7.5 - t_{(15,\ 0.005)}\frac{0.8}{\sqrt{16}} < \mu < 7.5 + t_{(15,\ 0.005)}\frac{0.8}{\sqrt{16}}$$
$$7.5 - 2.947(0.2) < \mu < 7.5 + 2.947(0.2)$$
$$7.5 - 0.5894 < \mu < 7.5 + 0.5894$$

or

$$6.9106 < \mu < 8.0894$$

9-3 exercises

1. When the variance is unknown, the procedure for testing hypotheses about the population mean is usually different for small and large samples. Explain the difference.

2. A car leasing company has been using tires with an average life of 30,000 miles. The company is willing to pay more for better tires. A tire vendor claims that his tires average more than 30,000 miles. A sample of 16 tires provided by this vendor yields an average of 31,000 miles and a standard deviation of 2000 miles. It is assumed that the life length of tires is normally distributed. Is there sufficient evidence to support the vendor's claim at $\alpha = 0.025$?

3. The federal Food and Drug Administration is conducting a test to determine whether a new medicine has the undesirable side effect of elevating body temperature. It is understood that the temperature of the human body is normally distributed with a mean of 98.6 degrees Farenheit. The new medicine is administered to nine patients, temperatures are taken, and a sample mean of 99 degrees and a standard deviation of 0.36 degrees are obtained. Should the company be per-

mitted to market the new drug if the level of significance is specified at 0.01?

4. A certain kind of screw produced by an automatic machine should average three inches in length. It is suspected, however, that the machine is no longer functioning properly, and that the screws it produces are, on the average, either longer or shorter than three inches. Suppose that a sample of 25 screws yields an average length of 2.9 inches and a standard deviation of 0.25 inches. At $\alpha = 0.10$ shall we conclude that the machine is functioning improperly? What is the 90 percent confidence interval for the true average length of the screws now that the sample average is 2.9 inches?

5. A production process is considered to be functioning properly when the average amount of instant coffee that is packed in a jar is six ounces. A random sample of 16 jars is selected; the sample average is found to be 6.1 ounces, with a standard deviation of 0.2 ounces. The level of significance is specified at 0.05.

   a) Is the process functioning properly?
   b) What are the 95 percent confidence limits for μ, the true average, in the light of the sample information?

6. In a sample of 25 light bulbs, it is found that the mean service life is 1580 hours, and that the sample variance is 14,400 hours.

   a) Test the hypothesis $\mu = 1600$ hours against the alternative $\mu \neq 1600$ with α specified at 0.01.
   b) Find the 99 percent confidence limits for μ.

7. From past records it is known that the average weight of hogs six months after birth is 100 pounds or less. A new diet fed to a group of 16 hogs since birth is supposed to increase the weight gain.

   a) Design a decision rule for rejecting the null hypothesis at a significance level of 1 percent.
   b) If the average weight of the sample of 16 hogs is found to be 107 pounds, and the sample standard deviation is 18 pounds, what will be your decision according to the rule in (a)?

8. The average weight of army recruits is believed to be normally distributed with a mean of 160 pounds. In a random sample of 25 recruits, the mean is 150 pounds and the standard deviation is 20 pounds.

   a) Test the null hypothesis against the alternative that the average weight of the most recent army recruits is different from 160 pounds at $\alpha = 0.02$.
   b) Find the 98 percent confidence interval for the true mean.

**9-4
testing the
difference
between
two means**

When two populations are both normally—or approximately normally—distributed, and when at least one sample size is small (less than 30), we use the *t* test to make decisions about differences between the population means. In using the *t* tests, however, we must remember that procedures are different for independent and dependent samples.

independent samples

Two samples are referred to as *independent* if the observations in one sample are not in any way related to the observations in the other. When testing about the difference between two population means on the basis of independent samples, we must make one additional assumption: that the two populations have an identical variance. The reason why this assumption has to be made is not a matter of concern here; we need only to point out that if the two population variances are identical, that is, $\sigma_1^2 = \sigma_2^2 = \sigma^2$, the variance of the difference between two sample means will become

$$\sigma_{\overline{D}}^2 = \frac{\sigma^2}{n_1} + \frac{\sigma^2}{n_2} = \sigma^2 \left(\frac{1}{n_1} + \frac{1}{n_2} \right)$$

It is then necessary to estimate only σ^2 for obtaining $s_{\overline{D}}^2$, the estimated variance of $\overline{D}$. Since the best estimator of σ^2 is s^2, we have

$$s_{\overline{D}}^2 = s^2 \left(\frac{1}{n_1} + \frac{1}{n_2} \right) \tag{9-3}$$

Our task of estimating $\sigma_{\overline{D}}^2$ is now reduced to finding the expression for s^2. It can be expected that the estimation of σ^2 will be improved if s^2 is computed by pooling the data in both the samples rather than by using either sample alone. The pooled estimator s^2 is obtained as follows:

$$s^2 = \frac{(n_1 - 1)s_1^2 + (n_2 - 1)s_2^2}{n_1 + n_2 - 2}$$

where s_1^2 and s_2^2 are the sample variances as defined in equation 5-10. Substituting this expression for s^2 in equation 9-3, and then taking the square root, we obtain the standard error of the difference between two sample means as follows:

$$s_{\overline{D}} = \sqrt{\frac{(n_1 - 1)s_1^2 + (n_2 - 1)s_2^2}{n_1 + n_2 - 2} \left(\frac{1}{n_1} + \frac{1}{n_2} \right)} \tag{9-4}$$

This quantity is the denominator of the T ratio used for testing the hypothesis about the difference between two population means. Accordingly, the test statistic used in testing such a hypothesis becomes

$$T = \frac{\overline{X}_1 - \overline{X}_2}{\sqrt{\dfrac{(n_1 - 1)s_1^2 + (n_2 - 1)s_2^2}{n_1 + n_2 - 2} \left(\dfrac{1}{n_1} + \dfrac{1}{n_2}\right)}} \qquad (9\text{-}5)$$

which is then compared with the critical t value for the α level of significance and $n_1 + n_2 - 2$ degrees of freedom.

EXAMPLE 9-5   An educator wants to know whether two different teaching methods have identical effects on learning. Two classes of students are selected at random and exposed to the two different methods. A comprehensive standard examination is then administered to the two classes to determine the effectiveness of the two methods. The relevant data are shown below.

	Class I	Class II
Sample size	$n_1 = 18$	$n_2 = 12$
Average test score	$\overline{X}_1 = 85$	$\overline{X}_2 = 80$
Sample variance	$s_1^2 = 36$	$s_2^2 = 34$

Assuming that the test scores for all potential students taught by each method are distributed normally, and have an identical variance, test the null hypothesis that the two teaching methods are equally effective at (a) $\alpha = 0.01$ and (b) $\alpha = 0.05$.

This is obviously a two-tailed test since the alternative hypothesis must be that the two teaching methods have different effectiveness. Thus the two hypotheses involved in this test are

$$H_0: \mu_1 - \mu_2 = 0 \quad \text{and} \quad H_1: \mu_1 - \mu_2 \neq 0$$

The test statistic T is computed by applying equation 9-5 as follows:

$$T = \frac{85 - 80}{\sqrt{\dfrac{17(36) + 11(34)}{18 + 12 - 2}\left(\dfrac{1}{18} + \dfrac{1}{12}\right)}} = \frac{5}{\sqrt{\dfrac{986}{28}\left(\dfrac{2}{36} + \dfrac{3}{36}\right)}}$$

$$= \frac{5}{\sqrt{\left(\dfrac{493}{14}\right)\left(\dfrac{5}{36}\right)}} = \frac{5}{\sqrt{\dfrac{2465}{504}}} = \frac{5}{\sqrt{4.89}} = \frac{5}{2.2113} = 2.26$$

a) At $\alpha = 0.01$ and $\nu = 28$, the critical t values are ± 2.763. Accordingly, the decision rule is

Reject H_0 if $T \geq 2.763$ or $T \leq -2.763$

Since the computed T value is 2.26, we do not reject H_0 at the 1 percent level.

b) At $\alpha = 0.05$ and $\nu = 28$, the critical *t* values are ± 2.048. Thus the decision rule is

Reject H_0 if $T \geq 2.048$ or $T \leq -2.048$

Since the computed T value is larger than 2.048, we reject H_0 at the 5 percent level.

The standard error of $\overline{D}$, as defined in equation 9-4, together with the appropriate *t* values, must be used for constructing the confidence interval for δ, the true difference between two population means, when both the samples are small and independent. The $(1 - \alpha)$ confidence interval for δ is expressed below.

$$\overline{D} - t_{(\nu, \alpha/2)} s_{\overline{D}} < \delta < \overline{D} + t_{(\nu, \alpha/2)} s_{\overline{D}} \tag{9-6}$$

EXAMPLE 9-6 Referring to example 9-5 above, determine the 95 percent confidence interval for the true difference between the two population means on the basis of the difference between the two sample means.

Since in the previous example, $\overline{D} = 5$, $t_{(\nu, \alpha/2)} = t_{(28, \ 0.025)} = \pm 2.048$, and $s_{\overline{D}} = 2.2113$, by applying equation 9-6 the 95 percent confidence interval for δ is constructed as follows:

$$5 - 2.048(2.2113) < \delta < 5 + 2.048(2.2113)$$

or

$$0.47 < \delta < 9.53$$

dependent samples

In this chapter and the last, we have repeatedly stated that the tests about the difference between two population means apply only when the samples are independent. It happens frequently, however, that observations are sampled in pairs, and that each observation in one sample is paired or related in some way with an observation in another sample. The two observations in each matched pair are said to be *dependent* because knowing one helps to reduce the error of predicting the other. For instance, if a student can do well in one subject, say, English, he will very likely do well in another subject, say, history. Thus, knowing his performance in English would help in predicting his performance in history.

It can be mathematically demonstrated that when samples are dependent, the *variance of the difference* between two sample means *is no longer equal to the sum of the variances* of the sample means. A new procedure must be found to handle such problems. We must derive a new random variable—the difference between two observations of each matched pair—and then calculate the mean and standard deviation of the differences. Let D be the difference between two observations of each matched pair; then the mean of n observations for D is

$$\overline{D} = \frac{\sum\limits_{i=1}^{n} D_i}{n} \tag{9-7}$$

the variance of D as defined in equation 5-8 is

$$\hat{s}_D^2 = \frac{\Sigma D^2}{n} - \overline{D}^2$$

The standard error of $\overline{D}$ as an estimator of $\sigma_{\overline{D}}$ is

$$s_{\overline{D}} = \frac{\hat{s}_D}{\sqrt{n-1}} \tag{9-8}$$

and finally the test statistic T becomes

$$T = \frac{\overline{D}}{s_{\overline{D}}} \tag{9-9}$$

Table 9-1   Performance in English and history classes by 11 students

English X_1	History X_2	Difference $D = X_1 - X_2$	Square of the difference (D^2)
88	85	3	9
93	96	−3	9
75	65	10	100
68	73	−5	25
98	88	10	100
58	65	−7	49
79	80	−1	1
83	75	8	64
88	93	−5	25
52	48	4	16
67	70	−3	9
Total		$\Sigma D = 11$	$\Sigma D^2 = 407$

EXAMPLE 9-7   We want to find out whether a student's performance in English is, on the average, different from his performance in history. Suppose further that a sample of 11 students is selected and their grades for these two subjects are obtained. (The grades are recorded in the first two columns of table 9-1.) Shall we conclude that the average performance in English is the same as the average performance in history at $\alpha = 0.02$?

Let the average performance in English be designated by μ_1 and that in history by μ_2. The hypotheses involved in this test are

$$H_0: \mu_1 - \mu_2 = 0 \quad \text{and} \quad H_1: \mu_1 - \mu_2 \neq 0$$

At $\alpha = 0.02$ and $\nu = 10$, the critical *t* values are ± 2.764. Thus the decision rule for this test is

$$\text{Reject } H_0 \text{ if } T \geq 2.764 \text{ or } T \leq -2.764$$

For $n = 11$, $\Sigma D = 11$, and $\Sigma D^2 = 407$ we have the mean of D

$$\overline{D} = \frac{11}{11} = 1$$

and the variance of D

$$\hat{s}_D^2 = \frac{407}{11} - 1^2 = 37 - 1 = 36$$

Consequently, by equation 9-8, the estimated standard error of $\overline{D}$ becomes

$$s_{\overline{D}} = \sqrt{\frac{36}{11-1}} = \sqrt{3.6} = 1.8974$$

Therefore, the test statistic T is calculated by equation 9-9 as follows:

$$T = \frac{1}{1.8974} = 0.527$$

which is greater than -2.764 but smaller than 2.764. We conclude that the observed average difference of 1 point is not significantly different from 0 and can be attributable to chance. In other words, we do not reject the null hypothesis that $\mu_1 = \mu_2$.

9-4 exercises

1. When the population variances are unknown, the procedure for testing hypotheses about the difference between two population means is usually different for small and large sample cases. Explain the difference.

2. When the population variances are unknown, the procedure for testing hypotheses about the difference between two population means is different for independent and dependent samples. Explain the difference.

3. Two different automobile engines are tested to determine whether they exhibit differences in pollution control. In a 16 day test of engine

I, the measurements indicated an average pollution index of 60 and a standard deviation (s_1) of 9; in a 16 day test of engine II, the measurements indicated an average pollution index of 55 and a standard deviation (s_2) of 9. The measurements are believed to have a normal distribution, and the two samples are obviously independent. Is there sufficient evidence that engine II has better pollution control than engine I at $\alpha = 0.05$?

4. Two machines produce identical bolts. The lengths of the bolts produced by the two machines are believed to be normally distributed and to have the same variance. It is now suspected that the average length of bolts produced by one machine is no longer equal to that produced by the other. Two independent samples are taken and the relevant sample data are shown below.

	Machine I	*Machine II*
Sample size	$n_1 = 8$	$n_2 = 10$
Sample average in inches	$\overline{X}_1 = 2.6$	$\overline{X}_2 = 2.5$
Sample variance in inches	$s_1^2 = 0.0055$	$s_2^2 = 0.0046$

a) Do the data show that there is a significant difference between the two sample means at $\alpha = 0.01$?
b) Find the 99 percent confidence interval for the true difference between the two population means.

5. We wish to determine whether grade-point averages (GPA) differ for boys and girls. It is assumed that the GPA is normally distributed with an identical variance for both sexes. Two independent samples of five students each yield the following observations.

GPA for boys: 2.9 3.1 2.7 3.3 3.0
GPA for girls: 3.6 2.8 3.6 3.2 2.8

a) Using $\alpha = 0.05$, test the hypothesis that the mean GPA for boys is the same as the mean GPA for girls against the alternative that the two means are different.
b) Find the 95 percent confidence limits for the true difference between the two population means.

6. A test has been designed to measure the speaking vocabulary of a child. We want to find out whether the average test score of children from one ethnic group is different from the average score for those from another ethnic group. It is believed that the test scores are normally distributed and the population variances for the two groups are identical. The test was given to two independent and randomly selected samples of four-year-old children. The following data were collected.

	Group I	Group II
Sample size	$n_1 = 10$	$n_2 = 8$
Sample mean	$\overline{X}_1 = 95$	$\overline{X}_2 = 97$
Sample variance	$s_1^2 = 40$	$s_2^2 = 36$

    a) Using $\alpha = 0.05$, test the hypothesis that the speaking vocabulary of the group I children is the same as that of the group II children against the alternative that they are different.

    b) Find the 95 percent confidence limits for the true difference.

7. We want to determine whether a class of 16 students can do equally well in English and mathematics. The test scores listed below are not independent.

Students	English	Mathematics
A	84	84
B	55	57
C	85	90
D	98	97
E	80	74
F	55	53
G	80	75
H	64	63
I	91	90
J	85	82
K	90	88
L	94	98
M	75	77
N	86	90
O	91	85
P	92	86

    a) Test the hypothesis that the population mean score in English is the same as that in mathematics against the alternative that they are different at $\alpha = 0.05$.

    b) Establish the 95 percent confidence interval for the true difference.

8. The management of a large company wishes to know whether a training program has increased the productivity of its workers. Let D be the difference between the hourly output before and after the training program was instituted. From a sample of 25 workers it is found that the mean of D is 26 and the standard deviation $s = 40$. Does the training program have any favorable effect on the productivity of the workers at $\alpha = 0.05$?

9. The productivity levels of a worker before and after a strike are believed to be dependent events. The weekly outputs of 10 workers before and after a strike are listed below. Test the null hypothesis that productivity remains unchanged against the alternative that it has increased at two levels of significance: 0.025 and 0.10.

Worker	Output before strike (B)	Output after strike (A)	Difference ($A - B$)
I	85 units	91 units	6
II	90	91	1
III	88	87	−1
IV	79	88	9
V	95	98	3
VI	86	94	6
VII	89	93	4
VIII	85	80	−5
IX	91	90	−1
X	84	90	6

the χ^2 test

Another frequently used probability distribution is the χ^2 (chi square) distribution. The χ^2 distribution has very broad applications, and in general usefulness ranks second to the normal distribution.

The χ^2 function, like the t function, has only one parameter, the number of degrees of freedom, which is also equal to $n - 1$. There is, however, one important difference between the two functions: the t distribution is always symmetrical around the center at $t = 0$, whereas the χ^2 distribution is *positively skewed,* or skewed to the right, especially when the number of degrees of freedom is small. As ν increases, the distribution becomes less skewed and rapidly approaches symmetry. In figure 10-1, for example, the curve for $\nu = 15$ is almost symmetrical. For a large ν, the χ^2 distribution is approximately normal; in any event, it is continuous and unimodal.

Like the areas under the z and t curves, the area under the χ^2 curve represents probability. A table of probabilities similar to the standard normal z table can be constructed for any χ^2 distribution. However, since there is a different χ^2 distribution for each possible value of ν, and since a full page is required for each distribution table, a comprehensive χ^2 table similar to the z table would run as many pages as there are values of ν.

Fortunately, in most cases, a χ^2 table, like the one reproduced in appendix G, is sufficient. The most commonly used probabilities are listed in the table head (the top row). They are represented by the upper tail area shown in the diagram at the top of the table. Each subsequent row presents the corresponding chi-square values for a particular distribution with the degrees of freedom shown in the left-hand column. Thus, each table entry for a given ν and a given probability shows the chi-square value that cuts off the *upper,* or *right-hand,* tail under the curve. For instance, the chi-square values that cut off the upper 1, 5, 95, and 99 percent of the distribution with 10 degrees of freedom are 23.2093, 18.3070, 3.9403, and 2.55821, respectively.

The χ^2 test can be applied to many kinds of problems, including testing hypotheses about the variance of a population. In the preceding

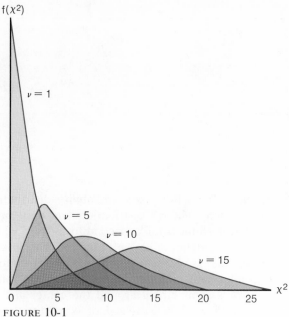

FIGURE 10-1
Chi-square distributions with 1,
5, 10, and 15 degrees of freedom

two chapters, we have been concerned with problems involving measures of central tendency. Our attention was particularly focused upon problems concerning whether a population mean is equal to a certain value and whether two population means are identical. Quite frequently, however, we are confronted with situations that call for a test of the variability of a distribution. The χ^2 test is especially appropriate for this type of test.

Let the square of the English capital letter X, or X^2, designate the test statistic to be used throughout this chapter. It can be mathematically proved that the X^2 variable as defined by

$$X^2 = \frac{(n-1)s^2}{\sigma^2} \qquad (10\text{-}1)$$

is distributed as χ^2 with $(n-1)$ degrees of freedom. Here n is the sample size, s^2 is the sample variance computed by equation 5-10 and σ^2 is the hypothesized population variance. In the next section, we shall illustrate, with concrete examples, how equation 10-1 is used for testing hypotheses about a single population variance. (The testing of hypotheses about the equality of two population variances requires a different test statistic and a different probability distribution; it will be discussed in the next chapter.)

Another well-known application of the χ^2 distribution is the *goodness of fit* test. This is the test used to compare an *observed* frequency distribution with an *expected* frequency distribution. The test statistic used in testing the goodness of fit is the *sum of all the quotients obtained from dividing the squared difference between each observed and expected frequency by the expected frequency*. It has been found that such a test statistic is distributed approximately like the χ^2 variable.

Another technique that also involves the χ^2 distribution and is closely related to the test of goodness of fit is the test of *statistical independence*. In the last two chapters, we have been dealing with inferential problems involving only one, or one type of, variable. Numerous practical problems, however, involve more than one variable. For example, in table 4-1 (page 51), academic training is one variable, and job performance rating is another. If they are independent, then the fact that a person has received college training will not affect the probability that he will be accorded a "good" job performance rating. Thus, if two variables are independent, knowing one will not help predict the other. As another example, Scholastic Aptitude Test (SAT) scores and college grade-point averages (GPA) are two variables. If they are not independent, that is, if they are related in some way, then knowing a student's SAT score will help in predicting his GPA in college.

A test of statistical independence determines, on the basis of empirical evidence, whether any apparent dependence between two variables can be attributed to chance alone or is due to some underlying cause or causes.

The test of independence is quite similar to that of goodness of fit. It involves essentially the same kind of comparison between observed and expected frequencies. The test of independence, however, puts emphasis on the classification of variables; its primary purpose is to find out whether there is sufficient evidence of dependence between the classification of one variable and the classification of another.

This chapter will discuss three kinds of tests, all involving the χ^2 distribution. They are the tests to determine whether a single population variance is equal to a certain value, whether an observed frequency distribution is significantly different from the expected, or theoretical, distribution, and finally whether the classification of one variable is independent of the classification of another.

10-1 exercises

1. Explain the similarity and difference between the χ^2 and t distributions.

2. Recall that the table entries in appendix F are t values that can be

positive or negative. What are the table entries in appendix G? Can they be negative? Which tail areas under the χ^2 curve are represented by the probabilities listed in the top row of appendix G?

3. Given a χ^2 distribution with 20 degrees of freedom, find the χ^2 value that cuts each of the following areas under the curve.

   a) The upper 2.5 percent   d) The lower 5 percent
   b) The upper 10 percent   e) The lower 1 percent
   c) The upper 90 percent

4. Given a χ^2 distribution with 15 degrees of freedom, find the probability of each of the following events:

   a) X^2 is greater than or equal to 30.5779
   b) X^2 is greater than or equal to 24.9958
   c) X^2 is greater than or equal to 14.3389
   d) X^2 is less than 8.54676
   e) X^2 is less than 4.60092

5. We have, in preceding chapters, discussed the procedures of hypothesis testing about population means. Can we also test hypotheses about a population variance? How?

6. What is the test of goodness of fit?

7. What is the test of statistical independence?

10-2 tests concerning a single variance

The basic assumption underlying the discussion in this section is that the population from which we are sampling is normally distributed. It can be mathematically demonstrated that when the population is normally distributed, the test statistic

$$X^2 = \frac{(n-1)s^2}{\sigma^2}$$

is distributed as χ^2 with $(n-1)$ degrees of freedom. Accordingly, the X^2 value can be compared with the critical chi-square value from appendix G for the purpose of hypothesis testing.

Since n and σ^2 in the above ratio are constants, it is not difficult to visualize that the sampling distribution of s^2 is quite closely associated with the distribution of the X^2 variable. Thus, on the basis of s^2 we may test the hypothesis about a single population variance by applying the χ^2 distribution. For this purpose, the χ^2 test can be either one-tailed or two-tailed. If the level of significance (α) is specified and the number of degrees of freedom $(\nu = n - 1)$ is known, and if the competing hypotheses are

$$H_0 : \sigma^2 = \sigma_0^2 \quad \text{and} \quad H_1 : \sigma^2 > \sigma_0^2$$

where σ_0^2 is the known or hypothesized variance under H_0, then the decision rule will be

$$\text{Reject } H_0 \text{ if } X^2 \geq \chi^2_{(v, \alpha)}$$

Here $\chi^2_{(v, \alpha)}$ is the critical chi-square value that cuts off the *upper tail* equal to α for the distribution with $(n - 1)$ degrees of freedom. If the alternative hypothesis is $H_1: \sigma^2 < \sigma_0^2$, then the decision rule will be

$$\text{Reject } H_0 \text{ if } X^2 \leq \chi^2_{(v, 1-\alpha)}$$

where $\chi^2_{(v, 1-\alpha)}$ is the critical chi-square value that cuts off the *lower tail* equal to α for the distribution with $(n - 1)$ degrees of freedom. If the alternative hypothesis is $H_1: \sigma^2 \neq \sigma_0^2$, then the decision rule becomes

$$\text{Reject } H_0 \text{ if } X^2 \geq \chi^2_{(v, \alpha/2)} \text{ or } X^2 \leq \chi^2_{(v, 1-\alpha/2)}$$

which is based on the assumption that the significance level is divided equally among the upper and lower tails of the distribution.

As pointed out previously, the sample statistic s^2 is a good estimator of the population parameter σ^2. Further, s^2 is usually used as a basis for constructing a confidence interval for σ^2. The $(1 - \alpha)$ confidence interval for σ^2 is

$$\frac{(n-1)s^2}{\chi^2_{(v, \alpha/2)}} < \sigma^2 < \frac{(n-1)s^2}{\chi^2_{(v, 1-\alpha/2)}} \tag{10-2}$$

EXAMPLE 10-1   The variance of the breaking strength of a cable manufactured by a certain company was known to be 40,000 pounds. After a new manufacturing process was introduced, however, it was suspected that the variance of the breaking strength had increased. A sample of 10 cables selected at random shows that the variance of the breaking strength is 50,000 pounds. Shall we conclude that there is a significant increase in the variability if the level of significance is specified at $\alpha = 0.01$?

The null and alternative hypotheses involved in this test are

$$H_0: \sigma^2 = 40,000 \quad \text{and} \quad H_1: \sigma^2 > 40,000$$

At $\alpha = 0.01$ and 9 degrees of freedom, the critical chi-square value is 21.666. Thus the decision rule is

$$\text{Reject } H_0 \text{ if } X^2 \geq 21.666$$

Since

$$X^2 = \frac{(n-1)s^2}{\sigma^2} = \frac{9(50,000)}{40,000} = \frac{450,000}{40,000} = 11.25$$

which is smaller than the critical value, we do not reject H_0. That is, on the basis of the sample evidence, we have to say that the apparent increase in variability is not significant at the specified level. (Figure 10-2 shows the χ^2 distribution and the critical value.)

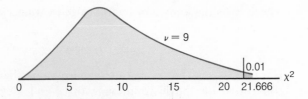

FIGURE 10-2
A χ^2 distribution with 9 degrees
of freedom

EXAMPLE 10-2   The variance of the heights of all college men in the
United States is known to be 30 inches. We believe that because of the
unique climatic conditions in California the variance of the heights of
male students in California colleges may be different from the national
variance. A random sample of 51 male California college students is
selected and measured; the sample variance (s^2) is found to be 25 inches.
Test the null hypothesis that the variance of the heights of California
students is the same as that of all students against the alternative that
they are different at $\alpha = 0.05$. Determine the 95 percent confidence interval for σ^2, the true variance of the heights of all male students in
California colleges.

The null and alternative hypotheses involved here are

$$H_0: \sigma^2 = 30 \quad \text{and} \quad H_1: \sigma^2 \neq 30$$

For a two-tailed test with $\alpha = 0.05$ and $\nu = n - 1 = 50$, the critical chi-
square values are

$$\chi^2_{(50, 0.025)} = 71.4202 \quad \text{and} \quad \chi^2_{(50, 0.975)} = 32.3574$$

Therefore the decision rule for this test is

Reject H_0 if $X^2 \geq 71.4202$ or $X^2 \leq 32.3574$

Since

$$X^2 = \frac{(n-1)s^2}{\sigma^2} = \frac{50(25)}{30} = \frac{1250}{30} = 41.67$$

which falls between the two critical values, we do not reject H_0. (Figure
10-3 shows the χ^2 distribution and the critical values.)

In light of the sample variance $s^2 = 25$, the 95 percent confidence
interval for σ^2 can be constructed by applying equation 10-2 as follows:

$$\frac{(51-1)25}{71.4202} < \sigma^2 < \frac{(51-1)25}{32.3574}$$

or

$$\frac{1250}{71.4202} < \sigma^2 < \frac{1250}{32.3574}$$

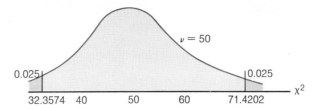

FIGURE 10-3
A χ^2 distribution with 50 degrees
of freedom

consequently,

$$17.5 < \sigma^2 < 38.6$$

That is, we are 95 percent confident that the true variance of the heights
of male students in California colleges is between 17.5 and 38.6 inches.

10-2 exercises

1. A group of 30 students is selected at random from a certain university,
   and the variance of their annual expenditures is found to be $10,000
   ($s^2 = \$10,000$). Do the annual expenditures of the students at this
   university have a significantly greater dispersion than those for all
   college and university students in the country? The variance of annual
   expenditures for all students is $9000, and the level of significance
   is 0.05.

2. It is suggested that after signing a labor contract, production workers
   will show a greater variation in hourly outputs than before the contract
   was signed. It is known that the variance of the hourly outputs before
   the labor agreement was $\sigma^2 = 80$. A random sample of 30 production
   workers is selected, and their hourly outputs after the labor agreement are obtained. It is found that the variance of the sample is 90
   ($s^2 = 90$). Shall we conclude that the dispersion of the hourly outputs
   has increased significantly at $\alpha = 0.05$?

3. The age distribution of a class of 30 freshmen has a standard deviation (s) of 5. Does the age distribution of this class differ significantly
   from that of all students, whose age distribution is known to have a
   standard deviation of 6, if this class is taken as a random sample and
   the level of significance is specified at 0.05? What are the 95 percent
   confidence limits for the true variance of the age distribution of all
   freshmen?

4. When a production process is functioning properly, the variance of
   the parts produced by this process is four inches. It is suggested that

the production process is now out of control. A sample of nine parts produced by the process is randomly selected and the following measurements are obtained.

$$9 \quad 10 \quad 12 \quad 13 \quad 12 \quad 8 \quad 6 \quad 11 \quad \text{and} \quad 9$$

a) Find the variance s^2.
b) Test the hypothesis that the production process is still functioning properly at 0.10.
c) Establish the 90 percent confidence interval for the true variance (σ^2) on the basis of the sample information.

5. The weight gain for all sheep during the first six months of their lives is known to have a variance of 100 pounds, or $\sigma^2 = 100$. It is suggested that for a certain breed of sheep the weight gain during this period is more uniform; that is, its variance is smaller than 100. A random sample of six sheep is taken, and their weight gains during the first six months after birth are shown as follows:

$$63 \quad 50 \quad 40 \quad 45 \quad 42 \quad \text{and} \quad 60$$

Using $\alpha = 0.10$, can we conclude that the dispersion of the weight gains of this breed of sheep is smaller than the norm?

6. It is suspected that the IQ's of college students are less variable than the IQ's of people in general, which are known to have a variance of 250 ($\sigma^2 = 250$). A random sample of IQ scores for 26 college students yields a variance of 220. Using $\alpha = 0.05$, test the hypothesis that the IQ's of college students have less variability than the IQ's of all people.

10-3 tests of goodness of fit

Up to now, problems of hypothesis testing and interval estimation have centered primarily on parameters, or summary population characteristics, such as the mean, proportion, difference between two means, and variance. In this section we will deal with a different type of problem. Using the χ^2 test of goodness of fit, we will learn how to make inferences about an entire population distribution on the basis of the distribution obtained in the sample.

In the goodness of fit test, the null hypothesis specifies some distribution, which may be uniform (all possible values of the random variable are equally probable), binomial, or any other. A sample is taken and the researcher uses the test to help him decide whether the sample distribution follows the distribution specified under the null hypothesis. The alternative hypothesis asserts, of course, that the sample is not selected from the population with the specified distribution.

The test involves the classification of sample data into a frequency distribution called the *observed frequencies*. These are compared with the *expected frequencies* that are derived from the specified theoretical distribution. If a certain test involves n observations which are classified into J classes, and if the letter O is used to denote the observed frequency and E the expected frequency, then the test statistic X^2 is defined by

$$X^2 = \Sigma \frac{(O - E)^2}{E} \qquad (10\text{-}3)$$

where the summation is over J classes. The variable X^2 as defined here is commonly used as a measure of the difference between the observed and the expected distributions. When the sample size is so large that no expected frequency is less than 5, X^2 is distributed approximately like chi-square with $(J - 1)$ degrees of freedom.

Observe that the numerator in equation 10-3 is the squared difference, which can only be a positive value. The greater the difference, the larger the X^2 value will be. Obviously, this is a *one-tailed test*. When the X^2 value is small, we say that the fit is good, and we do not reject the null hypothesis. We reject the null hypothesis only when the fit is bad, that is, only when the test statistic X^2 takes on a value equal to or larger than the critical chi-square value that cuts off the *upper* tail area equal to α of the distribution with $(J - 1)$ degrees of freedom.

EXAMPLE 10-3 We want to decide whether a cubic die is perfectly balanced. For this purpose the die is rolled 300 times; the outcomes of the rolls are noted and recorded. In this experiment there are of course six possible outcomes, namely, 1, 2, 3, 4, 5, and 6. If the observed frequencies for these six categories are 35, 40, 32, 60, 68, and 65, respectively, shall we conclude that the die is perfectly balanced at the 0.01 level?

Table 10-1 Calculation of X^2 in an experiment to compare observed with expected frequencies ($n = 300$)

Class	Observed frequencies (O)	Proba-bility	Expected frequencies (E)	$O - E$	$(O - E)^2$	$\dfrac{(O - E)^2}{E}$
1	35	$\frac{1}{6}$	50	-15	225	4.50
2	40	$\frac{1}{6}$	50	-10	100	2.00
3	32	$\frac{1}{6}$	50	-18	324	6.48
4	60	$\frac{1}{6}$	50	10	100	2.00
5	68	$\frac{1}{6}$	50	18	324	6.48
6	65	$\frac{1}{6}$	50	15	225	4.50
Total	300	1.00	300			25.96

The null hypothesis to be tested is that the die is perfectly balanced, while the alternative asserts that it is not. With α specified at 0.01 and the number of degrees of freedom $\nu = J - 1 = 6 - 1 = 5$, the critical chi-square value is found to be 15.0863. If the die is perfectly balanced, the probability for each side of the die to show up is $\frac{1}{6}$. With $n = 300$, the expected frequency for each class is $300(\frac{1}{6}) = 50$. The calculation of the test statistic is shown in table 10-1, from which we get

$$X^2 = \Sigma \frac{(O - E)^2}{E} = 25.96$$

which is greater than the critical chi-square value. Accordingly, we reject the null hypothesis that the die is perfectly balanced. The observed frequency distribution differs so much from the expected that it cannot be attributed to chance.

EXAMPLE 10-4   Let us consider a hypothetical example. Assume that researchers are currently studying the educational attainment of all eligible voters in the United States. Assume further that 20 years ago, all voters were classified into five mutually exclusive and collectively exhaustive categories of educational attainment: sixth grade or lower, 30 percent; between 7th and 9th grades, 35 percent; between 10th and 12th grades, 20 percent; undergraduate training, 10 percent; and graduate training, 5 percent. We want to decide whether voters today have the same pattern of distribution that they had 20 years ago. For this purpose, a sample of 1000 voters is randomly selected, and the observed frequencies for the five categories are found to be, respectively: 270, 315, 230, 125, and 60. If the level of significance is specified at 5 percent, can we conclude that the voters' distribution pattern has changed?

Table 10-2   Computation of X^2 for hypothetical educational attainment of voters in the United States ($n = 1000$)

Educational attainment	Observed frequencies (O)	Propor-tion	Expected frequencies (E)	$O - E$	$(O - E)^2$	$\dfrac{(O - E)^2}{E}$
Sixth grade or lower	270	0.30	300	−30	900	3.00
Between 7th and 9th grades	315	0.35	350	−35	1225	3.50
Between 10th and 12th grades	230	0.20	200	30	900	4.50
Undergraduate training	125	0.10	100	25	625	6.25
Graduate training	60	0.05	50	10	100	2.00
Total	1000	1.00	1000			19.25

This is a test of the null hypothesis that the distribution pattern of American voters in terms of educational attainment has not changed in the last 20 years against the alternative that it has changed significantly. This is again a one-tailed test because a significant change can result only in a sufficiently large X^2 value. With $\alpha = 0.05$ and the number of degrees of freedom $\nu = 5 - 1 = 4$, the critical chi-square value is 9.48773. The computation of the test statistic X^2 is shown in table 10-2, from which we get

$$X^2 = 19.25$$

which is much greater than the critical value. Accordingly, we reject the null hypothesis.

10-3 exercises

1. A random digit table is supposed to be unbiased; that is, each of the 10 digits should have the same probability to show up. To test whether or not this is actually the case, a sample of 100 digits is randomly selected and the following results are obtained.

Digit:	0	1	2	3	4	5	6	7	8	9	Total
No. of times each digit appears:	8	11	10	14	7	12	6	9	13	10	100

Shall we reject the hypothesis that the digits in the table are randomly arranged with $\alpha = 0.05$?

2. Suppose that the number of fatal accidents on the Los Angeles freeways during a certain week is distributed as follows:

Day	No. of fatal accidents
Sunday	28
Monday	12
Tuesday	10
Wednesday	7
Thursday	8
Friday	11
Saturday	24
Total	100

Test the hypothesis that Saturday and Sunday each accounts for 25 percent and that each of the five weekdays accounts for 10 percent of all the fatal accidents at $\alpha = 0.025$.

3. Four balanced coins are tossed together 160 times. The results are shown below.

No. of heads:	0	1	2	3	4	Total
Observed frequency:	16	35	55	48	6	160

At 0.05, test the null hypothesis that the four coins are all well balanced and randomly tossed.

4. Suppose that researchers want to find out if the distribution pattern of family income in the United States has changed significantly during the past five years. It is known that five years ago the percentage distribution of family income over the various income classes was as follows:

Income class	Percent of all families in the class
1) Under $3000	9
2) $3000–$4999	11
3) $5000–$6999	12
4) $7000–$9999	22
5) $10,000–$14,999	27
6) $15,000–$24,999	15
7) $25,000 and over	4
Total	100

A random sample of 1000 families is drawn, and the following distribution is obtained.

Income class:	1	2	3	4	5	6	7	Total
No. of families:	70	100	110	200	300	170	50	1000

At $\alpha = 0.05$, test the null hypothesis that the current pattern of family income distribution is not significantly different from that of five years ago.

5. In an experiment with peas, a biologist observed 186 tall and colorful, 66 tall and colorless, 54 dwarf and colorful, and 14 dwarf and colorless. According to Mendel's theory of heredity, one would expect the various categories to have the following proportions—$9:3:3:1$. Is there any evidence to support Mendel's theory at the 0.01 level of significance?

6. A new product has been developed by a manufacturing company. The company has used nationwide advertising to solicit inquiries about possible dealerships. The company has divided the country into 10 regions that are equal with respect to population and sales potential. Thus, the 10 regions are expected to produce equal numbers of inquiries. The actual results are shown as follows:

Region:	1	2	3	4	5	6	7	8	9	10	Total
No. of inquiries:	22	23	18	16	21	17	19	23	20	21	200

Test the null hypothesis that the 10 regions were equally responsive at $\alpha = 0.025$.

10-4
tests of
independence

The third type of χ^2 test we plan to consider is the test of *statistical independence*. This test involves making a decision about whether one variable is independent of another; therefore, the important problem here is to determine the existence of some relationship between two sets of attributes in a population. As usual, we make certain inferences about the population on the basis of sample observations.

The existence of some relationship in the sample, however, does not necessarily imply that such a relationship also exists in the population. The reason for this lies in the fact that even if two variables are independent in the population, any sample drawn at random from this population is likely to show, because of chance factors, that some relationship does exist. Therefore, the relationship shown in the sample must be striking enough to suggest that the same relationship actually exists in the population.

Suppose, for instance, that we want to decide whether a person's academic training (a variable) is related to his performance rating on the job (another variable). We classify academic training into two categories or attributes: "college-trained" and "not college-trained." Similarly, performance rating is classified as "good" and "poor." A test of independence for these two variables will show us whether college training, or the lack of it, is in any way related to job performance rating, good or poor. If they are related, we say that they are not independent, and vice versa. To test the null hypothesis that they are independent, we draw a random sample from the population of all those employed and classify our sample in a *contingency table*. On the basis of our findings we make a decision about the independence of academic training and job performance rating.

The χ^2 test of independence follows closely the methodology used in the test of goodness of fit. In fact, the two test statistics are quite similar. The procedure involves, first of all, a presentation of the sample data in a contingency table (see table 10-3). To illustrate the construction and use of a contingency table, let us suppose that for the purpose of testing whether academic training is related to performance rating, we obtain a sample of 100 people, of whom 40 are college-trained and 60 are not. An analysis of the data shows that among the 40 who are college-trained, 15 are rated good and 25 poor; and that among the 60 who are not college-trained, 15 are rated good and 45 poor. Thus the total numbers of those rated good and poor are 30 and 70, respectively.

Observe that table 10-3 involves two row categories and two column categories; this is why it is called a two-by-two, or 2×2, contingency table. In general, if there are J row and K column categories, it will be called a J-by-K, or $J \times K$, contingency table. Each individual cell in the table is the intersection of a particular row and a particular column. The number in each cell is the *observed* frequency for that cell. The numbers in the bottom and right-hand margins are referred to as *marginal totals*,

Table 10-3   Contingency table of observed frequencies for academic training and job performance rating

Performance rating	Academic training		Total
	College-trained	Not college-trained	
Good	15	15	30
Poor	25	45	70
Total	40	60	100

and the *grand total* in the sample appears in the lower right-hand corner of the table.

The χ^2 test will enable us to decide whether the two variables, academic training and job performance rating, are independent in the population. We know the observed frequencies, and for the χ^2 test, we must calculate the expected frequencies; the marginal totals and grand total are sufficient for us to achieve this purpose. The proportion of those college-trained is 40/100. Independence between the two variables implies that the proportion of college-trained among all those rated good will be the same as the proportion of college-trained among all those rated poor. In other words, independence suggests that 40/100 of those rated good are college-trained and also 40/100 of those rated poor are college-trained. Since altogether there are 30 people rated good, the expected frequency of the college-trained among the 30 rated good is

$$\frac{40}{100}(30) = 12$$

Similarly, the expected frequency of the college-trained among the 70 rated poor is

$$\frac{40}{100}(70) = 28$$

The proportion of those with no college training is 60/100. By similar reasoning, we find that the expected frequency of those with no college training among the 30 rated good is

$$\frac{60}{100}(30) = 18$$

and the expected frequency of those with no college training among the 70 rated poor is

$$\frac{60}{100}(70) = 42$$

The expected frequencies for the four cells just calculated are presented in table 10-4. Observe that the expected frequencies in any particular row or column add up to its marginal total.

In general, if we let R_j designate the marginal total of row j, and C_k designate the marginal total of column k in a contingency table with a total of n observations, then the expected frequency of the cell in row j and column k denoted by E_{jk} is

$$E_{jk} = \frac{C_k}{n} (R_j) \qquad (10\text{-}4)$$

Table 10-4 Contingency table of expected frequencies for academic training and job performance rating

Performance rating	Academic training		Total
	College-trained	Not college-trained	
Good	12	18	30
Poor	28	42	70
Total	40	60	100

Once the expected frequencies for all the cells are computed, we may then proceed to determine whether the observed frequencies differ significantly from them. The test statistic X^2 is computed by dividing each squared difference between an observed and expected frequency by the expected frequency, and then summing the quotients thus obtained. Thus we have

$$X^2 = \frac{(15-12)^2}{12} + \frac{(15-18)^2}{18} + \frac{(25-28)^2}{28} + \frac{(45-42)^2}{42}$$

$$= \frac{9}{12} + \frac{9}{18} + \frac{9}{28} + \frac{9}{42} = 0.75 + 0.5 + 0.32 + 0.22 = 1.79$$

In general, if there are J rows and K columns in the contingency table, the test statistic X^2 is computed as follows:

$$X^2 = \sum_{j=1}^{J} \sum_{k=1}^{K} \frac{(O_{jk} - E_{jk})^2}{E_{jk}} \qquad (10\text{-}5)$$

where the subscripts j and k designate any particular row and column, respectively.

These double summations can be interpreted as follows: First, let $j = 1$; then we take the summation of all the quantities in the first row. This is to say that only the right-hand summation sign is involved for computing the sum of the quantities in row 1. After this, we let $j = 2$, and take the summation of all the quantities in the second row. Proceed in this manner until the sum for the last row, $j = J$, is obtained. The overall sum as expressed by the double summation is the grand total of all the sums of the rows thus obtained.

The computed value of the test statistic is then compared with the critical chi-square value from appendix G. If the computed value is equal to or greater than the critical value, we shall reject the null hypothesis of independence. For obtaining the critical chi-square value we need to know the level of significance and the number of degrees of freedom. In a 2×2 contingency table, we have freedom to guess the number in only *one* of the four cells, because once the frequency in one of the four cells is known the frequencies in the other three cells are determined. Thus the number of degrees of freedom in a 2×2 contingency table is 1. In general, if there are J rows and K columns in a contingency table, the number of degrees of freedom is $(J - 1)(K - 1)$ because one degree of freedom is lost in the rows and again in the columns, and the total number of degrees of freedom is equal to the number of degrees of freedom of the rows multiplied by the number of degrees of freedom of the columns.

With the number of degrees of freedom equal to 1, and the level of significance specified at 0.05, the critical chi-square value is 3.84146. Since the computed X^2 value is less than this critical value, we do not reject the null hypothesis that the job performance rating is independent of academic training.

EXAMPLE 10-5 Suppose that we want to decide whether a person's racial background is related to his preference for certain brands of cigarettes. Assume that three brands, I, II, and III, are involved and the racial background of the people is classified into three categories, Whites, Blacks, and Others. A random sample of 200 smokers is selected and classified in a contingency table (table 10-5). If the level of significance is specified at 0.025, shall we reject the null hypothesis that the two variables, racial background and cigarette-brand preference, are independent?

Table 10-5 Contingency table of observed frequencies for racial background and preference for certain cigarette brands

Brand preference	Racial background			Total
	Whites	Blacks	Others	
I	20	10	20	50
II	20	20	3	70
III	20	30	30	80
Total	60	60	80	200

As a first step, we compute the expected frequency for each cell of the contingency table. For the cell formed by row I and column Whites, the marginal total for the row is 50 and the marginal total for the column is 60. By equation 10-4, the expected frequency for the cell of row I and column Whites is

$$\frac{60}{200}(50) = 15$$

By the same procedure, the expected frequencies for the other cells are computed. Table 10-6 shows the expected frequencies.

Table 10-6 Contingency table of expected frequencies for racial background and preference for certain cigarette brands

Brand preference	Racial background			Total
	Whites	Blacks	Others	
I	15	15	20	50
II	21	21	28	70
III	24	24	32	80
Total	60	60	80	200

By equation 10-5, the test statistic is computed as follows:

$$X^2 = \frac{(20-15)^2}{15} + \frac{(10-15)^2}{15} + \frac{(20-20)^2}{20}$$
$$+ \frac{(20-21)^2}{21} + \frac{(20-21)^2}{21} + \frac{(30-28)^2}{28}$$
$$+ \frac{(20-24)^2}{24} + \frac{(30-24)^2}{24} + \frac{(30-32)^2}{32}$$
$$= \frac{25}{15} + \frac{25}{15} + \frac{0}{20} + \frac{1}{21} + \frac{1}{21} + \frac{4}{28} + \frac{16}{24} + \frac{36}{24} + \frac{4}{32}$$
$$= \frac{50}{15} + \frac{2}{21} + \frac{4}{28} + \frac{52}{24} + \frac{4}{32}$$
$$= \frac{50(224) + 2(160) + 4(120) + 52(140) + 4(105)}{3360}$$
$$= \frac{19,700}{3360} = 5.86$$

Since there are three rows and three columns in the contingency table, the number of degrees of freedom is $(3-1)(3-1) = 4$. The critical chi-square value that cuts off the upper 2.50 percent of the distribution with 4 degrees of freedom is 11.1433. The computed X^2 value of 5.86 is smaller than the critical value. Consequently, we do not reject the null hypothesis that brand preference of potential cigarette buyers is independent of their racial background.

10-4 exercises

1. We wish to determine whether academic achievement at the community-college level is related to continued training at a four-year college or university. A sample of 1000 recent community-college

227

graduates is randomly selected; and their classification in terms of continued training (T), no further training (T'), and average grades (A, B, and C or lower) during their first two years of college is shown as follows:

Continued training	Average grades			Total
	A	B	C or lower	
T	90	150	160	400
T'	110	250	240	600
Total	200	400	400	1000

Test the null hypothesis that continued training in a four-year college or university is independent of the average grades earned in a community college at α 0.05.

2. A survey is conducted to determine whether performance rating on the job is independent of academic achievement at college. A sample of 100 employees is randomly selected and the classification of these employees in a three-by-three table is shown below.

Performance rating	Academic achievement at college			Total
	A	B	C or lower	
Excellent	10	5	5	20
Average	20	12	8	40
Poor	20	13	7	40
Total	50	30	20	100

With the level of significance level specified at 0.01, shall we conclude that performance rating on the job is not related to academic achievement at college?

3. A manufacturing company is conducting a study to determine whether the sex of potential customers is associated with model preference. Three models of a particular product are considered. A random sample of 1000 potential buyers is selected and interviewed, and the observed frequencies are shown in the following table.

Sex	Model preference			Total
	I	II	III	
Male	100	100	200	400
Female	300	150	150	600
Total	400	250	350	1000

Test the hypothesis that model preference is not associated with sex at $\alpha = 0.025$.

4. Suppose that all 100 United States senators were polled to ascertain their attitude toward capital punishment. The senators are classified according to their party affiliation and opinion on the issue. The results of the poll are tabulated as follows:

Party affiliation	Attitude toward capital punishment			Total
	Favor	Opposed	No opinion	
Democrats	22	33	5	60
Republicans	18	17	5	40
Total	40	50	10	100

Test the null hypothesis that the senators' opinion on capital punishment is independent of party affiliation at $\alpha = 0.01$.

5. A drug manufacturer was interested in testing the effectiveness of a new cold vaccine over a specific time period. A random sample of 50 people was selected. They were divided into two groups: One group consisted of 30 people who received the vaccine, and the other group of 20 people did not receive the vaccine. The results were:

Effectiveness	Vaccine	No vaccine	Total
Cold	10	10	20
No cold	20	10	30
Total	30	20	50

Is the vaccine effective at a significance level of 0.05?

6. A psychologist conducted an experiment to determine whether students' achievement is related to the teaching method used in a particular subject. Three teaching methods are involved, I, II, and III, and students' achievement is classified as A, B, or C. The results were:

Achievement	Teaching methods			Total
	I	II	III	
A	5	20	5	30
B	15	15	10	40
C	5	15	10	30
Total	25	50	25	100

Test the null hypothesis that students' achievement is not related to the methods of teaching at $\alpha = 0.01$.

11-1
introduction

Methods of statistical inference about a single mean, about the difference between two means, and about a single variance have already been discussed. In this chapter, we will learn how to conduct the test about the difference between two variances, and how to test hypotheses about the difference between three or more means.

To perform these tests, it is necessary to introduce another probability distribution, the *F distribution*, named after R. A. Fisher, who first developed it. The random variable *F* is the ratio of the *unbiased estimates of two population variances*. That is,

$$F = \frac{s_1^2}{s_2^2} \qquad (11\text{-}1)$$

The *F ratio* is another test statistic that is used primarily for testing whether the variance of one population is equal to that of another. There are many situations that require such a test. For instance, a researcher in education may want to compare two methods of teaching reading in elementary schools: the phonics approach and the "look-say" approach. Two groups of pupils are randomly selected. One group is taught by one method and one group by the other. After a period of time, a standard reading test is administered to the two groups. The two sample variances are then computed (by equation 5-10) to test the null hypothesis that the two populations of test scores have equal variability.

When the sample variances are equal, we will not, of course, reject the null hypothesis that the two population variances are identical. In most cases, however, the obtained variances are somewhat different. The *F* test enables us to decide whether the difference between the two sample variances is due to chance factors or is large enough (at the specified α level) for us to reject the null hypothesis.

The *F* distribution is used whenever a test about two variances is conducted. Thus, a brief discussion of the nature of the *F* distribution is in order. Unlike the *t* and the chi-square distributions, each of which has only one parameter, the *F* distribution has two parameters: the numbers of degrees of freedom generally designated ν_1 and ν_2. A particular

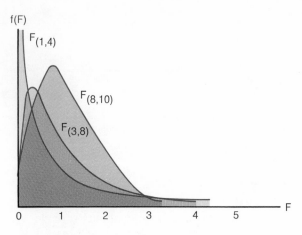

FIGURE 11-1
F distributions with specified
degrees of freedom

distribution of F, or the particular shape of an F curve, is completely determined when ν_1 and ν_2 are known. The parameter ν_1 is associated with the numerator of the F ratio, namely, s_1^2, the unbiased estimator of the variance σ_1^2 of population I. The variance s_1^2 is obtained from a sample of n_1 observations taken randomly from population I. Since the sum of squared deviations from the sample mean for s_1^2 has $(n_1 - 1)$ degrees of freedom, the parameter ν_1 is equal to $(n_1 - 1)$.

Similarly, ν_2 is associated with the denominator of the F ratio, namely s_2^2, the unbiased estimator of the variance σ_2^2 of population II. The variance s_2^2 is obtained from a sample of n_2 observations taken randomly from population II. Since the sum of squared deviations involved in s_2^2 has $(n_2 - 1)$ degrees of freedom, the second parameter ν_2 is equal to $(n_2 - 1)$. Consequently, for each different combination of the two numbers of degrees of freedom, $n_1 - 1$ and $n_2 - 1$, there is a different F distribution. Figure 11-1 shows a few F distributions with specified degrees of freedom, where the first subscript of F is ν_1 and the second is ν_2. This practice is followed throughout the discussion in this chapter.

We shall not be concerned about the density function of F, which is rather complex. All we need to point out is that the test statistic F assumes only *nonnegative* values, because both the numerator and the denominator of the F ratio are variances, which can not be negative values. Theoretically, the F value ranges from zero to infinity. In practical applications, however, we consider 1 as the lower limit of the F value because we may always use the larger sample variance as the numerator so that the F ratio would not be less than 1. The upper limit of the F value seldom runs more than a few digits in most situations.

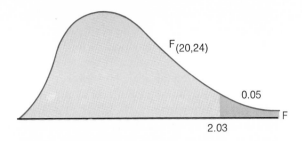

FIGURE 11-2
The F distribution with $\nu_1 = 20$
and $\nu_2 = 24$ showing the F value
cutting off the upper 5 percent
of the distribution

As mentioned above, the shape of the F curve depends upon the two numbers of degrees of freedom, ν_1 and ν_2. When the numbers of degrees of freedom are small, the F curve is skewed to the right. When one or both of the values increase, the F curve tends to become more symmetrical. The curves in figure 11-1 demonstrate this tendency.

Since the F distribution has two parameters ν_1 and ν_2, the F table is much more involved than the t and the chi-square tables. Thus, the F table has to be presented in an even more condensed form. (See appendix H, page 322.) One set of numbers of degrees of freedom forms the top row and the other forms the left-hand column; and a whole page is required to present the critical F values for each possible level of significance. The F table found in any statistics text therefore presents only the most commonly used α values, such as 0.10, 0.05, 0.025, and 0.01. For any particular value of α, the top row of the table presents the specified values of ν_1, the degrees of freedom associated with the numerator of the F ratio; the left-hand column presents the specified values of ν_2, the degrees of freedom associated with the denominator. Each cell entry in the table is a *critical F value*, usually designated $F_{(\alpha; \ \nu_1, \ \nu_2)}$, which cuts off the upper tail equal to α of the distribution with ν_1 and ν_2 degrees of freedom.

Figure 11-2 shows an F distribution for $\nu_1 = 20$ and $\nu_2 = 24$. The shaded area under the upper tail of the curve constitutes 5 percent of the distribution. The critical F value that cuts off the upper 5 percent of the distribution for 20 and 24 degrees of freedom is found, from the "five percent points" table in Appendix H, to be 2.03. Observe that 2.03 is the entry in the cell formed by the column headed 20 (ν_1) and the row marked 24 (ν_2). More formally,

$$F_{(0.05; \ 20, \ 24)} = 2.03$$

The F table is constructed on the basis that the numerator of the F ratio is larger than the denominator. Thus the F values are usually greater than 1, and approach 1 when both ν_1 and ν_2 approach infinity. Under the null hypothesis that two population variances are equal, the variances of any two samples taken respectively from the two populations are *expected* to be equal also. However, even if the null hypothesis is true, because of the random nature of sampling the two sample variances may very likely differ from each other. The larger the difference between the two sample variances, the greater the magnitude by which the test statistic F will rise above 1. The null hypothesis is rejected only when the value of the test statistic F is sufficiently larger than 1.

The F distribution has another important application. Recall that in chapters 8 and 9 the methods of testing hypotheses about the difference between two population means were discussed. Frequently, however, we want to decide whether three or more populations have identical means. The F distribution is often applied to the test about the equality between three or more population means. Such a test is usually conducted by analyzing the variability of the means of the samples that are taken from the populations under consideration. This procedure is generally referred to as the *analysis of variance,* which we shall consider after we examine the more basic technique of hypothesis testing about the equality of two population variances.

11-1 exercises

1. Explain the similarities and differences between the χ^2 and F distributions.

2. Suppose that a certain sample of 10 observations ($n_1 = 10$) taken from one population yields a variance s_1^2, and another sample of 15 observations ($n_2 = 15$) taken from a different population yields a variance s_2^2, and $F = s_1^2/s_2^2$. What are the values of ν_1 and ν_2 associated with this F ratio?

3. Why does the F ratio assume only nonnegative values?

4. Why do we consider 1 as the lower limit of the F ratio in practical applications?

5. What method do we usually use to test hypotheses about the equality between three or more means?

6. Find the following critical F values from appendix H:

   a) $F_{(0.10;\ 10,\ 8)}$ d) $F_{(0.05;\ 24,\ 20)}$
   b) $F_{(0.10;\ 8,\ 10)}$ e) $F_{(0.025;\ 20,\ 24)}$
   c) $F_{(0.05;\ 20,\ 24)}$ f) $F_{(0.01;\ 20,\ 24)}$

7. Given $F = s_1^2/s_2^2$, where s_1^2 is the variance of 25 observations and

s_2^2 is the variance of 20 observations. Find the critical F value that cuts off the

    a) upper 10 percent of the F curve

    b) upper 5 percent of the F curve

    c) upper 2.5 percent of the F curve

    d) upper 1 percent of the F curve

8. Suppose that a certain sample of 16 observations ($n_1 = 16$) taken from one population yields a variance $s_1^2 = 50$, and another sample of 11 observations ($n_2 = 11$) taken from a different population yields a variance $s_2^2 = 70$. What are the values of v_1 and v_2 according to appendix H? What is the critical F value that cuts off the

    a) upper 5 percent of the F distribution with v_1 and v_2 degrees of freedom?

    b) upper 1 percent of the same F distribution?

    c) Do we reject the null hypothesis that the two population variances are equal at $\alpha = 0.05$?

11-2 tests about the difference between two variances

As already pointed out, the F ratio as defined in equation 11-1 is used primarily for tests about the difference between two population variances. The example in the previous section should have made the need for such a statistical technique readily apparent. This technique has also been applied to many other problem situations, such as comparing the precision of one measuring device with that of another, the variability in grading practices of one teacher with that of another, and the consistency of one production process with that of another. In this section we shall illustrate the procedures involved in making such comparisons.

It is a fact that when two independent samples are randomly selected from two normal populations with identical variances, that is, $\sigma_1^2 = \sigma_2^2$, the ratio s_1^2/s_2^2 has a probability distribution known as the F distribution. Under the null hypothesis $\sigma_1^2 = \sigma_2^2$, the ratio s_1^2/s_2^2 (with the numerator larger than the denominator) is not expected to be much greater than 1. If it is, in fact, not much greater than 1, it would indicate that the two population variances are equal. On the other hand, if the ratio is sufficiently greater than 1, it would indicate a real or statistically significant difference between the two population variances. How large must the ratio be in order for us to reject the null hypothesis that two population variances are equal? The answer to this question depends on the critical F value, which is dictated by the level of significance α and the numbers of degrees of freedom v_1 and v_2.

EXAMPLE 11-1 Suppose that we want to determine whether test scores in reading are more variable when reading is taught by the phonics method than by the "look-say" method. Let the scores for all the elementary school students who may be taught by the phonics method be called population I, with the variance σ_1^2; and the scores of all those who may be taught by the look-say method be called population II, with the variance σ_2^2. The null hypothesis is that the two methods have essentially the same effect on the variability of the test scores, that is, $\sigma_1^2 = \sigma_2^2$.

Two independent samples are randomly drawn from the two populations, which are assumed to be normally distributed. A sample of 25 observations ($n_1 = 25$) taken from population I yields an unbiased estimate $s_1^2 = 40$, and another sample of 30 observations ($n_2 = 30$) taken from population II yields an unbiased estimate $s_2^2 = 25$. If we choose the level of significance at 0.05, may we conclude that test scores have a greater variability when reading is taught by the phonics approach than by the look-say approach?

The two competing hypotheses involved here are

$$H_0 : \sigma_1^2 = \sigma_2^2 \quad \text{and} \quad H_1 : \sigma_1^2 > \sigma_2^2$$

This is obviously a one-tailed test. With $\alpha = 0.05$ and $\nu_1 = 25 - 1 = 24$ and $\nu_2 = 30 - 1 = 29$, we find (in appendix H) the critical F value to be 1.90. Thus the decision rule is

$$\text{Reject } H_0 \text{ if } \frac{s_1^2}{s_2^2} \geq 1.90$$

Since the observed F ratio is $40/25 = 1.60$, which is smaller than the critical value, we conclude that the test scores for students taught by the phonics method are *not significantly* more variable than those for the students taught by the look-say method.

A sensible question we might raise here is: What would be our conclusion if it turns out that the variance for the group taught by the phonics method is *smaller than* the variance for the group taught by the look-say method? When such a situation happens, we do not proceed further with the test, since we could not conceivably reject the null hypothesis in such a situation.

When s_1^2 is larger than s_2^2, we will have no difficulty testing the null hypothesis $\sigma_1^2 = \sigma_2^2$ against the alternative $\sigma_1^2 > \sigma_2^2$. We simply divide the larger sample variance by the smaller one, and then compare the resulting ratio with the critical F value from the F table. However, the practice of treating s_1^2 as the numerator of the F ratio is appropriate only if it is the larger of the two sample variances in a one-tailed test such as the one in example 11-1. When two variances are computed from two samples taken from two populations that are hypothesized to have equal variances, s_1^2 can be greater than, equal to, or smaller than s_2^2. When s_1^2

is smaller than s_2^2, we want to determine whether the difference between them is attributable to chance or reflects the fact that σ_1^2 is *smaller than* σ_2^2.

The above discussion inevitably leads to the question: Is it possible to conduct a two-tailed test with the F distribution? The answer to this question is Yes. When we want to test the null hypothesis that two population variances are equal against the alternative that they are different, that is, $\sigma_1^2 \neq \sigma_2^2$, we divide the level of significance α by two, use $\alpha/2$ as the significance level, and conduct the test as if it were a one-tailed test. If the variance s_1^2 of a sample of n_1 observations taken from population I is greater than s_2^2 of a sample of n_2 observations taken from population II, and so $\nu_1 = n_1 - 1$ and $\nu_2 = n_2 - 1$, the decision rule will be

$$\text{Reject } H_0 \text{ if } s_1^2/s_2^2 \geq F_{(\alpha/2;\, \nu_1,\, \nu_2)}$$

On the other hand, if s_1^2 turns out to be smaller than s_2^2, then the test statistic becomes s_2^2/s_1^2 and the decision rule becomes

$$\text{Reject } H_0 \text{ if } s_2^2/s_1^2 \geq F_{(\alpha/2;\, \nu_2,\, \nu_1)}$$

where ν_2 designates any number in the top row and ν_1 any number in the left-hand column of the F table.

EXAMPLE 11-2   Two different machines, I and II, are used to produce identical screws, which are supposed to be three inches in length. Owing to a number of factors involved in the production process, the screws produced may vary slightly from the standard three inches. It is further suspected that the variability of the screws produced by machine I (population I) differs significantly from that of the screws produced by machine II (population II). Two random samples of 25 and 31 screws, respectively, are selected from populations I and II, and they yield the two variances $s_1^2 = 0.5$ inches and $s_2^2 = 0.4$ inches. Test the null hypothesis that the variability of population I is equal to that of population II against the alternative that they are different at $\alpha = 0.05$.

Let σ_1^2 be the variance of population I and σ_2^2 the variance of population II. The hypotheses to be tested in this case are

$$H_0: \sigma_1^2 = \sigma_2^2 \quad \text{and} \quad H_1: \sigma_1^2 \neq \sigma_2^2$$

Since this is a two-tailed test, the level of significance α is divided by 2, and 0.05/2, or 0.025, is used as the significance level for the one-tailed test. The larger of the two sample variances is $s_1^2 = 0.5$ computed from 25 observations; consequently, $\nu_1 = 25 - 1 = 24$. The smaller variance s_2^2 is computed from 31 observations; thus $\nu_2 = 31 - 1 = 30$. The critical F value, or $F_{(0.025;\, 24,\, 30)}$, is found in appendix H to be 2.14. Therefore the decision rule is

$$\text{Reject } H_0 \text{ if } s_1^2/s_2^2 \geq 2.14.$$

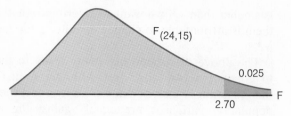

FIGURE 11-3
The *F* distribution with $\nu_1 = 24$
and $\nu_2 = 15$ showing the *F* value
cutting off the upper 2.5 percent
of the distribution

Since $s_1^2/s_2^2 = 0.5/0.4 = 1.25$, which is smaller than the critical value, we do not reject the null hypothesis.

EXAMPLE 11-3 Suppose that we wish to decide whether or not the variability in the intelligence of college men is equal to that for college women. Let σ_1^2 be the variance of the IQ's of all college men (population I) and σ_2^2 be the variance of the IQ's of all college women (population II). An intelligence test is administered to two groups of college students selected at random. One group consists of 16 male students ($n_1 = 16$), and their IQ's have a variance $s_1^2 = 200$; the other group consists of 25 female students ($n_2 = 25$), and their IQ's have a variance $s_2^2 = 600$. We know that the IQ's are normally distributed. If the level of significance is specified at $\alpha = 0.05$, do the sample variances differ significantly from each other?

In this example, the competing hypotheses are

$$H_0: \sigma_1^2 = \sigma_2^2 \quad \text{and} \quad H_1: \sigma_1^2 \neq \sigma_2^2$$

Since this is also a two-tailed test, the level of significance is again divided by 2. The degrees of freedom associated with the larger sample variance s_2^2, which becomes the numerator of the test statistic, is $n_2 - 1 = 25 - 1 = 24$; the degrees of freedom associated with the smaller sample variance s_1^2, which becomes the denominator of the test statistic, is $n_1 - 1 = 16 - 1 = 15$. Accordingly, the critical *F* value, or $F_{(0.025;\ 24,\ 15)}$, is found to be 2.70. The decision rule is therefore

$$\text{Reject } H_0 \text{ if } s_2^2/s_1^2 \geq 2.70$$

Since $s_2^2/s_1^2 = 600/200 = 3.00$, which is larger than the critical *F* value, we reject the null hypothesis. Figure 11-3 shows the *F* distribution and the critical region for this test.

Example 11-3 illustrates the procedure of conducting a two-tailed test by using the one-tailed test method. This is done by using the larger of the two sample variances as the numerator of the test statistic, and the degrees of freedom associated with the larger variance is considered as ν_1 in the F table, while the degrees of freedom associated with the smaller variance is considered as ν_2. At the same time, the level of significance is reduced to $\alpha/2$.

11-2 exercises

1. Suppose that the raw materials supplied by two vendors are compared. The two vendors seem to provide materials that are normally distributed with the same average, but the variability of the materials is a matter of concern. A sample of 16 lots from vendor I yields a variance of 150 ($s_1^2 = 150$), while a sample of 21 lots from vendor II gives a variance of 225 ($s_2^2 = 225$). Test the null hypothesis that their true variances are equal against the alternative that they are different at $\alpha = 0.05$.

2. A manufacturer of cotton cord claims that the cord his company produces has a smaller variability in breaking strength than nylon cord. The breaking strength of each cord is believed to be normally distributed. A random sample of 41 lengths of cotton cord is selected, and it yields a variance of 65 pounds ($s_1^2 = 65$). Another sample of 25 lengths of nylon cord is randomly drawn, and it gives a variance of 78 pounds ($s_2^2 = 78$). Shall we conclude that the cotton cord indeed has a smaller variability in breaking strength at $\alpha = 0.05$?

3. A psychologist wishes to determine whether the intelligence of girls is more variable than that of boys. It is known that the IQ's of both boys and girls are normally distributed. Suppose that a random sample of the IQ's of 61 girls provides a variance of $s_1^2 = 240$, and a random sample of the IQ's of 61 boys provides a variance $s_2^2 = 200$. At $\alpha = 0.01$, test the null hypothesis that the variability of girls' IQ's is the same as that of boys' against the alternative that the former is greater than the latter.

4. Two methods for teaching reading are applied to two randomly selected groups of nine-year-old children. It is desired to determine whether the results of the two methods in terms of the scores of a standard reading test have the same variability. Suppose that the following data are obtained:

	Method I	Method II
Sample size	$n_1 = 25$	$n_2 = 30$
Sample variance	$s_1^2 = 108$	$s_2^2 = 95$

With the level of significance at 0.05, shall we conclude that the test scores of the two groups have the same population variance?

5. Two machines, I and II, are used to produce identical bolts, the lengths of which are believed to be normally distributed. A random sample of 41 bolts produced by machine I yields a variance $s_1^2 = 0.5$ inches, while a random sample of 61 bolts produced by machine II yields a variance $s_2^2 = 0.3$ inches. Test the null hypothesis that the bolts produced by the two machines have identical variability against the alternative that they have different variability at $\alpha = 0.10$.

6. Let X and Y be two random variables. Two samples of 31 observations each yield two variances $s_X^2 = 49$ and $s_Y^2 = 25$. Test the hypothesis that the variance of X is significantly greater than that of Y at $\alpha = 0.05$.

11-3 analysis of variance

The discussion in the preceding two sections has made it possible for us to consider another type of problem: deciding whether differences between *three or more* sample means reflect actual differences in the means of the populations from which the samples are taken or whether they can be attributed to chance. For instance, an agronomist may want to know whether there is any difference in the average acre yield of soybeans if three different fertilizers are used in three identical plots of land; a motorist may wish to determine whether there is any difference in the average mileage obtained with several brands of gasoline; a business manager may want to find out whether there is any difference in the average service life of several kinds of light bulbs; and an educational psychologist may want to determine whether there is any difference in the effectiveness of several methods of teaching reading in the elementary schools.

In chapter 9 we used the t test to decide whether an observed difference between two sample means can be attributed to chance or whether it is indicative of an actual difference between the means of the two populations from which the samples are taken. If more than two samples are involved, however, the t test is of little use. For one thing, the t test is quite impractical to apply when the number of samples involved is large. Suppose that a certain problem involves five samples which yield five means: $\overline{X}_1$, $\overline{X}_2$, $\overline{X}_3$, $\overline{X}_4$, and $\overline{X}_5$. It will require the following 10 t tests to compare the 10 pairs of sample means.

X_1 and X_2	X_2 and X_3	X_3 and X_4	X_4 and X_5
X_1 and X_3	X_2 and X_4	X_3 and X_5	
X_1 and X_4	X_2 and X_5		
X_1 and X_5			

If more sample means are to be compared, more t tests will have to be performed: a total of six samples, for instance, would increase the number of pairs to 15.

Furthermore, when a sample mean is paired with each of many other sample means and a t test is performed for each pair of means, it is very likely that the t test will show significant differences for some pairs of sample means, even if the samples were in fact drawn from the *same* population. Thus, whenever three or more sample means are to be compared with one another, the t test becomes inappropriate; some other technique must be used for testing the hypothesis that the means are computed from samples taken from populations with an identical mean (μ). Such a technique is generally known as the *analysis of variance*.

analysis of variance—the rationale

Suppose that three different methods are used in teaching elementary statistics to three groups of randomly selected students. At the end of the course, the students are given a standard test. For the sake of simplicity, let us assume that each group consists of only five students. Their test scores are shown below.

Method I	Method II	Method III
74	84	83
78	77	85
73	79	86
73	79	87
72	81	89
Total 370	400	430

The means of these three samples, each with five observations, are respectively

$$\overline{X}_1 = 74, \quad \overline{X}_2 = 80, \quad \text{and} \quad \overline{X}_3 = 86$$

The question we wish to answer here is: Shall we, on the basis of these sample data, reject the null hypothesis that the three teaching methods are equally effective? Let μ_1, μ_2, and μ_3 be, respectively, the means of test scores earned by all potential students taught by the three methods, then the null hypothesis to be tested is $\mu_1 = \mu_2 = \mu_3$ against the alternative that they are *all unequal*.

Obviously, if the sample means differ very little from one another, it indicates that the null hypothesis is true. On the other hand, if the sample means differ very much from one another, it would tend to support the alternative hypothesis. Our task is therefore to identify a device for measuring the differences between the $\overline{X}$'s to determine whether the differences are *sufficiently* large, and the most commonly used and familiar measure is, of course, the standard deviation or variance.

For computing the variance of the $\overline{X}$'s, we need the grand mean of the three samples. The grand mean of all the samples is equal to the mean of the sample means since the samples are of equal size.

$$\text{Grand mean} = \frac{\overline{X}_1 + \overline{X}_2 + \overline{X}_3}{3} = \frac{74 + 80 + 86}{3} = \frac{240}{3} = 80$$

By equation 5-10, the variance of the sample means, or $s_{\overline{X}}^2$, is found as follows:

$$s_{\overline{X}}^2 = \frac{(74 - 80)^2 + (80 - 80)^2 + (86 - 80)^2}{3 - 1} = \frac{36 + 0 + 36}{2} = \frac{72}{2} = 36$$

Here the subscript $\overline{X}$ is used to show that this variance measures the variability of the sample means.

The method of analysis of variance is based on the *assumption that the populations from which the various samples are selected are normal or approximately normal with the same variance* σ^2. This implies that when the null hypothesis is true, all the populations from which we are sampling are really a single population with a mean μ and a variance σ^2. Thus, under the condition that the null hypothesis is true, we can consider the above three samples of five observations each as being taken from the *same population*, and use $s_{\overline{X}}^2$ as the estimator of $\sigma_{\overline{X}}^2$, which is the variance of the means of all possible samples of five observations each from the same population.

Equation 6-8 shows that when a sample is computed from n observations, then

$$\sigma_{\overline{X}}^2 = \frac{\sigma^2}{n}$$

That is, the variance of the sample mean $\overline{X}$ is equal to the variance of the population divided by the sample size. Since $s_{\overline{X}}^2$ is the unbiased estimator of $\sigma_{\overline{X}}^2$, and since $s_{\overline{X}}^2 = 36$, we obtain the following relationship.

$$s_{\overline{X}}^2 = 36 = \frac{\text{estimated } \sigma^2}{n}$$

Since n is the number of observations in each of the samples, the population variance σ^2 is estimated as

$$ns_{\overline{X}}^2 = 5(36) = 180$$

Thus, we have converted the variance for the means into an estimate of the variance of one single population. This estimate of σ^2, which is derived entirely from three sample means, is based on $3 - 1 = 2$ degrees of freedom. In general, if K samples are involved, the estimate of σ^2 derived from K sample means will be based on $K - 1$ degrees of freedom. Let us designate this estimate of σ^2 by s_B^2, that is, $s_B^2 = ns_{\overline{X}}^2$; the subscript B suggests that this estimate of σ^2 is based entirely on the dispersion *between* the sample means.

Suppose that our null hypothesis is false, which implies that the teaching methods do have different effects on the test scores. The three sample means would tend to show greater differences if the null hypothesis is, in fact, false. When the differences between the sample means are large, the variance of the sample means, or $s_{\bar{X}}^2$, will be large. Consequently, the estimate of σ^2, which is equal to $ns_{\bar{X}}^2$, will be large. Our original problem of deciding whether the differences between the three means are *too large* to be attributed to chance is shifted to "whether $s_{\bar{X}}^2$ is *too large*," and finally to "whether the estimate s_B^2 is *too large*."

This is done because there is a method for testing whether s_B^2 is sufficiently large; it is the F test. Since the test statistic involving the F distribution is the ratio of two variances, or two unbiased estimators of the same population variance σ^2, we need to identify the other estimate of σ^2 for testing our null hypothesis.

There is one condition that must be met with respect to the nature of the second estimate of σ^2. It is that the estimate must not be affected if the null hypothesis is false, that is, if there are in fact differences between the population means. Let s_1^2, s_2^2, and s_3^2 be the variances of the three samples, respectively. Then each of these three variances cannot possibly be affected by differences between the sample means no matter how large those differences happen to be, and each may therefore be used as an estimate of σ^2. Since the means of the three samples are, respectively, 74, 80, and 86, the three estimates are computed as follows:

$$s_1^2 = \frac{(74-74)^2 + (78-74)^2 + (73-74)^2 + (73-74)^2 + (72-74)^2}{5-1}$$

$$= \frac{(0)^2 + (4)^2 + (-1)^2 + (-1)^2 + (-2)^2}{4} = \frac{0+16+1+1+4}{4} = \frac{22}{4} = 5.5$$

$$s_2^2 = \frac{(84-80)^2 + (77-80)^2 + (79-80)^2 + (79-80)^2 + (81-80)^2}{5-1}$$

$$= \frac{(4)^2 + (-3)^2 + (-1)^2 + (-1)^2 + (1)^2}{4} = \frac{16+9+1+1+1}{4} = \frac{28}{4} = 7$$

and

$$s_3^2 = \frac{(83-86)^2 + (85-86)^2 + (86-86)^2 + (87-86)^2 + (89-86)^2}{5-1}$$

$$= \frac{(-3)^2 + (-1)^2 + (0)^2 + (1)^2 + (3)^2}{4} = \frac{9+1+0+1+9}{4} = \frac{20}{4} = 5$$

Although each of these three variances is an unbiased estimate of σ^2, the estimation would be greatly improved if all the information available in the three samples is used. Thus, we would have a better estimate if the average of all three variances is used as the estimate. Let us use s_W^2 to designate such an estimate of σ^2, where the subscript W suggests that this estimate is based entirely on the dispersion *within* each and all the samples. Then we have

$$s_W^2 = \frac{s_1^2 + s_2^2 + s_3^2}{3} = \frac{5.5 + 7 + 5}{3} = \frac{17.5}{3} = 5.83$$

The total degrees of freedom for s_W^2 is the sum of the numbers of degrees of freedom for all the sample variances. Since in this example the number of observations in each of the three samples is the same, the number of degrees of freedom for each sample variance is $n - 1 = 5 - 1 = 4$. As a result, the total degrees of freedom for s_W^2 is $3(4) = 12$. In general, if there are K samples, the first sample has n_1 observations, the second sample has n_2 observations, . . . , and the Kth sample has n_K observations, and $n_1 + n_2 + . . . + n_K = N$, the total number of observations in all the K samples. Then, the total degrees of freedom for s_W^2 is $(n_1 - 1) + (n_2 - 1) + . . . + (n_K - 1) = (n_1 + n_2 + . . . + n_K) - K = N - K$. In this example, $N = 15$, $K = 3$, the degrees of freedom is therefore $15 - 3 = 12$, the same number as obtained above.

The value of s_B^2, the first estimate of σ^2, will tend to be large if the null hypothesis is false. The value of s_W^2, the second estimate of σ^2, will not be influenced by whether the null hypothesis is true or false. Since the expected value of the unbiased estimate s_B^2, as well as that of s_W^2, is σ^2 when the null hypothesis is true; and s_B^2 tends to be large when the null hypothesis is not true, the test statistic F is formed with s_B^2 as the numerator. That is,

$$F = \frac{s_B^2}{s_W^2} \tag{11-2}$$

which has $K - 1$ and $N - K$ degrees of freedom. When the null hypothesis is false, this ratio tends to be large. Thus the test about the differences between three or more means is always a *one-tailed* test.

Let us specify the level of significance at 0.05. Since in the above example the degrees of freedom $v_1 = 2$ and $v_2 = 12$, and $F_{(0.05;\ 2,\ 12)} = 3.89$, the decision rule is

Reject H_0 if $s_B^2/s_W^2 \geq 3.89$

Our observed F ratio is $180/5.83 = 30.9$. Since this observed ratio is larger than the critical F value, we conclude that the three teaching methods have significantly different effects on the test scores for the students. That is, we *reject* the null hypothesis that the three teaching methods are equally effective.

the computation process

The foregoing discussion explains, by illustration, how the technique of analysis of variance is used for testing whether the differences between three sample means can be attributed to chance or whether they are indicative of real differences in the population means. Our purpose now is to develop a general computation procedure to deal with problems of this kind.

Suppose that a certain problem involves testing about the differences between the means of K samples. There are n_1 observations in sample 1, n_2 in sample 2, . . . , and n_K in sample K. Let X_{ki} designate each observation; the subscript k designates the kth sample, and the subscript i designates the ith observation in each sample. Then the sample data will appear in a matrix form as shown in table 11-1.

Table 11-1   Tabulation of sample data for analysis of variance

	Sample 1	Sample 2	. . . .	Sample K
	X_{11}	X_{21}		X_{K1}
	X_{12}	X_{22}		X_{K2}
	.	.		.
	.	.		.
	.	.		.
	X_{1n_1}	X_{2n_2}		X_{Kn_K}
Sample size n_1		n_2	. . . .	n_K
Sample total T_1		T_2	. . . .	T_K
Total number of observations:			$n_1 + n_2 + \ldots + n_K = N$	
Grand total of values:			$T_1 + T_2 + \ldots + T_K = T$	

In table 11-1 the sample total T_1 is the sum of values in Sample 1. Symbolically,

$$T_1 = \sum_{i=1}^{n_1} X_{1i} = X_{11} + X_{12} + \ldots + X_{1n_1}$$

The other T's can be similarly obtained. The grand total T is expressed by the use of double summations, namely,

$$T = \sum_{k=1}^{K} \sum_{i=1}^{n_k} X_{ki}$$

The total value T is obtained by first holding $k = 1$; then the summation gives the value T_1. Next holding $k = 2$, we obtain the sum T_2, and so on, until finally we hold $k = K$ to obtain the sum T_K. The summation of the T_K's gives the grand total value T.

Recall that in equation 11-2 the test statistic was defined as the ratio of the variance *between* the sample means to the variance *within* each and all the samples, or s_B^2/s_W^2. The denominator of each of these two variances is its degrees of freedom. The numerator of $s_B{}^2$ is the sum of squared deviations of all the sample means from the grand mean of all the samples. It is commonly known as the *sum of squares between*, or *SSB*. It can be demonstrated mathematically that

$$SSB = \sum_{k=1}^{K} \frac{T_k{}^2}{n_k} - \frac{T^2}{N} \qquad (11\text{-}3)$$

Similarly, the numerator of $s_W{}^2$ is the sum of squared deviations of all the observations from the corresponding sample means for all the samples. It is commonly known as the *sum of squares within*, or *SSW*. It can be shown that

$$SSW = \sum_{k=1}^{K} \sum_{i=1}^{n_k} X_{ki}{}^2 - \sum_{k=1}^{K} \frac{T_k{}^2}{n_k} \qquad (11\text{-}4)$$

The *sum of squares total*, or *SST*, is the sum of squared deviations of all the observations from the grand mean. It is equal to the sum of *SSB* and *SSW*; that is,

$$SST = SSB + SSW \qquad (11\text{-}5)$$

Once *SSB* and *SSW* are computed and each is divided by its degrees of freedom, we are ready to compute the test statistic. The procedure is summarized in table 11-2.

Table 11-2 Summary table for an analysis of variance

Source of variation	Sum of squares (SS)	Degrees of freedom	Variance	F ratio
Between groups (B)	$SSB = \sum_{k=1}^{K} \frac{T_k{}^2}{n_k} - \frac{T^2}{N}$	$\nu_1 = K - 1$	$s_B{}^2 = \dfrac{SSB}{K-1}$	$s_B{}^2 / s_W{}^2$
Within groups (W)	$SSW = \sum_{k=1}^{K} \sum_{i=1}^{n_K} X_{ki}{}^2 - \sum_{k=1}^{K} \frac{T_k{}^2}{n_K}$	$\nu_2 = N - K$	$s_W{}^2 = \dfrac{SSW}{N-K}$	
Total (T)	$SST = \sum_{k=1}^{K} \sum_{i=1}^{n_K} X_{ki}{}^2 - \frac{T^2}{N}$	$N - 1$		

an example

In order to verify that the computation procedure outlined in table 11-2 will result in the same answer as the procedures illustrated in "Analysis of Variance—The Rationale," we shall use the same sample data in the following example.

EXAMPLE 11-4   By the procedure summarized in table 11-2, we shall test the null hypothesis that the three methods of teaching elementary statistics are equally effective against the alternative that they are not. The original sample data, the square of each observed value, and the various totals are shown in table 11-3.

Table 11-3 Test scores resulting from three teaching methods in elementary statistics

Method I		Method II		Method III	
Raw score X_1	Square X_1^2	Raw score X_2	Square X_2^2	Raw score X_3	Square X_3^2
74	5476	84	7056	83	6889
78	6084	77	5929	85	7225
73	5329	79	6241	86	7396
73	5329	79	6241	87	7569
72	5184	81	6561	89	7921
$n_1 = 5$		$n_2 = 5$		$n_3 = 5$	$N = 15$
$T_1 = 370$		$T_2 = 400$		$T_3 = 430$	$T = 1200$

The values for SSB and SSW can now be computed from the data in table 11-3. Thus we have

$$SSB = \sum_{k=1}^{K} \frac{T_k^2}{n_k} - \frac{T^2}{N} = \frac{370^2}{5} + \frac{400^2}{5} + \frac{430^2}{5} - \frac{1200^2}{15}$$

$$= \frac{136,900}{5} + \frac{160,000}{5} + \frac{184,900}{5} - \frac{1,440,000}{15}$$

$$= 96,360 - 96,000 = 360$$

and

$$SSW = \sum_{k=1}^{K} \sum_{i=1}^{n_k} X_{ki}^2 - \sum_{k=1}^{K} \frac{T_k^2}{n_k}$$

$$= 5476 + 6084 + 5329 + 5329 + 5184$$
$$+ 7056 + 5929 + 6241 + 6241 + 6561$$
$$+ 6889 + 7225 + 7396 + 7569 + 7921 - 96,360$$
$$= 96,430 - 96,360 = 70$$

The variance measuring the dispersion between the sample means is therefore

$$s_B^2 = \frac{SSB}{K-1} = \frac{360}{2} = 180$$

and the variance measuring the dispersion within each and all the samples is

$$s_W^2 = \frac{SSW}{N-K} = \frac{70}{12} = 5.83$$

Table 11-4   Summary table of the analysis of variance

Source of variation	Sum of squares (SS)	Degrees of freedom	Variance	F ratio
Between groups (B)	$SSB = 360$	2	180	$180/5.83 = 30.9$
Within groups (W)	$SSW = \ \ 70$	12	5.83	
Total (T)	$SST = 430$	14		

The analysis is summarized in table 11-4. The computed *F* ratio 30.9 is greater than the critical value 3.89. Accordingly, the null hypothesis is rejected at $\alpha = 0.05$.

11-3 exercises

1. Three randomly selected groups of chickens are fed three different diets. Each group consists of five chickens. Their weight gains during a specified period of time are as follows:

Diet I	Diet II	Diet III
4	3	6
4	4	7
7	5	7
7	6	7
8	7	8

Use $\alpha = 0.05$ to test the null hypothesis that the three diets have the same effect on the weight gain of chickens against the alternative that they have different effects.

2. Suppose that four groups of five students each used four different methods of programmed learning to study statistics. We want to determine whether there is a significant difference in the results of the four methods. A standard test is administered to the four groups and is graded on an eight-point scale. The following results are obtained.

Method I	Method II	Method III	Method IV
1	1	3	5
2	2	5	5
3	5	5	6
4	5	6	7
5	7	6	7

Test the null hypothesis that the four means of test scores are equal against the alternative they are all unequal at $\alpha = 0.025$.

3. A manufacturing company has four identical machines in a particular production process. Each machine is operated by a different worker. A sample of products produced during a five-hour period is taken

from each machine, and the number of defective parts produced each hour is obtained. The results are as follows:

Machine I	Machine II	Machine III	Machine IV
10	7	2	3
9	7	3	3
9	8	3	6
9	8	3	6
8	5	4	7

Using $\alpha = 0.01$, test the null hypothesis that the machines produce the same average of defective parts per hour against the alternative that the four averages are different.

4. Three fertilizers, I, II, and III, were used on three plots of land. Plot I grows five strawberry plants, plot II four plants, and plot III six plants. The yield of each strawberry plant is measured and recorded. The results are shown below.

Fertilizer I	Fertilizer II	Fertilizer III
3	2	4
4	3	5
4	4	6
5	5	6
7		7
		8

Test the null hypothesis that the mean yields of the three plots are the same against the alternative that they are different at (a) $\alpha = 0.01$ and (b) $\alpha = 0.10$.

5. Consider four brands of light bulbs. Suppose that we want to know whether or not the four brands have the same average service life. A random sample of three bulbs from each brand is selected and tested. The total numbers of service hours (in thousands) of the four kinds of bulbs are found as follows:

$$T_1 = 50 \qquad T_2 = 40 \qquad T_3 = 40 \qquad T_4 = 50$$

and the grand total is $T = 50 + 40 + 40 + 50 = 180$. The sum of squares, or $\Sigma\Sigma X^2$, is 2778.

Test the null hypothesis that the average service lives are equal against the alternative that they are different at $\alpha = 0.05$.

6. Suppose that the employees of a large corporation are classified into three age groups. Management wants to find out whether the three groups have the same level of morale in terms of a scale used for morale measurement. A random sample of four employees is selected from each group, and a morale index is obtained for each employee. The total indexes for the three groups are shown below.

$$T_1 = 30 \qquad T_2 = 40 \qquad T_3 = 35$$

The grand total $T = 30 + 40 + 35 = 105$. The sum of squares, or $\Sigma\Sigma X^2$, is 947.

Test the null hypothesis that the three age groups have the same average index of morale against the alternative that they have different average indexes at $\alpha = 0.01$.

linear regression

and correlation

12-1 introduction

By and large our discussions so far have been concerned with problems involving only a single variable. Our attention now turns to problems involving a pair of variables. There are many situations in which we are confronted with paired measurements and must decide whether a relationship exists between the two. For instance, we may be interested in the relationship between heights and weights of college men, between students' grade-point averages (GPA) in high school and their GPAs in college, between employees' scores on an aptitude test and their performance on the job, between the amount of precipitation and the yield of wheat. If there is a relationship between the two variables under consideration, we may also want to know the strength of that relationship or dependence. Further, we may want to predict the value of one variable from the value of the other.

The methods dealing with this type of problem are known as the *regression and correlation techniques*. If only two variables are involved, the technique is referred to as *simple* regression or correlation. When three or more variables are involved, the term *multiple* is used.

Regression technique refers to the procedures involved in deriving an equation for the purpose of estimation or prediction. The variable to be estimated or predicted is called the *dependent* variable; and the other variable, the one that provides the basis for the estimation, is called the *independent* variable. In a simple regression problem, there is only one independent variable and one dependent variable. Multiple regression involves *two or more* independent variables and one dependent variable. (In this chapter, we will consider simple regression equations only.) A simple regression equation involving X as the independent variable and Y as the dependent variable is usually referred to as a *regression of Y on X*. (The term regression will be clarified further at the end of the next section.)

Furthermore, our interest here is limited to linear regression only. By linear regression we mean a relationship that can be represented

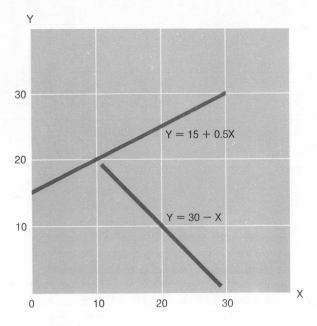

FIGURE 12-1
Two linear equations

graphically by a straight line. (Frequently, a regression equation is *curvilinear*; that is, its graphical representation is a curve rather than a straight line.) Figure 12-1 shows two regression lines. The equation represented by the upward sloping line is

$$Y = 15 + 0.5X$$

while the equation represented by the downward sloping line is

$$Y = 30 - X$$

In general, a simple regression line takes the form

$$Y = a + bX \qquad (12\text{-}1)$$

where a is the Y intercept, that is, the point where the line and the Y axis intersect, and b is the *slope* of the line, which is the *change in Y for each unit change in X*. The task of deriving a regression equation involves the computation of the values for a and b.

When the measurements for two variables are analyzed and a regression equation is set up, we then try to determine how closely the variables are related. This requires a technique known as *correlation analysis*, which deals with the measurement of the closeness of the relationship between the two variables involved in the regression equation. Sometimes, we may wish to measure the closeness of the relationship between two variables without deriving a regression equation. Other times, we may want to develop a regression equation without conducting a correlation analysis. In short, each type of analysis can be conducted independently and each serves its separate purpose.

In the next section, we shall discuss the method of deriving a simple linear regression equation; the method of measuring the closeness of the relationship between the two variables involved in the regression equation will be considered in the third section of the chapter.

12-1 exercises

1. Why do we usually include correlation analysis in the study of the technique of regression? Can these two subjects be studied separately?

2. From the following equations identify the simple, linear, multiple, and curvilinear regressions:

   a) $Y = 2 + 3X$ c) $Y = 10 + 2X_1 + 3X_2 + 4X_3$

   b) $Y = 1000 - 5000X$ d) $Y = 2 + 3X + X^2$

3. What are some of the purposes of regression and correlation analyses?

4. What is meant by "regression of Y on X"? Which is the independent and which is the dependent variable?

5. For the equation $Y = a + bX$, show graphically the meaning of a and b.

6. For the equation $Y = a + bX$, show graphically each of the following and describe the relationship between X and Y:

   a) $a = 2$ and $b = 0$ d) $a = -2$ and $b = 2$

   b) $a = 0$ and $b = 1$ e) $a = 4$ and $b = -2$

   c) $a = 2$ and $b = 1$

7. Give an example to show paired observations for two variables.

12-2 linear regression

The primary objective in ascertaining the relationship between two variables is to make more accurate predictions. If a relationship has been established between scores on two variables, then knowing the score on one variable will help us predict the score on the other variable. For instance, if we know the relationship between achievement-test scores and grade-point averages in college, we will be able to make a more accurate estimation of a student's college performance on the basis of his score in the achievement test. Similarly, if we know the relationship between the budget of the federal government and the number of single-family residential units built in the country for a given year, a knowledge

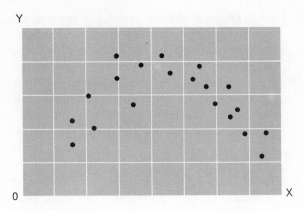

FIGURE 12-2
A scatter diagram showing a non-
linear relationship between X and Y

of the federal budget will enable us to make a better prediction of the
number of single-family housing units to be constructed.

It is important to remember that relationships between two variables
are not necessarily *cause-and-effect*, or *causal*, relationships. A causal
relationship implies that the independent variable is the cause and the
dependent variable is the effect. It is apparent that a causal relationship
exists between the amount of fertilizer used and the amount of wheat
harvested from a given acreage. However, it is ludicrous to say that a
student's score in an achievement test is a "cause" of his grade-point
average in college. Likewise, it is imprudent to surmise that a cause-and-
effect relationship exists between the federal budget and the number of
single-family housing units built throughout the country. It may be that
some third variable such as intelligence is the causal factor for the
relationship between achievement test scores and grade-point averages
in college, and some other economic or political factor is responsible for
the relationship between the federal budget and the housing units mentioned above. Keep in mind that some relationships are causal, while
others may be *consequential* or *functional*. A consequential relationship
is a relationship existing between two variables that may be caused by,
or attributable to, a third variable. A functional relationship exists when
variables are related in some functional way such as shown in equation
12-1. Our primary concern is with the relationships themselves and
not with the reasons they exist or the circumstances surrounding them.

the scatter diagram

As a first step in deriving a regression equation, the collected sample
data are usually plotted on a scatter diagram. (See figure 12-2.) A scatter
diagram provides a visual picture of the kind of relationship involved

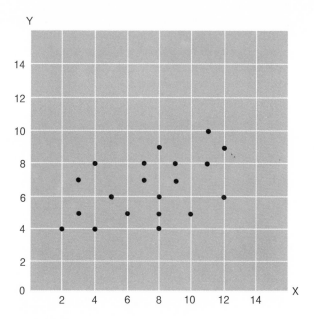

FIGURE 12-3
A scatter diagram showing GPA in
high school (X) and college (Y)
for 20 students (source: table 12-1)

and suggests the type of equation that will best fit the data. The usual way
to construct a scatter diagram is to have the independent variable X
scaled on the horizontal axis and the dependent variable Y on the vertical
axis; thus a two-dimensional plane is formed by X and Y. Each pair of
observations of X and Y (X, Y) is represented by a point in the plane.
Looking at figure 12-2, we can see that a linear regression equation does
not best fit the data represented by those points. It is not the best choice
because when the values of the independent variable X are small, an
increase in the value of X is accompanied by an increase in the value of
Y, whereas for large values of X an increase in X is accompanied by a
decrease in Y. It would appear that a bell-shaped curve would fit the data
best.

Suppose that we wish to determine the relationship between high
school and college GPA. It seems apparent that if a student's GPA in
high school is high, he is likely to do well in college. For the sake of
simplicity, let us assume that a scale ranging from 2 to 12 is used in both
high schools and colleges. Let X and Y designate a student's GPA in
high school and college, respectively. Suppose that a random sample of
20 students is taken, and the 20 pairs of observations are shown in table
12-1. Our objective is to see if we can predict the value of Y from the
known value of X. Let us first plot the data in a scatter diagram; figure
12-3 shows the scatter diagram for the data in table 12-1.

The scatter diagram in figure 12-3 clearly suggests that there exists a *positive* linear relationship between X and Y; that is, students' GPA in college tends to *vary directly* with their GPA in high school, and a straight line would result if we trace along the center of the diagram from the lower left to the upper right.

Table 12-1   High school and college GPAs
for 20 students

Students	High school GPA (X)	College GPA (Y)
1	3	5
2	2	4
3	4	4
4	12	9
5	11	8
6	8	9
7	9	7
8	7	8
9	6	5
10	5	6
11	4	8
12	8	4
13	3	7
14	12	6
15	9	8
16	8	5
17	11	10
18	7	7
19	8	6
20	10	5

Once we have decided that a linear relation exists between the two variables, we are faced with the task of finding the particular line that can provide a good fit to the points on the diagram. The crudest way of drawing a line through those points is the so-called freehand method. It involves the drawing of a straight line freehand through or near the points on the scatter diagram so that the number of points falling below the line is approximately equal to those above the line. Obviously, such a line has its drawback since it lacks any immediate assurance that the fit is the "best" and cannot be improved.

the method of least squares

There is, of course, no limit to the number of straight lines that could be drawn on any scatter diagram. Obviously, many of the lines would not fit the data and must be disregarded, while others may appear to fit the points very well. We need only one line, however; our primary objective now is to select the line that best fits the data. What do we mean by "best"? What criterion do we use in the selection of the "best" line? If all the points on the diagram fall on a line, that line certainly would be the best-fitting line. Such a situation, however, will seldom occur. Since the points are usually scattered as in figure 12-3, we need first to identify the criterion by which the best-fitting line can be determined.

The criterion that is most commonly used is known as the *least squares,* and the method of fitting a straight line to paired observations on the basis of this criterion is known as the *method of least squares*. Briefly,

> The least squares criterion requires that the line designed to fit the points on the scatter diagram must be such that the sum of the squares of the vertical distances from the points to the line is the smallest possible.

Let us consider this idea in greater detail. There are 20 points ($n = 20$) in figure 12-3; each point represents an ordered pair of observations. The 20 ordered pairs are (X_1, Y_1), (X_2, Y_2), . . . , (X_{20}, Y_{20}). The equation of the line we wish to fit to the 20 points is equation 12-1; in other words,

$$Y = a + bX$$

Thus, there are two values of Y for each of the values of X: one is the observed value of Y paired with the observed value of X, and the other is the Y coordinate of the corresponding point on the line, which is obtained by substituting the given observed value of X into the above equation. For X_1, the observed value Y_1 is one value of Y, and the Y coordinate of the point on the line (or $a + bX_1$) is the other; for X_2, the observed value Y_2 is one value of Y, and the Y coordinate of the point on the line (or $a + bX_2$) is the other; . . . and finally for X_{20}, the observed value Y_{20} is one value of Y, and the Y coordinate of the point on the line (or $a + bX_{20}$) is the other. Since there are 20 points, we have 20 vertical deviations between the two Y values for the 20 observed X values. They are $Y_1 - (a + bX_1)$, $Y_2 - (a + bX_2)$, . . . , and $Y_{20} - (a + bX_{20})$; these vertical deviations are shown in figure 12-4.

More generally speaking, if there are n pairs of observations in the sample, the least squares criterion requires that

$$\Sigma [Y - (a + bX)]^2$$

must be the minimum sum. Thus, any line that minimizes this quantity is called the *least squares line*.

This explains why the least squares fit is generally considered the "best" fit; it minimizes the sum of the squared deviations from the points to the line. In the regression line

$$Y = a + bX$$

the values of X and Y are given in the sample. The Y-intercept a and the slope b, which is also known as the *regression coefficient,* are the *unknowns* and have to be computed from the sample data. It is apparent that any particular line is determined by the values of a and b, and the best-fitting least squares line can be obtained only if the *correct* values for a and b are selected.

It can be proved mathematically that the values of a and b which make the sum of squares of the deviations as small as possible are obtained by solving simultaneously the following two equations. They are generally referred to as *normal equations:*

$$\Sigma\, Y = na + b\, \Sigma\, X \qquad\qquad (12\text{-}2)$$

$$\Sigma\, XY = a\, \Sigma\, X + b\, \Sigma\, X^2 \qquad\qquad (12\text{-}3)$$

Here n is the number of pairs of observations. With the exception of a and b, all the quantities in the two equations can be computed from the data in table 12-1. The values of the two unknowns a and b can be calculated from these quantities.

Let us proceed to solve the equations by first constructing a table for computing the various quantities in equations 12-2 and 12-3. Table 12-2 shows the computation process.

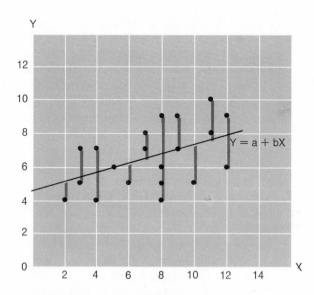

FIGURE 12-4
A diagram showing the vertical distance between each point and the line (source: table 12-1)

Table 12-2 Computation for regression of college GPA on high school GPA for 20 students

Students	High school GPA (X)	College GPA (Y)	XY	X²	Y²
1	3	5	15	9	25
2	2	4	8	4	16
3	4	4	16	16	16
4	12	9	108	144	81
5	11	8	88	121	64
6	8	9	72	64	81
7	9	7	63	81	49
8	7	8	56	49	64
9	6	5	30	36	25
10	5	6	30	25	36
11	4	8	32	16	64
12	8	4	32	64	16
13	3	7	21	9	49
14	12	6	72	144	36
15	9	8	72	81	64
16	8	5	40	64	25
17	11	10	110	121	100
18	7	7	49	49	49
19	8	6	48	64	36
20	10	5	50	100	25
$n = 20$	$\Sigma X = 147$	$\Sigma Y = 131$	$\Sigma XY = 1012$	$\Sigma X^2 = 1261$	$\Sigma Y^2 = 921$

The bottom row of table 12-2 provides all the quantities needed to solve equations 12-2 and 12-3. (The column totals can be accumulated directly; it is not necessary to make all the calculations if a desk calculator or some other computing equipment is used. The total in the last column, or ΣY^2, is included for later use.) Substituting these quantities into the two normal equations, we obtain

$$\text{I} \quad 131 = 20a + 147b$$
$$\text{II} \quad 1012 = 147a + 1261b$$

The remaining task is to solve these two equations for a and b. We can do this by the *method of elimination*. We can eliminate a by multiplying the terms on both sides of the first equation by 147, multiplying the terms on both sides of the second equation by 20, and then subtracting the first equation from the second. Thus we have

$$\text{I} \quad 19,257 = 2940a + 21,609b$$
$$\text{II} \quad 20,240 = 2940a + 25,220b$$

and by subtraction

$$983 = 3611b$$

Therefore,

$$b = \frac{983}{3611} = 0.2722$$

Substituting this value of b into the first equation, we get

$$131 = 20a + 147(0.2722)$$
$$20a = 131 - 40.0134 = 90.9866$$

Thus,

$$a = \frac{90.9866}{20} = 4.55$$

Consequently, the least squares equation is

$$Y = 4.55 + 0.27X$$

The least squares equation can also be obtained by the use of formulas for a and b. The formulas for a and b are actually derived by solving equations 12-2 and 12-3 algebraically by the method of elimination. Accordingly, we have

$$a = \frac{(\Sigma X^2)(\Sigma Y) - (\Sigma X)(\Sigma XY)}{n(\Sigma X^2) - (\Sigma X)^2} \qquad (12\text{-}4)$$

$$b = \frac{n(\Sigma XY) - (\Sigma X)(\Sigma Y)}{n(\Sigma X^2) - (\Sigma X)^2} \qquad (12\text{-}5)$$

If we use these two formulas for computing the values for a and b, we will have

$$a = \frac{(1261)(131) - (147)(1012)}{20(1261) - (147)^2} = \frac{165,191 - 148,764}{25,220 - 21,609} = \frac{16,427}{3611} = 4.55$$

$$b = \frac{20(1012) - (147)(131)}{20(1261) - (147)^2} = \frac{20,240 - 19,257}{25,220 - 21,609} = \frac{983}{3611} = 0.27$$

Observe that these formulas yield the same values for a and b as before. The line thus obtained may be plotted on the scatter diagram to show its fit to the observed data. The line $Y = a + bX$ drawn in figure 12-4 is actually the least squares line $Y = 4.55 + 0.27X$ obtained above. Accordingly, the sum of squares of the vertical distances from the points to the line as shown on the diagram is smaller than it would be if any other line were used.

Once the equation of the least squares line is determined, we can predict a student's academic performance in college on the basis of his performance in high school. Suppose that a student's GPA in high school is 10; we can predict his GPA in college by substituting $X = 10$ into the above equation as follows:

$$Y = 4.55 + 0.27(10) = 7.25$$

The predicted GPA 7.25 is what can be *expected* if his high school GPA is 10. It is quite likely, not only possible, that different students with the same GPA, say, 10, may earn quite different GPA's in college. This being the case, any predictions based on least squares lines should be regarded as averages. For this reason, the least squares line is often referred to as the *conditional mean*—any point on the line is the mean of all potential Y values associated with (on the condition of) a given pre-determined value of X.

As mentioned before, a least squares line $Y = a + bX$ is a *regression* of Y on X. We may now explain why the term regression is used to name such a line. The term was first introduced by Sir Francis Galton in the nineteenth century. He found, in his study of the relationship between the heights of fathers and sons, that tall fathers were likely to have tall sons and short fathers were likely to have short sons. However, the mean height of the sons of tall fathers was lower than the mean height of their fathers, and the mean height of the sons of short fathers was higher than the mean height of their short fathers. Galton referred to this tendency to return to the mean height of all men as regression. It is therefore not surprising that the concepts of conditional mean and regression line are used interchangeably.

The regression line as obtained above is a *sample* regression line, since it is based on a sample of 20 observations. It is obvious that if we draw a different sample of 20 students, we would very likely obtain different values for a and b and thus have a different regression line. Let the *true*, or *population*, *regression line* be expressed by the following equation

$$Y = \alpha + \beta X \tag{12-6}$$

Here α is the Y intercept, and β is the slope, of the true regression line. We use the value of a as an estimate of α and the value of b as an estimate of β. In short, the line $Y = a + bX$ is used as an estimate of the line $Y = \alpha + \beta X$. An analysis of the goodness of this estimate would involve a great deal of work. The problem of interval estimation of β involves the correlation coefficient, and will not be discussed in this text.

12-2 exercises

1. A life insurance company wishes to determine the relationship between sales experience and sales volume. A random sample of nine salesmen is selected, and their years of experience (X) and current annual sales (Y) are found to be:

$$X: 1 \quad 2 \quad 3 \quad 4 \quad 5 \quad 6 \quad 7 \quad 8 \quad 9$$
$$Y: 2 \quad 1 \quad 3 \quad 3 \quad 4 \quad 5 \quad 6 \quad 5 \quad 7$$

(in \$100,000)

a) Construct a scatter diagram and plot the regression line of Y on X on the diagram.

b) Estimate the volume of sales per year for a salesman who has had 10 years of sales experience.

2. An experiment is conducted to determine the relationship between rainfall and wheat yield. Suppose that the following data are obtained.

Rainfall in inches: X 1 2 3 4 5 5 6 7 8 9
Wheat yield in bushels: Y 1 3 2 5 5 4 7 6 9 8

a) Fit a least square line to the data with X as the independent variable, and then plot the line on a scatter diagram.

b) Estimate the wheat yield if the rainfall is 10 inches.

3. A record of maintenance cost is kept on six identical machines of different ages. Management wants to determine whether there is a functional relationship between machine age (X) and maintenance cost (Y). The following data are obtained.

Machine	X	Y
1	2	$ 70
2	1	40
3	3	100
4	2	80
5	1	30
6	3	100

Find the regression equation with X as the independent variable and Y the dependent variable. What would be the maintenance cost for a four-year old machine?

4. The record of a school district shows the following data for teachers who voluntarily resign.

Number of years in service (X)	Number who voluntarily resigned (Y)
15	10
9	16
13	14
11	15
12	15

Derive a regression equation, and find the number of teachers that would resign among those who have had 14 years of service.

5. Consider 10 college students. Their grade-point averages in high school and in college are shown below.

High school GPA (X)	College GPA (Y)	X^2	Y^2	XY
2.5	2.2	6.25	4.84	5.50
2.6	3.0	6.76	9.00	7.80
2.8	2.4	7.84	5.76	6.72
3.0	2.7	9.00	7.29	8.10
3.2	3.0	10.24	9.00	9.60
3.4	3.0	11.56	9.00	10.20
3.5	3.2	12.25	10.24	11.20
3.6	3.5	12.96	12.25	12.60
3.8	3.4	14.44	11.56	12.92
3.8	3.6	14.44	12.96	13.68
Total 32.2	30.0	105.74	91.90	98.32

Derive the regression equation for the above data with X as the independent variable.

6. Determine the regression of Y on X by using the following data:

$$n = 10 \quad \Sigma X = 550 \quad \Sigma Y = 680$$
$$\Sigma XY = 45,900 \quad \Sigma X^2 = 38,500 \quad \Sigma Y^2 = 56,000$$

12-3 simple correlation

A legitimate question may now be raised: How accurately can we predict the value of Y by means of the least squares equation? The amount of accuracy in the prediction depends on the *closeness of the relationship* between X and Y, which is also known as the *strength of correlation* between the two variables. When the correlation is insignificant, little accuracy is obtained by using the least squares line to make predictions; but when the correlation is strong, that is, when the least squares line is close to all the points on the diagram, great accuracy will be achieved by using the line for the purpose of estimation and prediction.

The usual measure of the strength of correlation based on a sample of n pairs of observations is the *correlation coefficient*, commonly designated as r. In this section, we shall attempt first to analyze the concept of a correlation coefficient, and then, on the basis of the sample correlation coefficient r, to test the hypothesis about the population correlation coefficient ρ (the Greek letter rho).

the coefficient of correlation

The accuracy of predictions about the values of Y will naturally be greater if the variation among the Y values is smaller. For the time being, let us assume that we are making predictions about the Y value without

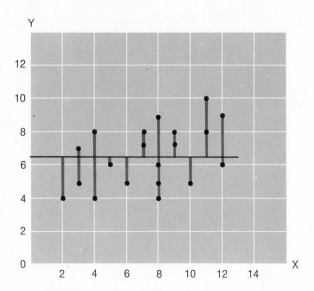

FIGURE 12-5
Deviations of Y values from the
mean ($\overline{Y}$) (source: table 12-1)

reference to the value of X. Such predictions become more precise as the variance of Y becomes smaller. The variance of Y varies directly with the deviations of the Y values from the sample mean $\overline{Y}$. Such deviations for the data in table 12-1 are shown in figure 12-5. A close examination of these deviations reveals immediately that they are in general greater than the deviations from the least squares line shown in figure 12-4. Since the deviations of the Y values from the least squares line are smaller, we have reason to expect that it is more accurate to predict the value of Y by means of the least squares line than without the aid of such a line. In other words, the estimation of the Y value will be more precise if it is made on the condition that X is given than without reference to X.

This is to say that the variability of the random variable Y is reduced by the regression of Y on X. Thus, the total variability is divided into two parts: *the amount that has been eliminated by the regression line* and *the amount that remains* in spite of the regression line. Let Y_c be the computed Y value, or $Y_c = a + bX$; we get the partition of the total variability of Y in terms of the sum of squared deviations as follows:

$$\Sigma(Y - \overline{Y})^2 \quad = \quad \Sigma(Y_c - \overline{Y})^2 \quad + \quad \Sigma(Y - Y_c)^2 \qquad (12\text{-}7)$$

Total variation | Variation eliminated | Variation remaining
by regression of Y on X

Apparently, the more variation that is eliminated by the regression line, the closer the relationship between X and Y will be, and the more precise the estimation of the Y value will become. Usually, the variability that is eliminated is expressed in terms of proportions. Dividing each side of equation 12-7 by $\Sigma(Y - \overline{Y})^2$, we have

$$\frac{\Sigma(Y - \overline{Y})^2}{\Sigma(Y - \overline{Y})^2} \quad \frac{\Sigma(Y_c - \overline{Y})^2}{\Sigma(Y - \overline{Y})^2} + \frac{\Sigma(Y - Y_c)^2}{\Sigma(Y - \overline{Y})^2}$$

Consequently, we have the *proportional reduction* in the variability of Y by the regression of Y on X expressed as follows:

$$\frac{\Sigma(Y_c - \overline{Y})^2}{\Sigma(Y - \overline{Y})^2} = \frac{\Sigma(Y - \overline{Y})^2}{\Sigma(Y - \overline{Y})^2} - \frac{\Sigma(Y - Y_c)^2}{\Sigma(Y - \overline{Y})^2} \tag{12-8}$$

The proportion on the left-hand side of this equation is commonly designated by r^2. Thus we have

$$r^2 = 1 - \frac{\Sigma(Y - Y_c)^2}{\Sigma(Y - \overline{Y})^2} \tag{12-9}$$

which is the proportion of the total variability of Y *accounted for*, or *explained by*, its relationship with X, and is generally known as the *coefficient of determination*. The square root of r^2, or r, is the *coefficient of correlation*. That is,

$$r = \sqrt{1 - \frac{\Sigma(Y - Y_c)^2}{\Sigma(Y - \overline{Y})^2}} \tag{12-10}$$

EXAMPLE 12-1   Using the data from table 12-1, determine the strength of the relationship between X and Y by computing the coefficient of correlation r.

For computing the value of r by equation 12-10, we need the two sums of squared deviations: $\Sigma(Y - \overline{Y})^2$ and $\Sigma(Y - Y_c)^2$. From the last section, we know that $\overline{Y} = 131/20 = 6.55$ and $Y_c = 4.55 + 0.27X$. The computation of these sums of squares is shown in table 12-3. If the sums of squares are used in equation 12-10, we obtain

$$r = \sqrt{1 - \frac{49.5759}{62.95}} = \sqrt{1 - 0.787544} = \sqrt{0.212456} = 0.461$$

Since $r = 0.461$ and $r^2 = 0.2125$, we say that 21.25 percent of the total variability of Y is accounted for by its relationship with X, or by the regression of Y on X. Of course, the greater this percentage, the stronger the relationship between X and Y will be.

Observe that by equation 12-10, the obtained r invariably has a *positive sign*. However, the sign of r is always the same as that of b in the least squares equation. If b has a negative sign, then the sign of r must be changed to negative. Since in this example b is positive, no change is necessary.

Table 12-3 Computation of sums of squared deviations from the data in table 12-1

X	Y	$Y_c = 4.55 + 0.27X$	$(Y - Y_c)$	$(Y - Y_c)^2$	$(Y - \overline{Y}) =$ $(Y - 6.55)$	$(Y - \overline{Y})^2 =$ $(Y - 6.55)^2$
3	5	5.36	−0.36	0.1296	−1.55	2.4025
2	4	5.09	−1.09	1.1881	−2.55	6.5025
4	4	5.63	−1.63	2.6569	−2.55	6.5025
12	9	7.79	1.21	1.4641	2.45	6.0025
11	8	7.52	0.48	0.2304	1.45	2.1025
8	9	6.71	2.29	5.2441	2.45	6.0025
9	7	6.98	−0.02	0.0004	0.45	0.2025
7	8	6.44	1.56	2.4336	1.45	2.1025
6	5	6.17	−1.17	1.3689	−1.55	2.4025
5	6	5.90	0.10	0.0100	−0.55	0.3025
4	8	5.63	2.37	5.6169	1.45	2.1025
8	4	6.71	−2.71	7.3441	−2.55	6.5025
3	7	5.36	1.64	2.6896	0.45	0.2025
12	6	7.79	−1.79	3.2041	−0.55	0.3025
9	8	6.98	1.02	1.0404	1.45	2.1025
8	5	6.71	−1.71	2.924	−1.55	2.4025
11	10	7.52	2.48	6.1504	3.45	11.9025
7	7	6.44	0.56	0.3136	0.45	0.2025
8	6	6.71	−0.71	0.5041	−0.55	0.3025
10	5	7.25	−2.25	5.0625	−1.55	2.4025

Sum of squares: $\Sigma(Y - Y_c)^2 = 49.5759$ $\qquad$ $\Sigma(Y - \overline{Y})^2 = 62.9500$

The preceding procedure of computing r is used for the primary purpose of illustrating the meaning of the correlation coefficient. It shows that the correlation coefficient is the square root of the proportional reduction in the variability of Y by the regression of Y on X. The greater the reduction, the larger the value of r will become, and the closer the relationship between X and Y will be.

In actual practice, however, we seldom use the above procedure to compute the value of r. Instead, the following formula is more frequently used.

$$r = \frac{n(\Sigma XY) - (\Sigma X)(\Sigma Y)}{\sqrt{n(\Sigma X^2) - (\Sigma X)^2} \sqrt{n(\Sigma Y^2) - (\Sigma Y)^2}} \tag{12-11}$$

The various sums in this formula are all provided in table 12-2. If we use these sums in equation 12-11, we have

$$r = \frac{20(1012) - (147)(131)}{\sqrt{20(1261) - (147)^2}\ \sqrt{20(921) - (131)^2}}$$

$$= \frac{20,240 - 19,257}{\sqrt{25,220 - 21,609}\ \sqrt{18,420 - 17,161}}$$

$$= \frac{983}{\sqrt{3611}\ \sqrt{1259}}$$

$$= \frac{983}{\sqrt{4,546,249}}$$

$$= \frac{983}{2132.2}$$

$$= 0.461$$

which is identical with the result obtained earlier. This latter method is preferred not only because it is easier but also because it has the advantage of automatically giving r the correct sign (either plus or minus).

If the correlation (relationship, or association) between X and Y is strong, most of the variability of Y can be accounted for by the relationship with X, and r will be close to 1 or -1. The value of r ranges from -1 to $+1$ since it is defined as plus or minus the square root of a *proportion*. If $r = 1$ or -1, it means that a *perfect fit* exists; that is, the observed points all fall on the regression line. If the correlation between X and Y is weak, very little of the variability of Y can be accounted for by its relationship with X, and r will be close to 0. When $r = 0$, it indicates no correlation; that is, none of the variability of Y can be accounted for by its relationship with X.

When r is between 0 and 1, there exists a *positive correlation* between the two variables X and Y. If X and Y have a positive correlation, the values of X and Y tend to move in the same direction; the value of one variable tends to vary directly with that of the other. On the other hand, when r is between -1 and 0, there exists a *negative correlation* between X and Y. Then the values of X and Y tend to move in the opposite direction; when one increases, the other tends to decrease, and vice versa.

test of hypothesis

Frequently, we may want to decide whether a correlation coefficient (r) is large enough to suggest the existence of a correlation between paired values in the population or whether the obtained r can be attributed to chance. In other words, we may want to test the hypothesis that the population correlation coefficient ρ is equal to zero against the alternative that it is not.

Under the assumption that the two variables involved are each normally or approximately normally distributed, we can apply the t distribution to test the hypothesis about the population correlation coefficient ρ. The test statistic T used for such a test is defined as

$$T = r \sqrt{\frac{n-2}{1-r^2}} \qquad (12\text{-}12)$$

which is distributed as t with $(n-2)$ degrees of freedom.

EXAMPLE 12-2   On the basis of the r value obtained in example 12-1, test the hypothesis that the population correlation coefficient ρ is equal to zero at the 0.05 significance level.

This is obviously a two-tailed test, and the hypotheses to be tested are

$$H_0 : \rho = 0 \quad \text{and} \quad H_1 : \rho \neq 0$$

Since the level of significance is specified at 0.05 and the number of degrees of freedom is $20 - 2 = 18$, from appendix F the critical t values are found to be ± 2.101. The decision rule therefore is

$$\text{Reject } H_0 \text{ if } T \geq 2.101 \text{ or } T \leq -2.101$$

In example 12-1 we found that the correlation coefficient is 0.461. Using this value in equation 12-12, we have

$$T = 0.461 \sqrt{\frac{20-2}{1-(0.461)^2}} = 0.461 \sqrt{\frac{18}{1-0.2125}} = 0.461 \sqrt{\frac{18}{0.7875}}$$
$$= 0.461 \sqrt{22.857142} = 0.461(4.78) = 2.204.$$

which is larger than the critical value 2.101. As a result, we reject H_0. The obtained value $r = 0.461$ cannot be attributed to chance and there is a correlation between X and Y at the specified significance level, 0.05.

12-3 exercises

1. Referring to problem 1, exercises 12-2, do the following.
   a) Find the correlation coefficient r.
   b) Test the null hypothesis that no relation exists between the length of sales experience and the volume of sales at $\alpha = 0.05$.

2. Referring to problem 2, exercises 12-2, do the following.
   a) Find the correlation coefficient r.
   b) Test the null hypothesis that no relation exists between rainfall and wheat yield at $\alpha = 0.02$.

3. Referring to problem 3, exercises 12-2, do the following.
   a) Find the correlation coefficient r.
   b) Test the null hypothesis that maintenance cost is not related to machine age at $\alpha = 0.05$.

4. Referring to problem 4, exercises 12-2, do the following.

   a) Find the correlation coefficient r.

   b) Test the null hypothesis that the number of teachers who voluntarily resign has no relation to years of service at $\alpha = 0.01$.

5. Use the data in problem 5, exercises 12-2 to compute the correlation coefficient r.

6. Use the data in problem 6, exercises 12-2 to compute the correlation coefficient r.

13-1
introduction

Most of the techniques of hypothesis testing discussed in the preceeding chapters are *parametric*, because they are concerned with population parameters such as the mean, the difference between two means, and the population correlation coefficient. There are many situations, however, to which the parametric methods cannot be applied. For instance, if a committee is called upon to evaluate and rate the competence or qualifications of 10 candidates for a position, or if several judges are asked to rank the relative merits of five finalists in a beauty contest, the measurements obtained will be of a completely different nature. Therefore, whenever observations or measurements are made in terms of *ranks* or *orders*, *nonparametric methods*, which are often referred to as *order techniques*, become appropriate, because these techniques are not concerned with population parameters.

Furthermore, parametric methods are based on the assumption that the population distribution is known. Recall that when testing hypotheses about the mean, the variance, or the differences between means or variances, we invariably assumed a normal or an approximately normal distribution of the population from which the samples are taken. Likewise, when testing the hypothesis that the population correlation coefficient is zero, we assumed that the population distribution of each of the two variables is normal or roughly so. Nonparametric methods, however, are *distribution free;* that is, they involve no assumption about the pattern of the underlying population distribution. For instance, when testing hypotheses about *statistical independence* between two types of variables, as we did with the contingency tables in chapter 10, we did not stipulate that the variables involved must be normal, or must have any other shape of distribution. Thus, the χ^2 test of statistical independence is a distribution-free method.

The two terms "nonparametric" and "distribution free" are not, however, synonymous; "nonparametric" is used to describe a test that involves no specific population parameters, while "distribution free" refers to a test that involves no assumption about the shape of the population distribution. In spite of this difference, the term "nonparametric methods" is commonly used to refer to tests involving either or both of these two conditions.

In many situations the assumption of a normal population distribution is not justified. To replace the traditional parametric tests, numerous statistical tests that involve an unknown distributional pattern have been developed. A nonparametric procedure must be used whenever the population distribution is unknown or known to be not normal. This does not mean, however, that nonparametric methods are preferred under all circumstances. In fact, when the normality actually exists, the use of a nonparametric procedure would be a gross mistake. A nonparametric test yields less precise results, and therefore is a weaker tool than a parametric technique. Thus, a nonparametric method is used only if the researcher concludes, after careful deliberations, that a parametric test is not applicable.

Nonparametric methods require fewer assumptions, and are easier to explain, understand, and use. They are especially useful where the observations are susceptible to ordering but not to measurements on a quantitative scale.

Numerous nonparametric methods have been developed in recent decades. We shall discuss only a few such techniques in this chapter: the sign test, the rank sum test, and the rank correlation test.

13-1 exercises

1. Cite two parametric methods of hypothesis testing and explain why they are referred to as parametric.

2. Cite one nonparametric method of hypothesis testing and explain why it is referred to as nonparametric.

3. Nonparametric statistics is also known as order statistics. Why?

4. Explain why nonparametric techniques are often referred to as distribution free.

5. Explain the relationship between the two concepts "nonparametric" and "distribution free." Are they synonymous?

6. Name the advantages and disadvantages of nonparametric methods.

13-2 the sign test

Recall that when samples are small, and the t test is used to determine whether two population means are equal, we must assume that each of the two populations is normal or approximately normal and that the variances of the two populations are identical. In many instances, however, either or both of these assumptions may be false and the t test cannot be used. Renouncing these assumptions requires rephrasing our hypothesis. Rather than hypothesizing that $\mu_1 = \mu_2$, we hypothesize that the two populations have an identical distribution. Observe that the two hypotheses are not exactly the same since two populations may have different distributions and yet possess the same mean. For testing the hypothesis that two population distributions are identical, we use a non-parametric procedure known as the *sign test*.

Suppose that A and B are two random variables, and we hypothesize that the population distribution of A is the same as that of B. For testing this hypothesis we draw two samples of equal size: one for A and the other for B. The observations in the two samples may be matched pairs such as a worker's hourly output before a strike (B) and his hourly output after the strike (A). This "before" and "after" type of data is similar to the data of dependent samples involving the t test (example 9-7, page 205). The sign test is based on the signs, negative or positive, of the differences between the observations of matched pairs without regard to the magnitude of the differences. Thus, wherever the normality assumption is not warranted and the t test should not be used, the sign test provides a convenient substitute for the t test involving dependent samples.

Even in studies involving two independent samples, or unpaired data, the sign test can also be used. When two samples are selected from two populations, I and II, there are many situations in which the observation of I is independent of II, or I and II are not in any way related. For instance, suppose that two machines are used to produce identical parts. The number of defective parts produced by the first machine in any day (A) cannot be dependent on the number of defective parts produced by the second machine during the same day (B). The sign test can be used to decide whether the population of A, the number of defective parts produced daily by machine I, is identical with the population of B, the number of defective parts produced daily by machine II.

When the null hypothesis that the distributions of A and B are identical is true, we can expect that the differences between A and B, or $A - B$, will have the same number of *positive* signs as *negative* signs. That is, the proportion of positive (or negative) signs is expected to be 50 percent, since the null hypothesis implies that the population proportion of either sign is 0.5, or $p = 0.5$. Accordingly, the procedure of the sign test is similar to the parametric test involving the null hypothesis that $p = 0.5$.

When the difference between A and B, or $A - B$, is zero, it is disregarded and the number of pairs is then reduced. The total number of differences with either positive or negative signs is taken as n, a quantity required in the test.

EXAMPLE 13-1 Suppose that we wish to decide whether a certain diet is effective in reducing weight. Twenty persons are put on the diet. Let the weights *before* and *after* the diet be designated by B and A

respectively. Table 13-1 lists the values of B and A for those 20 persons as well as the signs of the differences between A and B, or $A - B$.

Table 13-1 Weights of 20 persons before and after a diet

Person	Weight before (B)	Weight after (A)	Signs (A − B)
A	188	185	−
B	185	189	+
C	180	190	+
D	230	215	−
E	196	165	−
F	185	186	+
G	198	185	−
H	215	213	−
I	240	243	+
J	160	135	−
K	187	180	−
L	213	200	−
M	210	212	+
N	179	169	−
O	199	200	+
P	256	223	−
Q	215	201	−
R	169	175	+
S	217	205	−
T	175	180	+

Test the null hypothesis that the diet is not effective against the alternative that it is effective at $\alpha = 0.0577$.

If the diet is not effective in reducing weight, we should expect *no more* persons to lose weight than to gain it. In other words, our null hypothesis simply implies that a negative sign is as likely to occur as a positive sign. This is equivalent to saying that the null hypothesis is $H_0 : p = 0.5$, where p is the probability of getting a negative sign. If the diet is effective in reducing weight, we should expect more persons to lose than to gain weight, and there should be more negative than positive signs. Consequently, the alternative hypothesis is $H_1 : p > 0.5$. At $\alpha = 0.0577$, the critical value for the binomial random variable X, the number of negative signs, is 14, because in appendix B we find that for $n = 20$ and $p = 0.5$, the cumulative probability for X equal to or less than 13 is 0.9423. Thus the probability of getting 14 or more negative signs for $n = 20$ and $p = 0.5$ is $1 - 0.9423 = 0.0577$. Thus, the decision rule is

Reject H_0 if $X \geq 14$

Since the observed number of negative signs is 12, which is smaller than the critical value, we do not reject the null hypothesis. That is, the diet, on the basis of the experiences of those 20 persons, is not significantly effective in reducing weight.

As discussed in section 8-5, if the sample size is sufficiently large, the method of normal approximation may be used for testing hypothesis about a population proportion p. When $p = 0.5$, any sample size greater than 10 would be considered large enough to invoke this method. As n increases, the binomial random variable X will become more and more normal with a mean np and a variance $np(1 - p)$. Accordingly, the test statistic Z becomes

$$Z = \frac{X - np}{\sqrt{np(1 - p)}}$$

We may, if we wish, use the method of normal approximation to test the hypothesis in example 13-1 that the diet is not effective, since the sample size $n = 20$ is large enough for this purpose.

In the next example we shall use data that are not of the "before" and "after" type, and shall conduct the sign test by the normal approximation method.

EXAMPLE 13-2   Two automatic machines are used to produce identical screws. Let A be the number of defective screws produced by the first machine in any hour, and B the number of defective screws produced by the second machine in the same hour. The outputs of the two machines in an 18-hour period are inspected, and the number of defective screws produced in each hour by each machine, together with the signs of differences between A and B, are presented in table 13-2. Test the null hypothesis that the population distribution of A is the same as that of B against the alternative that they are different at $\alpha = 0.05$.

Because there are two zeros among the 18 differences, the total number of signs of the differences is 16, or $n = 16$. When the two population distributions are identical, a positive sign is expected to be as likely to occur as a negative sign, and the proportion of positive (or negative) signs can be expected to be 0.5, or $p = 0.5$. Let X be the number of positive signs of the differences, or $A - B$. Since $n = 16$ and $p = 0.5$, we have the mean of X

$$\mu = np = 16(0.5) = 8$$

and the standard deviation of X

$$\sigma = \sqrt{np(1 - p)} = \sqrt{16(0.5)(0.5)} = 2$$

This is obviously a two-tailed test, since the null hypothesis will be

Table 13-2 Number of defective screws produced per hour by two machines for 18 hours

Hours	Defective screws by first machine (A)	Defective screws by second machine (B)	Signs (A − B)
1	12	10	+
2	8	12	−
3	15	12	+
4	8	7	+
5	16	16	0
6	20	22	−
7	18	20	−
8	19	10	+
9	15	12	+
10	27	25	+
11	11	16	−
12	24	21	+
13	17	17	0
14	19	15	+
15	13	17	−
16	9	10	−
17	11	10	+
18	26	7	+

rejected if the positive signs are either too many or too few. Thus the hypotheses to be tested in this case are

$$H_0: \mu = 8 \quad \text{and} \quad H_1: \mu \neq 8$$

Applying the standard normal distribution (appendix E), we have, with $\alpha = 0.05$, 1.96 and −1.96 as the critical values. Consequently, the decision rule is

Reject H_0 if $Z \geq 1.96$ or $Z \leq -1.96$

From table 13-2 we obtain 10 positive signs, or $X = 10$. Thus the test statistic Z becomes

$$Z = \frac{X - np}{\sqrt{np(1 - p)}} = \frac{10 - 8}{\sqrt{16(0.5)(0.5)}} = \frac{2}{2} = 1$$

Since the observed value for Z falls between the two critical values, we do not reject H_0. The number of defective screws produced per hour by the first machine is not significantly different from the second machine.

The principal drawback of the sign test lies in the fact that it ignores completely the magnitude of the difference between paired values. A different sign test was developed by Frank Wilcoxon in 1945, and the revised test is called Wilcoxon's signed-rank test, which takes into consideration the magnitude of the difference between paired values. Wilcoxon's test is presented in the exercises below.

13-2 exercises

1. Refer to example 13-1. Use the normal approximation method to test the null hypothesis that the diet is not effective in reducing weights against the alternative that it is at $\alpha = 0.05$.

2. Two production lines are used to produce electric fuses. Let X be the number of defective fuses produced by one line in any hour, and Y be the number of defective fuses produced by the other line in the same hour. The outputs of the two production lines in a 16 hour period are inspected, and the following information is obtained.

Hour:	1	2	3	4	5	6	7	8
X:	20	10	61	18	25	28	22	20
Y:	21	25	64	25	21	22	30	29

Hour:	9	10	11	12	13	14	15	16
X:	18	15	23	25	18	5	30	31
Y:	16	10	18	35	34	25	19	13

By normal approximation test the null hypothesis that the distribution of X is the same as that of Y against the alternative that they are different at $\alpha = 0.01$.

3. Twelve pairs of hogs, paired by birth date, are randomly selected. They are divided into two groups; one hog from each pair is placed in group I, and the other hog is placed in group II. Two different diets are used to feed the two groups; and after a period of time their weight gains are recorded as follows:

Pair	Weight gain, group I (A)	Weight gain, group II (B)	Sign of difference $(A - B)$
1	17	14	+
2	17	21	−
3	21	36	−
4	18	20	−
5	17	24	−
6	14	12	+
7	15	15	0
8	19	28	−
9	16	16	0
10	17	16	+
11	16	21	−
12	12	20	−

Does the above information present sufficient evidence to indicate that the diet for group I is less effective than the diet for group II at $\alpha = 0.0547$?

4. An experiment is designed to compare the effectiveness of two methods, I and II, of teaching reading. The subjects are 15 sets of identical twins. One child is randomly selected from each set of twins and assigned to class I; the other is assigned to class II. The two classes are taught with the two methods, respectively. The test scores in a standard reading comprehension examination are shown below.

Twin pair:	1	2	3	4	5	6	7	8
Test score, class I (A):	70	95	80	73	61	96	66	75
Test score, class II (B):	75	92	71	63	64	84	51	78

Twin pair:	9	10	11	12	13	14	15
Test score, class I (A):	90	77	89	90	90	75	67
Test score, class II (B):	84	59	78	74	97	83	58

a) Use the sign test to determine whether the two teaching methods are equally effective against the alternative that method I is more effective than method II at $\alpha = 0.0593$.

b) Use normal approximation to test the null hypothesis that the two teaching methods are equally effective against the alternative that they are not at $\alpha = 0.05$.

5.

Hog pairs	Weight gains from the two diets				Signed rank	
	X	Y	$X - Y$	Rank	Negative	Positive
1	17	14	+3	4		+4
2	17	21	−4	5	−5	
3	21	36	−15	10	−10	
4	18	20	−2	2.5	−2.5	
5	17	24	−7	7	7	
						+2.5
6	14	12	+2	2.5		
7	19	28	−9	9	−9	
8	17	16	+1	1		+1
9	16	21	−5	6	−6	
10	12	20	−8	8	−8	
Total					−47.5	+7.5

The signed-rank test developed by Frank Wilcoxon takes into consideration the magnitude of the difference between paired values. The test requires the ranking of all the *absolute* differences between paired values from the smallest to the largest. The smallest absolute difference is assigned the rank 1, the next smallest the rank 2, and so on. Any pair with a difference 0 may be disregarded. Since the differences are ranked without regard to their signs, a difference of -1 or $+1$ is assigned the same rank. So is a difference of -2 or $+2$, and so forth. Once the differences are ranked, the sign of each difference is affixed to its rank. The sums of the *negative* and *positive ranks* are then computed. The computation of the two sums for the data in problem 3 (page 277) is shown in the table above with the two cases of zero difference ignored.

Observe that the difference -2 for the fourth pair is tied with the difference $+2$ for the sixth pair; they are tied for the second and the third ranks. Each is assigned the rank 2.5, the average of the ranks 2 and 3.

The random variable involved in the signed-rank test is usually designated by T (not to be confused with the test statistic T used in chapter 9). When the number of pairs, or n, is eight or larger, the variable T is distributed approximately normally with the mean

$$E(T) = \frac{n(n + 1)}{4}$$

and the standard deviation

$$\sigma_T = \sqrt{\frac{n(n + 1)(2n + 1)}{24}}$$

Thus, the Z score is computed as follows:

$$Z = \frac{T - E(T)}{\sigma_T}$$

Either the absolute value of the negative sum or the positive sum can be used as the value of T, and the test can be one tailed or two tailed. If the alternative hypothesis is true, Y is expected to be greater than X. Consequently, the difference $X - Y$ is expected to have a negative sign, and the absolute value of the negative sum is expected to be greater than the average of the two absolute sums. Let T assume the value 47.5; the decision rule for a one-tailed test with $\alpha = 0.05$ is

Reject H_0 if $Z \geq 1.645$

Since

$$E(T) = \frac{10(10 + 1)}{4} = \frac{110}{4} = 27.5$$

$$\sigma_T = \sqrt{\frac{10(10 + 1)(2(10) + 1)}{24}} = \sqrt{\frac{110(21)}{24}} = \sqrt{\frac{2,310}{24}}$$
$$= \sqrt{96.25} = 9.81$$

and

$$Z = \frac{47.5 - 27.5}{9.81} = \frac{20}{9.81} = 2.04$$

which is greater than the critical value, we reject H_0. That is, the weight gain for hogs fed with diet II is in general greater than that with diet I.

Referring to problem 4 above, use the signed-rank test to determine whether the two teaching methods are equally effective against the alternative that they are not equally effective at $\alpha = 0.05$.

6. Refer to problem 2 above. Use the signed-rank test to show whether the distribution of X is the same as that of Y with the level of significance specified at $\alpha = 0.01$.

13-3
the rank
sum test

In testing hypotheses about the differences between three or more population means (the *analysis of variance*), we assumed that the populations involved were normally distributed and that the variances of the various populations were identical. When these assumptions are met, we can apply the F test to decide whether the differences between the sample means can be attributed to chance or are indicative of real differences in the populations means. When these assumptions are not warranted, however, and the F test cannot be used to handle this type of problem, a nonparametric method known as the *rank sum test* provides a convenient substitute.

The rank sum test that is especially suitable for testing whether the various population distributions are identical is called the Kruskal-Wallis test, named after the two men who published the method in 1952. It is also referred to as the H test, because the test statistic is designated by H. The method does not require any assumption except that the populations involved are continuous. The null hypothesis to be tested by the Kruskal-Wallis method is that the various populations have identical distributions, whereas the alternative hypothesis asserts that they do not.

Suppose that a certain problem involves K populations, from which K samples are randomly selected. The samples are not necessarily of equal size; the first sample may have n_1 observations, the second may have n_2 observations, and so forth. To perform the test, all the observations in the K samples are combined and then ranked from the lowest to the highest. The lowest score is assigned rank 1, the second lowest score rank 2, ..., and the highest score rank N, which is the total number of observations in all the samples. The sum of the ranks for each sample is then calculated, and is denoted by R_k for $k = 1, 2, \ldots, K$. To conduct such a test, each of the samples must have more than five observations; then the test statistic H defined as

$$H = \frac{12}{N(N+1)} \sum_{k=1}^{K} \frac{R_k^2}{n_k} - 3(N+1) \tag{13-1}$$

is distributed approximately as χ^2 with $(K - 1)$ degrees of freedom. As usual, the null hypothesis that the distributions of the K populations are identical is tested by comparing the value of the test statistic H with the critical χ^2 value at the specified level of significance. The null hypothesis is rejected only if the value of H is *equal to or larger than* the critical χ^2 value.

EXAMPLE 13-3 Suppose that three different methods, A, B, and C, are used to teach English to three groups I, II, and III, of randomly selected adult immigrants. After a period of time, a standard examination is administered to all three groups, which consist of 9, 10, and 10 students, respectively. The test scores are shown in table 13-3.

Table 13-3 Test scores and ranks for three groups taught by methods A, B, and C

Group I		Group II		Group III	
Test score (by A)	Rank	Test score (by B)	Rank	Test score (by C)	Rank
93	23.5	89	19	78	9
77	8	90	20	80	11
93	23.5	85	15	75	6
79	10	76	7	81	12
92	22	84	14	91	21
99	28	95	26	88	18
98	27	82	13	86	16
71	3	72	4	94	25
87	17	73	5	69	2
		68	1	100	29
$R_1 = 162$		$R_2 = 124$		$R_3 = 149$	

Test the null hypothesis that the populations of test scores relative to the three different methods have an identical distribution against the alternative that the three distributions are different at $\alpha = 0.05$.

In this example, $n_1 = 9$, $n_2 = 10$, and $n_3 = 10$, and the total number of observations is $N = n_1 + n_2 + n_3 = 9 + 10 + 10 = 29$. We combine these 29 observations and then rank them from the lowest to the highest. The ranks from 1 to 29 are shown in the columns headed "rank" in table 13-3. If two or more scores are identical, the average of the ranks is used for each. For instance, the score 93 appears twice and is between the 22nd and the 25th ranks; thus the score 93 is assigned the average of the two ranks 23 and 24, namely, 23.5. The sum of the ranks for each group is computed; we have

$$R_1 = 162 \qquad R_2 = 124 \qquad R_3 = 149$$

Placing the values of N, the n's, and the R's into equation 13-1 we obtain

$$H = \frac{12}{29(29+1)} \left[\frac{(162)^2}{9} + \frac{(124)^2}{10} + \frac{(149)^2}{10} \right] - 3(29+1)$$

$$= \frac{12}{870} \left(\frac{26,244}{9} + \frac{15,376}{10} + \frac{22,201}{10} \right) - 90$$

$$= \frac{1}{72.5} (2916 + 1537.6 + 2220.1) - 90$$

$$= \frac{6673.7}{72.5} - 90 = 92.05 - 90 = 2.05$$

Since in appendix G the critical χ^2 value for $K - 1 = 3 - 1 = 2$ degrees of freedom and $\alpha = 0.05$ is found to be 5.99146, which is greater than the computed H value, we do not reject the null hypothesis. The three teaching methods A, B, and C do not have significantly different effects upon the test scores of the students.

The rank sum test is always a one-tailed test; the null hypothesis is rejected only if the computed H value is sufficiently large. When the ranks are rather evenly distributed among the groups, which results in a relatively small value for H, it indicates that the population distributions are not different. On the other hand, when the high ranks are concentrated in one or a very few groups, while the other groups consist mainly of low ranks, making the H value relatively large, it reflects the fact that the distributions of the populations are different. Thus, we shall reject the null hypothesis if the H value is sufficiently large, because it indicates that the population distributions differ from one another.

13-3 exercises

1. Suppose that 20 plots of nearly equal fertility are planted with three different varieties of grain. It is assumed that the distribution of the varieties among the 20 plots is made at random. The yields in bushels per acre are shown below.

Variety I: 20 21 22 23 24 26 28
Variety II: 19 20 23 18 25 13 12
Variety III: 15 20 16 17 14 27

Does a significant difference in yields between the three varieties exist at $\alpha = 0.05$?

2. Suppose that we wish to decide whether or not four brands of automobile tires have the same average service life. A random sample of

five tires of each brand is selected and tested. The following mileage figures are obtained.

Brand I	Brand II	Brand III	Brand IV
35,000	34,500	34,000	34,600
37,000	33,000	33,500	38,000
37,500	34,300	36,000	36,400
32,000	32,500	37,400	40,000
38,500	31,000	36,700	39,100

Is there sufficient evidence to indicate that the four brands of tires have different average service lives at $\alpha = 0.05$?

3. A large corporation has classified its employees into three tenure groups. Management wants to find out whether morale is the same for all three tenure groups. Ten employees are randomly selected from each group, and their morale is measured on a 40-point scale. The following results are obtained.

Group I: 21 23 25 27 31 33 30 35 13 39
Group II: 20 22 26 28 30 32 19 34 36 38
Group III: 15 16 17 24 18 29 30 14 37 24

Can we conclude that the three tenure groups have equal morale at $\alpha = 0.01$?

4. The final grades earned by five students in economics, statistics, and psychology are given below. At the 0.05 level of significance, test the hypothesis that the five students have the same level of proficiency in the three subjects.

Economics	Statistics	Psychology
98	90	99
86	91	89
78	95	80
84	97	88
87	93	96

5. Three different treatments are used to destroy bacteria in a certain product. The bacteria count is recorded after each product unit is subjected to one of the treatments. The following results are obtained.

Treatment I: 110 108 105 98 102 111 131 106
Treatment II: 120 109 115 104 122 119 123 117
Treatment III: 130 125 135 133 128 103 100 127

Are the three treatments equally effective in destroying bacteria at $\alpha = 0.025$?

6. Consider the following four samples of five observations each.

Sample I	Sample II	Sample III	Sample IV
101	102	105	106
103	104	107	108
110	109	111	112
99	98	115	114
97	96	113	116

Do these four samples come from the same population—that is, from populations with the same mean—at $\alpha = 0.05$?

13-4 the rank correlation test

For testing hypotheses about the correlation coefficient, it was assumed that the two variables are normal or approximately normal. When the truth of such an assumption is questionable, a nonparametric technique may be used. The nonparametric method that tests the correlation between two variables is known as the *rank correlation test*. Several rank correlation methods have been developed. In this section, we shall discuss only one of these methods; it is known as Spearman's method named after the man who proposed it in 1904.

The Spearman rank correlation method is the oldest and for many years the most widely used of all nonparametric methods. It requires no assumption about the distributions of the underlying populations. In the application of this method, we need only to arrange the sample observations in rank orders. The relationship between the variables is determined on the basis of the rank of each variable. The method has the distinct advantage of circumventing the elaborate computation process required in the calculation of the correlation coefficient r. Furthermore, it can be applied to situations where numerical measurements are difficult or impossible to obtain. For instance, it may be very difficult, if not impossible, to measure the morale of each employee in a department; but it is a relatively easy job to assign ranks to the morale of employees.

Suppose that we wish to determine the relationship between two variables, X and Y. The observations for X and Y are taken in pairs; for example, a person's height and weight or two judges' ratings of each contestant in a beauty pageant. Let us first consider a situation in which X and Y take on numerical scores. As an initial step, the scores for each of the two variables are ranked, with the lowest score assigned rank 1, the next lowest rank 2, and so on. (Or rank 1 may be assigned to the highest score, rank 2 to the next highest score, and so on, for each of the two variables.) For the sake of clarity, the scores (and ranks) of one of the two variables are arranged in an ascending (or descending) order. The rank correlation method calls for, among other things, the computation of the *sum* of the *squared differences* between the corresponding ranks for all the pairs. Let D be the difference between the paired ranks, n the number of pairs in the sample, and R the rank correlation coefficient. Then R is defined as follows:

$$R = 1 - \frac{6 \sum\limits_{i=1}^{n} D_i^2}{n(n^2 - 1)} \tag{13-2}$$

Recall that in chapter 12 the test for significance of the correlation coefficient r was made by computing the test statistic T by equation 12-12. The test as presented then was based on the assumption that the samples were drawn from normal distributions. The test for significance of the rank correlation coefficient R requires no such assumption, and the test statistic T is computed in the same way as equation 12-12, with R replacing r.

$$T = \frac{R \sqrt{n-2}}{\sqrt{1 - R^2}} \tag{13-3}$$

where the number of degrees of freedom is also $(n - 2)$.

Table 13-4 Computation of rank correlation between test scores in English and mathematics

Students	Scores in English		Scores in mathematics		Differences between ranks D	D^2
	X	Rank	Y	Rank		
A	61	1	63	2	−1	1
B	64	2	62	1	1	1
C	68	3	69	4	−1	1
D	69	4	65	3	1	1
E	76	5	78	7	−2	4
F	78	6	70	5	1	1
G	82	7	75	6	1	1
H	84	8	90	11	−3	9
I	86	9	85	9	0	0
J	90	10	84	8	2	4
K	97	11	95	12	−1	1
L	98	12	86	10	2	4

$$\Sigma D^2 = 28$$

EXAMPLE 13-4   Suppose that we want to decide whether college freshmen's performances in English (X) correlate with their performances in mathematics (Y). A sample of 12 freshmen who have taken both courses is randomly selected, and their test scores for the two courses are shown in the X and Y columns of table 13-4. Find (a) the rank correlation coefficient R, and (b) test the hypothesis that the correlation in the population is zero against the alternative that it is not zero at $\alpha = 0.01$.

a) Table 13-4 shows the ranks of X, which are arranged in ascending order, and also the ranks of Y. The squares of the differences between the paired ranks are listed in the last column, and the total of the squares is 28. Placing this quantity and the value of n into equation 13-2, we have

$$R = 1 - \frac{6(28)}{12(12^2 - 1)} = 1 - \frac{168}{1716} = 0.902$$

b) The null hypothesis to be tested here is that there is no correlation in the population and the alternative is that there is. This is obviously a two-tailed test, because we shall reject the null hypothesis if R is sufficiently close to either $+1$ or -1. With $\nu = 12 - 2 = 10$ and $\alpha = 0.01$, the critical t values taken from appendix F are ± 3.169. Thus, the decision rule is

Reject H_0 if $T \geq 3.169$ or $T \leq -3.169$

Since the value of T is

$$T = \frac{0.902\sqrt{12 - 2}}{\sqrt{1 - 0.902^2}} = \frac{0.902\sqrt{10}}{\sqrt{1 - 0.813604}} = \frac{0.902(3.1623)}{\sqrt{0.186396}} = \frac{2.8523946}{0.432}$$
$$= 6.603$$

which is larger than 3.169, we conclude that the observed rank correlation coefficient R is too large to be attributable to chance and that there is a correlation in the population.

Let us now consider the situation in which the sample observations themselves are in ranks. For instance, two executives may be required to rank several employees eligible for promotion in terms of their qualifications, or, two judges may be called upon to rank a number of applicants for certain prizes on the basis of their merits. This type of problem involves an analysis of two sets of rankings to determine the closeness of the dependence between ranks and to decide whether the dependence as shown in the sample is statistically significant.

EXAMPLE 13-5   Two coaches, I and II, are asked to rate the 11 players of a football club's defensive team. Let the ranks assigned by coaches I and II be designated by X and Y, respectively. The rankings are shown in columns X and Y of table 13-5. Find (a) the rank correlation coefficient R, and (b) test the null hypothesis that there is no correlation between X and Y in the population at $\alpha = 0.05$.

Table 13-5 Computation of a rank correlation coefficient for two rankings of a football team

Players	Ranks given by Coach I (X)	Ranks given by Coach II (Y)	D $(X - Y)$	D^2
A	1	3	−2	4
B	2	5	−3	9
C	3	1	2	4
D	4	2	2	4
E	5	6	−1	1
F	6	4	2	4
G	7	9	−2	4
H	8	10	−2	4
I	9	7	2	4
J	10	8	2	4
K	11	11	0	0
			$\Sigma D^2 = 42$	

a) In this sample $n = 11$ and the sum of squared differences between the paired ranks is 42. Using these quantities in equation 13-2, we obtain

$$R = 1 - \frac{6(42)}{11(11^2 - 1)} = 1 - \frac{252}{1320} = 0.81$$

b) This is again a two-tailed test. Since the number of degrees of freedom is $11 - 2 = 9$ and the level of significance is specified at 0.05, appendix F shows that the critical values are ± 2.262. The test statistic T is computed as follows:

$$T = \frac{0.81 \sqrt{11 - 2}}{\sqrt{1 - 0.81^2}} = \frac{0.81 \sqrt{9}}{\sqrt{1 - 0.6561}} = \frac{0.81(3)}{\sqrt{0.3439}} = \frac{2.43}{0.586} = 4.147$$

which is larger than the critical value 2.262. Accordingly, we conclude that the computed rank correlation coefficient R is too large to be attributable to chance and the correlation as shown in the sample is statistically significant.

13-4 exercises

1. In a beauty contest two judges are employed to rate 11 contestants. The ratings are shown below.

Contestant:	A	B	C	D	E	F	G	H	I	J	K
Judge I:	1	2	3	4	5	6	7	8	9	10	11
Judge II:	2	3	1	6	4	5	8	7	10	11	9

a) Compute the rank correlation coefficient R.

b) At $\alpha = 0.01$, can we conclude that there is significant correlation between the rankings of the two judges?

2. The heights and weights of 18 randomly selected college sophomores are as follows.

Sophomore:	1	2	3	4	5	6	7	8	9
Height (in.):	60	61	63	64	65	66	67	68	69
Weight (lb.):	125	120	124	135	150	145	140	160	155
Sophomore:	10	11	12	13	14	15	16	17	18
Height (in.):	70	71	72	73	74	75	76	77	78
Weight (lb.):	170	180	175	174	185	190	200	220	210

a) Compute the correlation coefficient R by equation 13-2.
b) Test the null hypothesis that no relation exists between the heights and weights of college sophomores at $\alpha = 0.02$.

3. A new administrative position is created in a certain institution. Six candidates are applying for the position. The vice-president and the personnel director of the institution are called upon to rank the six candidates, and the results are shown as follows.

	Ranks assigned by	
Candidates	vice-president	personnel director
A	1	2
B	2	1
C	3	3
D	4	5
E	5	6
F	6	4

a) Compute the rank correlation coefficient R.
b) Is there a significant correlation between the two rankings at $\alpha = 0.05$?

4. The final grades earned by 11 students in English and German are listed below.

Student:	A	B	C	D	E	F	G	H	I	J	K
English grade:	69	72	75	77	80	84	86	90	93	95	99
German grade:	75	70	80	79	85	90	83	89	91	98	95

a) Compute the rank correlation coefficient R.
b) Test the null hypothesis that the students' grades in English and German are not related to each other at $\alpha = 0.01$.

rudiments of modern

decision theory

As indicated at the beginning of this text, the emphasis in statistics has changed from the summarizing of numerical data to the testing of hypotheses. Like other disciplines, statistics has been changing, and new techniques and theories are constantly being developed to meet the needs and challenges of the times. Recently attempts have been made to include all the techniques of statistical decision making in an all-inclusive theory called *decision theory*. Decision theory covers not only the methods of statistical inference discussed in the last several chapters, but also decision rules not yet considered in this text. The primary objective of this chapter is to examine some of the popular decision rules that are integral parts of modern decision theory.

Thus far the decision problems we have encountered have involved a decision to accept or reject the hypothesized value of an unknown population parameter on the basis of sample information taken from the population. A level of significance is established, a decision rule is set up, and the null hypothesis is then rejected or accepted by comparing the test statistic with the critical value or values specified in the decision rule. If the null hypothesis is true (false), the probability of making the erroneous decision by rejecting (accepting) it is α (β), and the probability of making the correct decision by accepting (rejecting) it is $1 - \alpha$ ($1 - \beta$). The task of decision making is completed when the error probabilities for a given decision rule are determined, and a choice between the null and alternative hypotheses is made according to the rule.

In other words, our previous discussions were not concerned with the problem of *gains* or *losses* relative to correct or wrong decisions. They did not deal with the payoff of a decision. Decision theory involving payoffs is a more advanced subject, and a rather thorough treatment of the subject is beyond the scope of an elementary text. We will, however, review the rudiments of decision theory with the notion of payoff taken into consideration.

Generally speaking, the essence of decision making may be regarded as the selection between two or more alternative courses of action. In hypothesis testing, there are only two courses of action available to the decision maker, namely, *accept* or *reject* the null hypothesis. We now consider a problem situation in which there may be more than two available courses of action. It is assumed that the decision maker knows the possible *consequences* of each available course of action and that such consequences take the form of *payoffs*, or gains or losses. It is further assumed that the decision maker is a rational being; that is, he prefers greater gains or smaller losses, and his preference dictates his choice of a course of action.

There can be a number of consequences for each available course of action. This is the case because the consequences are the results not only of the decision maker's action, but also of factors beyond his control. Let us first use a familiar example to illustrate this point. In hypothesis testing, as we know, accepting the null hypothesis (H_0) can have two consequences: (1) a *correct decision* if H_0 is true and (2) a *type II error* if it is false. On the other hand, the two possible consequences of rejecting the null hypothesis are (1) a *type I error* if H_0 is true and (2) a *correct decision* if H_0 is false. Let A_1 denote the action to accept H_0, and A_2 denote the action to reject H_0. Also, let S_1 and S_2 designate the states of nature. Then we have a decision matrix like the one in table 14-1.

Table 14-1 A decision matrix

States of nature	Courses of action	
	A_1 (Accept H_0)	A_2 (Reject H_0)
S_1 (H_0 is true)	Correct decision	Type I error
S_2 (H_0 is false)	Type II error	Correct decision

When payoffs are taken into consideration in the decision process, the cell entries in the decision matrix will be *gains* or *losses*, which can be cash values. Suppose, for instance, that a decision problem involves three alternative courses of action and three states of nature. Then, there will be three consequences for each of the three alternative courses of action. Let C designate a consequence, and the first and second subscripts of C designate the state of nature and course of action, respectively; we will have a three-by-three decision matrix like the one shown in table 14-2. When a payoff forms a part of the framework in decision making, the C's in the decision matrix will be gains or losses.

The matrix formulation makes it easier for us to identify different types of decision situations. First, let us consider a situation in which we are perfectly sure what the state of nature is. If the decision matrix consists of only a single state of nature, it has only one row. In such a situation, the decision problem is a simple one; the decision maker just

Table 14-2   A decision matrix

States of nature	Courses of action		
	A_1	A_2	A_3
S_1	C_{11}	C_{12}	C_{13}
S_2	C_{21}	C_{22}	C_{23}
S_3	C_{31}	C_{32}	C_{33}

selects the course of action that will result in the most favorable consequence. This decision situation is generally referred to as *decision making under certainty*. We shall not be concerned with this type of decision situation because it presents very little statistical problem to the decision maker.

The decision situation that people face most frequently in everyday life is called *decision making under risk*. This is the situation in which there are two or more states of nature and the decision maker does not possess complete information about them. The information that is available about the states of nature is *probabilistic* at the best. This type of decision problem will be discussed in the next section.

Another kind of decision problem that people sometimes encounter is called *decision making under conflict*, which is dealt with in the *theory of games*. In this decision situation, there are two or more states of nature, which are under the control of an intelligent adversary. It is often found in business competition, bargaining, warfare, and the like. This type of problem will be briefly considered in the final section of this chapter.

14-1 exercises

1. Decision making involves the making of a choice between alternative actions. What are the alternatives available in hypothesis testing?

2. Decision making leads to consequences. What are the consequences of the available actions in hypothesis testing?

3. What is meant by payoffs in decision making? Why is it important that payoffs are taken into consideration by the decision maker?

4. Describe the essential features of the decision matrix.

5. What is meant by decision making under certainty?

6. Explain the circumstances in which decision making is conducted under risk.

7. Explain the circumstances in which decision making is conducted under conflict.

8. Define decision theory.

When we have to choose between several alternative courses of action in the face of uncertainty about the states of nature, one of several criteria may be used as a basis for our decision. Among the decision criteria to be considered in this section are the maximax and minimax rules. In addition, the Bayesian rule of mathematical expectation will be examined.

maximax rule

An adventurous and aggressive decision maker may choose to take the action that would result in the maximum payoff possible. A prisoner, for instance, may decide to try to escape, even though the chance of success is extremely small; to him, the possibility of success is worth the risk and the penalties of failure. A football team that is trailing late in the fourth quarter will very likely launch a passing attack; the cost of losing additional points is rather small in comparison with the possible reward of winning the game. People purchase lottery tickets even though the expected value of the ticket is negative; the prospect of becoming rich outweighs the price they have to pay for the ticket. The goal of the decision maker in each of these cases is to obtain the maximum possible payoff from the available courses of action; any decision maker who takes this approach is said to follow the *maximax rule*.

The term "maximax" is an abbreviation of the phrase "maximum of the maximums." Table 14-2 shows that for each action there are three possible payoffs corresponding to the three states of nature. Suppose that C_{21}, C_{32}, and C_{13} are the maximum payoffs under A_1, A_2, and A_3, respectively, and that C_{32} is the maximum among these three maximums. Then, according to this criterion, the decision maker will choose action A_2, because it will give him the maximum possible payoff.

Table 14-3 Earnings from selling refreshments under different weather conditions

States of nature	Decision		
	A_1	A_2	A_3
S_1	\$22	\$18	\$10
S_2	\$16	\$19	\$18
S_3	\$14	\$17	\$20

EXAMPLE 14-1 Suppose that three groups of vendors are hired to sell three different types of refreshments in a football stadium. Some vendors sell ice cream (A_1), some sell peanuts (A_2), and others sell coffee and doughnuts (A_3). John Smith, a college sophomore, has been hired as a vendor for the next game, and he is permitted to choose which type of refreshment to sell. Past records show that the earnings from A_1, A_2,

and A_3 depend on the state of nature, namely, weather conditions on the day the game is played. Let S_1 designate a sunny day, S_2 a cloudy day, and S_3 a rainy day. Table 14-3 shows the possible earnings from selling each type of refreshment under each state of nature. If the decision of John Smith is based on the maximax criterion, what type of refreshment should he choose to sell?

The maximum earnings of selling the three types of refreshments in table 14-3 are as follows:

A_1	A_2	A_3
$22	$19	$20

The maximum of these three maximums is $22. Consequently, according to the maximax rule, John Smith is going to sell ice cream, or take action A_1.

minimax rule

Minimax is just the opposite of maximax. Application of the minimax criterion requires a payoff table of losses instead of gains. The losses are the costs to be incurred or the damages to be suffered for each of the alternative actions and states of nature. The minimax rule minimizes the maximum possible loss for each course of action. Buying an automobile insurance policy costs money; but the premium to be paid is rather small compared with the liability that might be incurred as a result of a fatal accident. A basketball team that has absolutely no hope of winning the game should replace its super stars, because the damage of losing additional points is unimportant compared with the possibility of injuries to its best players. The discomfort resulting from an injection of a cold vaccine is negligible compared with the misery of catching cold.

The term "minimax" is an abbreviation of the phrase "minimum of the maximums." Suppose that the C's of table 14-2 are losses. Under each of the various actions there is a maximum loss, and the action that is associated with the *minimum of the various maximum losses* is the action to take according to the minimax rule.

EXAMPLE 14-2   Let us assume that a decision situation involves the same states of nature as in example 14-1 above, namely,

S_1: Sunny day S_2: Cloudy day S_3: Rainy day

The decision maker has three available actions.

A_1: Wear warm-day outfit including sun glasses.
A_2: Wear chilly-day outfit.
A_3: Wear rainy-day outfit including raincoat, boots, and rain hat.

Suppose that the losses resulting from wearing the improper outfit can be measured on a scale ranging from 0 to 9. (See entries in table 14-4.) What outfit should the decision maker wear, according to the minimax criterion?

Table 14-4 Losses from wearing different outfits under various weather conditions

States of nature	Decisions		
	A_1	A_2	A_3
S_1	0	2	5
S_2	4	1	3
S_3	9	7	4

Table 14-4 shows that the maximum losses incurred by the various actions are

$$
\begin{array}{ccc}
A_1 & A_2 & A_3 \\
9 & 7 & 5
\end{array}
$$

and the minimum among these three maximums is 5 under action A_3. Thus, according to the minimax criterion, the decision maker should wear the complete rainy-day outfit.

bayesian rule

In the foregoing discussion, we talked about states of nature, but we did not consider how probable they are. Yet this information is quite important to the decision maker. The probabilities of the states of nature are often called *personal* or *subjective probabilities*, because they usually reflect hunches or the intensity of belief on the part of the decision maker. Sometimes, however, such probabilities can be estimated on the basis of official records or personal experiences. The relevance of probabilities in decision making can be demonstrated by referring to example 14-1. If John Smith is perfectly certain that the day will be cloudy, he will surely take action A_2. On the other hand, if he is perfectly certain that the day will be rainy, he will take action A_3. The expression "He is perfectly certain" implies that the probability of a particular state of nature is 100 percent.

When the probabilities of the various states of nature are given, it is possible to calculate the mathematical expectation of gain or loss for each course of action. The decision maker then selects, from the available alternative actions, the action that leads to the *maximum expected gain* or *minimum expected loss*. As shown in equation 6-3, the expected value for each action is the sum of the products of the various values multiplied by the corresponding probabilities. Such an *expected-value criterion* is widely used and is generally referred to as the *Bayesian rule*.

EXAMPLE 14-3   Referring to example 14-1, let us assume that John Smith has done some research at the local weather bureau. According to available records, the three states of nature, S_1, S_2, and S_3, occur in the following proportions: 0.2, 0.4, and 0.4, respectively. On the assumption that these proportions are the probabilities for the various weather conditions of the day when the next game will be played, what would be John Smith's decision?

The computation of the expected earnings for the data in table 14-3 is presented in table 14-5. (The information in table 14-3 is reproduced in table 14-5 for the sake of easy reference.) The computation shows that the expected earnings of the three actions, A_1, A_2, and A_3, are 16.4, 18.0, and 17.2, respectively, and that the maximum of these is 18.0. Thus, according to the Bayesian criterion, John Smith will take action A_2, the one that gives him the maximum expected earnings.

Table 14-5   Computation of expected earnings for the data in table 14-3

Probabilities	States of nature	Earnings A_1 A_2 A_3	Probabilities × Earnings		
			A_1	A_2	A_3
0.2	S_1	22 18 10	0.2(22) = 4.4	0.2(18) = 3.6	0.2(10) = 2.0
0.4	S_2	16 19 18	0.4(16) = 6.4	0.4(19) = 7.6	0.4(18) = 7.2
0.4	S_3	14 17 20	0.4(14) = 5.6	0.4(17) = 6.8	0.4(20) = 8.0
Expected earnings:			16.4	18.0	17.2

EXAMPLE 14-4   Suppose that in example 14-2 above the probabilities of the three states of nature are also 0.2, 0.4, and 0.4. What action should the decision maker take if he follows the Bayesian rule?

The computation of the expected losses for the data in table 14-4 is presented in table 14-6. It shows that the expected losses of the three actions A_1, A_2, and A_3 are, respectively, 5.2, 3.6, and 3.8. Since the *minimum* of the expected losses is 3.6, the decision maker should, according to the Bayesian criterion, take action A_2.

Table 14-6   Computation of expected losses for the data in table 14-4

Probabilities	States of nature	Losses A_1 A_2 A_3	Probability × Loss		
			A_1	A_2	A_3
0.2	S_1	0 2 5	0.2(0) = 0	0.2(2) = 0.4	0.2(5) = 1.0
0.4	S_2	4 1 3	0.4(4) = 1.6	0.4(1) = 0.4	0.4(3) = 1.2
0.4	S_3	9 7 4	0.4(9) = 3.6	0.4(7) = 2.8	0.4(4) = 1.6
Expected loss:			5.2	3.6	3.8

1. Suppose that the management of a manufacturing company has to decide on an investment that will increase its capacity by 100,000 500,000 or 1,000,000 units. The future demand for its product may be brisk, average, or slack, which is the unknown state of nature. The business research department has provided the management with the following payoff (profit) table.

State of nature	Investment to increase capacity by		
	100,000	500,000	1,000,000
Brisk demand	$1,000,000	$3,000,000	$5,000,000
Average demand	$2,000,000	$2,500,000	$3,000,000
Slack demand	$2,500,000	$2,000,000	$1,000,000

a) By the maximax rule what decision should management choose?
b) If the probabilities for the three states of nature (brisk, average, and slack demand) are, respectively, 0.2, 0.2, and 0.6, what decision should management make according to the Bayesian criterion?

2. A decision maker has to decide whether he should purchase fire insurance. Only two alternatives are available to him: buy insurance and not buy insurance. There are two states of nature: fire and no fire. The payoff table of *losses* is shown below.

State of nature	Buy insurance	Not buy insurance
Fire	$52,000	$5,000,000
No fire	$50,000	0

a) What action should the decision maker take by the minimax rule?
b) If the probabilities for the two states of nature, fire and no fire, are respectively, 0.001 and 0.999, what action should the decision maker take by the expected-value criterion?

3. Consider the predicament of a jury. The jury has two alternatives: acquit and condemn. There are two states of nature: innocent and guilty. Suppose that the matrix of *losses* is given as follows:

State of nature	Condemn	Acquit
Guilty	1	5
Innocent	9	0

a) How, according to the minimax rule, should the jury rule?
b) If the person being prosecuted has an even chance of being guilty or innocent, what would be the sensible action for the jury to take by the minimum expected loss criterion?

4. An electronic company is capable of making one of three models of TV sets. The most expensive model will sell well when the economy is booming, the least expensive model is in bigger demand when the economy is in a recession, and the medium-priced model usually does well when the economy is stable. The payoff (profit) matrix for the company is given below (in millions of dollars).

State of the economy	Most expensive	Medium priced	Least expensive
Booming	10	7	5
Stable	6	8	6.5
Depressed	2	4	7

  a) By the maximax criterion, which model should the company produce?
  b) If the probabilities for a booming, stable, and depressed economy are, respectively, 0.3, 0.3, and 0.4, which model should the company decide to make?

14-3 game theory

One of the most interesting aspects of decision theory is that statistical decision problems can be viewed as games. As indicated earlier, game theory deals with those problem situations in which the decision maker is in conflict or competition with an intelligent adversary. The courses of action available to this intelligent opponent become the states of nature that face the decision maker. He realizes that the primary objective of his opponent is to inflict on him as much damage as possible, and with this in mind the decision maker formulates his decision strategies. Decision problems of this kind exist in many situations such as military conflict, business competition, and labor-management negotiations.

Frequently, however, there is no human opponent involved, and the decision maker does not possess even probabilistic information about the states of nature. Under such circumstances, we may still consider the situation as one of conflict, that is, a *game against Nature or Fate*. For the sake of simplicity, we shall, in this section, discuss a decision problem involving only two states of nature and two possible courses of action. The states of nature are assumed to be controlled by Fate, the opponent, who is trying to outsmart the decision maker. Therefore, the decision maker will try to minimize his maximum losses; that is, he will follow the minimax rule, a rather conservative strategy. We shall see shortly that this decision rule will provide the best solution possible even under the assumption that Fate is seeking to outwit him and to inflict the greatest amount of damage possible.

Suppose that a manufacturing company must decide whether to produce a deluxe or an economy model of a certain product. If the company decides to make the deluxe model (A_1), and there is prosperity (S_1), it will end up with a gain of $1,000,000. If the company decides to make the deluxe model, and there is a depression (S_2), there will be a loss of $600,000. If the company decides to make the economy model (A_2), and there is prosperity, it will suffer a loss of $400,000. If the company decides to make the economy model and there is a depression, it will gain $500,000. To the company, S_1 and S_2 are states of nature about which they have absolutely no information. To their opponent, Fate, S_1 and S_2 are considered as two courses of action, over which he has complete control. The decision matrix for the manufacturing company is presented in table 14-7.

Table 14-7 Decision matrix for the manufacturing company

Action of Fate	Action of the company	
	A_1 Make deluxe model	A_2 Make economy model
S_1 (Prosperity)	1,000,000	−400,000
S_2 (Depression)	−600,000	500,000

If the company takes action A_1, the worst possible result would be a loss of $600,000; and if the company takes action A_2, the worst possible result would be a loss of $400,000. The possible maximum loss of $400,000 is preferable to a possible maximum loss of $600,000. Accordingly, under the minimax rule, the company will elect to take action A_2. Keep in mind, however, the company's opponent, Fate, will try to make things as difficult as possible for it. Therefore, Fate will choose S_1 which will cause the company to lose $400,000 rather than gain $500,000. If the company anticipates Fate's response, it would shift to A_1 in order to win $1,000,000. If Fate thinks that the company will try to outwit him by taking action A_1, he will in turn try to outsmart the company by shifting to S_2, and thus cause the company to lose $600,000. This argument can continue indefinitely and the only way in which each party can avoid being outwitted by the other party is to use a *mixed strategy* by introducing an element of chance.

By mixed strategy, we mean selecting a course of action by the use of some random device. First, however, probabilities must be assigned to the various available actions. The probabilities to be assigned to A_1 and A_2 can be found on the basis that the best possible solution under the mixed strategy is one by which

The expected payoff, or the *value of the game*, from the decision maker's action must remain the same regardless of the action the opponent is going to take.

Let p be the probability of taking A_1, and $(1 - p)$ be the probability of taking action A_2. Then,

If S_1 occurs, the expected payoff $= 1,000,000p - 400,000(1 - p)$
and
If S_2 occurs, the expected payoff $= -600,000p + 500,000(1 - p)$

Since these two expected payoffs must be equal, we have

$$1,000,000p - 400,000(1 - p) = -600,000p + 500,000(1 - p)$$
$$1,000,000p - 400,000 + 400,000p = -600,000p + 500,000 - 500,000p$$
$$1,400,000p - 400,000 = -1,100,000p + 500,000$$
$$2,500,000p = 900,000$$
$$p = \tfrac{9}{25}$$

and

$$1 - p = \tfrac{16}{25}$$

Therefore, the probabilities of A_1 and A_2 are 9/25 and 16/25, respectively. Thus, the company may label 9 balls as A_1 and 16 balls as A_2, and then mix the 25 balls thoroughly in an urn. If a ball marked A_1 is drawn, the company will make the deluxe model; if a ball marked A_2 is drawn, the company will make the economy model. In so doing, the expected payoff to the company will be

If S_1 occurs, expected payoff $= 1,000,000(9/25) - 400,000(16/25)$
$$= 360,000 - 256,000 = 104,000$$
If S_2 occurs, expected payoff $= -600,000(9/25) + 500,000(16/25)$
$$= -216,000 + 320,000 = 104,000$$

Thus, regardless of the states of nature, the expected payoff always remains $104,000. Even though this expected payoff of $104,000 is far below the possible gain of $1,000,000 under S_1, and $500,000 under S_2, it is much better than the possible loss of $400,000 and $600,000 respectively.

14-3 exercises

1. Suppose that a petroleum company is playing some sort of game, with Nature as its opponent. The company has the choice of two different actions, to drill a new well or not to drill a new well, while Nature also has the choice of two different strategies, to provide oil or not to provide oil. Suppose further that the payoff matrix for the company is given below.

Nature	The petroleum company	
	Action A_1 (Drill)	Action A_2 (Not drill)
Strategy S_1 (Oil)	$4,000,000	$1,000,000
Strategy S_2 (No oil)	−$2,000,000	0

a) What are the action probabilities for A_1 and A_2—that is, the probabilities of taking action A_1 and A_2?

b) What is the expected payoff of the game?

2. David Jones has been hired to sell refreshments in a football stadium. He may sell either coffee or coke, but he has to make his choice known to the management one week before the day when the next game will be played. Obviously, he will do better selling coke if the weather is hot, and selling coffee if the weather is cool. From past experience, he constructs the following payoff matrix.

Nature	David Jones	
	Action A_1 (Sell coke)	Action A_2 (Sell coffee)
Strategy S_1 (Hot day)	$30	$20
Strategy S_2 (Cool day)	$15	$25

If David Jones is playing game against Nature, what is his expected payoff?

3. Suppose that there are two and only two sellers of a particular commodity. The net profit to seller A will be greater if he cuts his price while seller B keeps his price unchanged. On the other hand, the net profit to seller B will be greater if he cuts his price while seller A does not cut his. A payoff table of net profit for the two competitors is given below.

Competitor B	Competitor A	
	S_1 (Cut price)	S_2 (Not cut price)
S_1 (Cut price)	3	6
S_2 (Not cut price)	5	4

a) Find the probabilities that should be assigned to S_1 and S_2 by each of the two competitors.
b) What is the value of the game for A and for B?

4. The weather of a certain region is either rainy or not rainy. A professional golfer practices every day and often finds himself in the wrong outfit. He considers the weather a sort of Devil who tries to make his life as miserable as possible. He seems to have a knowledge of the loss or discomfort he has to suffer if he wears the wrong outfit. His loss table is given below.

Weather	Possible action	
	Wear rainy-day outfit S_1	Wear non-rainy-day outfit S_2
Rainy S_1	2	10
Not rainy S_2	4	0

a) What is the mixed strategy for the golfer?
b) What is the expected loss if the golfer uses this mixed strategy?

appendixes

answers to selected exercises

appendix A
binomial probability distributions

Table entries are values of $\binom{n}{x}p^x q^{n-x}$. Items omitted are less than .00005.

$n = 5$

x	.1	.2	.3	.4	.5	.6	.7	.8	.9
0	.5905	.3277	.1681	.0778	.0312	.0102	.0024	.0003	.0000
1	.3280	.4096	.3602	.2592	.1562	.0768	.0284	.0064	.0004
2	.0729	.2048	.3087	.3456	.3125	.2304	.1323	.0512	.0081
3	.0081	.0512	.1323	.2304	.3125	.3456	.3087	.2048	.0729
4	.0004	.0064	.0284	.0768	.1562	.2592	.3602	.4096	.3280
5	.0000	.0003	.0024	.0102	.0312	.0778	.1681	.3277	.5905

$n = 10$

x	.1	.2	.3	.4	.5	.6	.7	.8	.9
0	.3487	.1074	.0282	.0060	.0010	.0001	.0000	.0000	.0000
1	.3874	.2684	.1211	.0403	.0098	.0016	.0001	.0000	.0000
2	.1937	.3020	.2335	.1209	.0439	.0106	.0014	.0001	.0000
3	.0574	.2013	.2668	.2150	.1172	.0425	.0090	.0008	.0000
4	.0112	.0881	.2001	.2508	.2051	.1115	.0368	.0055	.0001
5	.0015	.0264	.1029	.2007	.2461	.2007	.1029	.0264	.0015
6	.0001	.0055	.0368	.1115	.2051	.2508	.2001	.0881	.0112
7	.0000	.0008	.0090	.0425	.1172	.2150	.2668	.2013	.0574
8	.0000	.0001	.0014	.0106	.0439	.1209	.2335	.3020	.1937
9	.0000	.0000	.0001	.0016	.0098	.0403	.1211	.2684	.3874
10	.0000	.0000	.0000	.0001	.0010	.0060	.0282	.1074	.3487

$n = 15$

x	.1	.2	.3	.4	.5	.6	.7	.8	.9
0	.2059	.0352	.0047	.0005	.0000	.0000	.0000	.0000	.0000
1	.3432	.1319	.0305	.0047	.0005	.0000	.0000	.0000	.0000
2	.2669	.2309	.0916	.0219	.0032	.0003	.0000	.0000	.0000
3	.1285	.2501	.1700	.0634	.0139	.0016	.0001	.0000	.0000
4	.0428	.1876	.2186	.1268	.0417	.0074	.0006	.0000	.0000
5	.0105	.1032	.2061	.1859	.0916	.0245	.0030	.0001	.0000
6	.0019	.0430	.1472	.2066	.1527	.0612	.0116	.0007	.0000
7	.0003	.0138	.0811	.1771	.1964	.1181	.0348	.0035	.0000
8	.0000	.0035	.0348	.1181	.1964	.1771	.0811	.0138	.0003
9	.0000	.0007	.0116	.0612	.1527	.2066	.1472	.0430	.0019
10	.0000	.0001	.0030	.0245	.0916	.1859	.2061	.1032	.0105
11	.0000	.0000	.0006	.0074	.0417	.1268	.2186	.1876	.0428
12	.0000	.0000	.0001	.0016	.0139	.0634	.1700	.2501	.1285
13	.0000	.0000	.0000	.0003	.0032	.0219	.0916	.2309	.2669
14	.0000	.0000	.0000	.0000	.0005	.0047	.0305	.1319	.3432
15	.0000	.0000	.0000	.0000	.0000	.0005	.0047	.0352	.2059

n = 20

x	.1	.2	.3	.4	.5	.6	.7	.8	.9
0	.1216	.0115	.0008	.0000	.0000	.0000	.0000	.0000	.0000
1	.2702	.0576	.0068	.0005	.0000	.0000	.0000	.0000	.0000
2	.2852	.1369	.0278	.0031	.0002	.0000	.0000	.0000	.0000
3	.1901	.2054	.0716	.0123	.0011	.0000	.0000	.0000	.0000
4	.0898	.2182	.1304	.0350	.0046	.0003	.0000	.0000	.0000
5	.0319	.1746	.1789	.0746	.0148	.0013	.0000	.0000	.0000
6	.0089	.1091	.1916	.1244	.0370	.0049	.0002	.0000	.0000
7	.0020	.0545	.1643	.1659	.0739	.0146	.0010	.0000	.0000
8	.0004	.0222	.1144	.1797	.1201	.0355	.0039	.0001	.0000
9	.0001	.0074	.0654	.1597	.1602	.0710	.0120	.0005	.0000
10	.0000	.0020	.0308	.1171	.1762	.1171	.0308	.0020	.0000
11	.0000	.0005	.0120	.0710	.1602	.1597	.0654	.0074	.0001
12	.0000	.0001	.0039	.0355	.1201	.1797	.1144	.0222	.0004
13	.0000	.0000	.0010	.0146	.0739	.1659	.1643	.0545	.0020
14	.0000	.0000	.0002	.0049	.0370	.1244	.1916	.1091	.0089
15	.0000	.0000	.0000	.0013	.0148	.0746	.1789	.1746	.0319
16	.0000	.0000	.0000	.0003	.0046	.0350	.1304	.2182	.0898
17	.0000	.0000	.0000	.0000	.0011	.0123	.0716	.2054	.1901
18	.0000	.0000	.0000	.0000	.0002	.0031	.0278	.1369	.2852
19	.0000	.0000	.0000	.0000	.0000	.0005	.0068	.0576	.2702
20	.0000	.0000	.0000	.0000	.0000	.0000	.0008	.0115	.1216

n = 30

x	.1	.2	.3	.4	.5	.6	.7	.8	.9
0	.0424	.0012	.0000	.0000	.0000	.0000	.0000	.0000	.0000
1	.1413	.0093	.0003	.0000	.0000	.0000	.0000	.0000	.0000
2	.2277	.0337	.0018	.0000	.0000	.0000	.0000	.0000	.0000
3	.2361	.0785	.0072	.0003	.0000	.0000	.0000	.0000	.0000
4	.1771	.1325	.0208	.0012	.0000	.0000	.0000	.0000	.0000
5	.1023	.1723	.0464	.0041	.0001	.0000	.0000	.0000	.0000
6	.0474	.1795	.0829	.0115	.0006	.0000	.0000	.0000	.0000
7	.0180	.1538	.1219	.0263	.0019	.0000	.0000	.0000	.0000
8	.0058	.1106	.1501	.0505	.0055	.0001	.0000	.0000	.0000
9	.0016	.0676	.1573	.0823	.0133	.0006	.0000	.0000	.0000
10	.0004	.0355	.1416	.1152	.0280	.0020	.0000	.0000	.0000
11	.0001	.0161	.1103	.1396	.0509	.0054	.0001	.0000	.0000
12	.0000	.0064	.0749	.1474	.0805	.0129	.0005	.0000	.0000
13	.0000	.0022	.0444	.1360	.1115	.0269	.0015	.0000	.0000
14	.0000	.0007	.0231	.1101	.1354	.0489	.0042	.0000	.0000
15	.0000	.0002	.0106	.0783	.1445	.0783	.0106	.0002	.0000
16	.0000	.0000	.0042	.0489	.1354	.1101	.0231	.0007	.0000
17	.0000	.0000	.0015	.0269	.1115	.1360	.0444	.0022	.0000
18	.0000	.0000	.0005	.0129	.0805	.1474	.0749	.0064	.0000

$$n = 30$$

x	.1	.2	.3	.4	.5	.6	.7	.8	.9
19	.0000	.0000	.0001	.0054	.0509	.1396	.1103	.0161	.0001
20	.0000	.0000	.0000	.0020	.0280	.1152	.1416	.0355	.0004
21	.0000	.0000	.0000	.0006	.0133	.0823	.1573	.0676	.0016
22	.0000	.0000	.0000	.0001	.0055	.0505	.1501	.1106	.0058
23	.0000	.0000	.0000	.0000	.0019	.0263	.1219	.1538	.0180
24	.0000	.0000	.0000	.0000	.0006	.0115	.0829	.1795	.0474
25	.0000	.0000	.0000	.0000	.0001	.0041	.0464	.1723	.1023
26	.0000	.0000	.0000	.0000	.0000	.0012	.0208	.1325	.1771
27	.0000	.0000	.0000	.0000	.0000	.0003	.0072	.0785	.2361
28	.0000	.0000	.0000	.0000	.0000	.0000	.0018	.0337	.2277
29	.0000	.0000	.0000	.0000	.0000	.0000	.0003	.0093	.1413
30	.0000	.0000	.0000	.0000	.0000	.0000	.0000	.0012	.0424

From National Bureau of Standards, *Tables of the Binomial Distribution*, Applied Mathematics Series, No. 6, 1950. With the permission of the publisher.

appendix B
cumulative probabilities for binomial distributions

$$CP(x) = \sum_{k=0}^{x} \binom{n}{k} p^k q^{n-k}$$

n = 5

x	.1	.2	.3	.4	.5	.6	.7	.8	.9
0	.5905	.3277	.1681	.0778	.0312	.0102	.0024	.0003	.0000
1	.9185	.7373	.5283	.3370	.1874	.0870	.0308	.0067	.0004
2	.9914	.9421	.8370	.6826	.5000	.3174	.1631	.0579	.0085
3	.9995	.9933	.9693	.9130	.8124	.6630	.4718	.2627	.0814
4	.9999	.9997	.9977	.9898	.9686	.9222	.8320	.6723	.4094
5	1.0000	1.0000	1.0000	1.0000	1.0000	1.0000	1.0000	1.0000	1.0000

n = 10

x	.1	.2	.3	.4	.5	.6	.7	.8	.9
0	.3487	.1074	.0282	.0060	.0010	.0001	.0000	.0000	.0000
1	.7361	.3758	.1493	.0463	.0108	.0017	.0001	.0000	.0000
2	.9298	.6778	.3828	.1672	.0547	.0123	.0015	.0001	.0000
3	.9872	.8791	.6496	.3822	.1719	.0548	.0105	.0009	.0000
4	.9984	.9672	.8497	.6330	.3770	.1663	.0473	.0064	.0001
5	.9999	.9936	.9526	.8337	.6231	.3670	.1502	.0328	.0016
6	1.0000	.9991	.9894	.9452	.8282	.6178	.3503	.1209	.0128
7	1.0000	.9999	.9984	.9877	.9454	.8328	.6171	.3222	.0702
8	1.0000	1.0000	.9998	.9983	.9893	.9537	.8506	.6242	.2639
9	1.0000	1.0000	.9999	.9999	.9991	.9940	.9717	.8926	.6513
10	1.0000	1.0000	1.0000	1.0000	1.0000	1.0000	1.0000	1.0000	1.0000

n = 15

x	.1	.2	.3	.4	.5	.6	.7	.8	.9
0	.2059	.0352	.0047	.0005	.0000	.0000	.0000	.0000	.0000
1	.5491	.1671	.0352	.0052	.0005	.0000	.0000	.0000	.0000
2	.8160	.3980	.1268	.0271	.0037	.0003	.0000	.0000	.0000
3	.9445	.6481	.2968	.0905	.0176	.0019	.0001	.0000	.0000
4	.9873	.8357	.5154	.2173	.0593	.0093	.0007	.0000	.0000
5	.9978	.9389	.7215	.4032	.1509	.0338	.0037	.0001	.0000
6	.9997	.9819	.8687	.6098	.3036	.0950	.0153	.0008	.0000
7	1.0000	.9957	.9498	.7869	.5000	.2131	.0501	.0043	.0000
8	1.0000	.9992	.9846	.9050	.6964	.3902	.1312	.0181	.0003
9	1.0000	.9999	.9962	.9662	.8491	.5968	.2784	.0611	.0022
10	1.0000	1.0000	.9992	.9907	.9407	.7827	.4845	.1643	.0127
11	1.0000	1.0000	.9999	.9981	.9824	.9095	.7031	.3519	.0555
12	1.0000	1.0000	.9999	.9997	.9963	.9729	.8731	.6020	.1840

$n = 15$

x	.1	.2	.3	.4	.5	.6	.7	.8	.9
13	1.0000	1.0000	1.0000	1.0000	.9995	.9948	.9647	.8329	.4509
14	1.0000	1.0000	1.0000	1.0000	1.0000	.9995	.9952	.9648	.7941
15	1.0000	1.0000	1.0000	1.0000	1.0000	1.0000	1.0000	1.0000	1.0000

$n = 20$

x	.1	.2	.3	.4	.5	.6	.7	.8	.9
0	.1216	.0115	.0008	.0000	.0000	.0000	.0000	.0000	.0000
1	.3918	.0691	.0076	.0005	.0000	.0000	.0000	.0000	.0000
2	.6770	.2060	.0354	.0036	.0002	.0000	.0000	.0000	.0000
3	.8671	.4114	.1070	.0159	.0013	.0000	.0000	.0000	.0000
4	.9569	.6296	.2374	.0509	.0059	.0003	.0000	.0000	.0000
5	.9888	.8042	.4163	.1255	.0207	.0016	.0000	.0000	.0000
6	.9977	.9133	.6079	.2499	.0577	.0065	.0002	.0000	.0000
7	.9997	.9678	.7722	.4158	.1316	.0211	.0012	.0000	.0000
8	.9999	.9900	.8866	.5955	.2517	.0566	.0051	.0001	.0000
9	1.0000	.9974	.9520	.7552	.4119	.1276	.0171	.0006	.0000
10	1.0000	.9994	.9828	.8723	.5881	.2447	.0479	.0026	.0000
11	1.0000	.9999	.9948	.9433	.7483	.4044	.1133	.0100	.0001
12	1.0000	1.0000	.9987	.9788	.8684	.5841	.2277	.0322	.0005
13	1.0000	1.0000	.9997	.9934	.9423	.7500	.3920	.0867	.0025
14	1.0000	1.0000	.9999	.9983	.9793	.8744	.5836	.1958	.0114
15	1.0000	1.0000	1.0000	.9996	.9941	.9490	.7625	.3704	.0433
16	1.0000	1.0000	1.0000	.9999	.9987	.9840	.8929	.5886	.1331
17	1.0000	1.0000	1.0000	1.0000	.9998	.9963	.9645	.7940	.3232
18	1.0000	1.0000	1.0000	1.0000	1.0000	.9994	.9923	.9309	.6084
19	1.0000	1.0000	1.0000	1.0000	1.0000	.9999	.9992	.9885	.8784
20	1.0000	1.0000	1.0000	1.0000	1.0000	1.0000	1.0000	1.0000	1.0000

$n = 30$

x	.1	.2	.3	.4	.5	.6	.7	.8	.9
0	.0424	.0012	.0000	.0000	.0000	.0000	.0000	.0000	.0000
1	.1837	.0105	.0003	.0000	.0000	.0000	.0000	.0000	.0000
2	.4114	.0442	.0021	.0000	.0000	.0000	.0000	.0000	.0000
3	.6475	.1227	.0093	.0003	.0000	.0000	.0000	.0000	.0000
4	.8246	.2552	.0301	.0015	.0000	.0000	.0000	.0000	.0000
5	.9269	.4275	.0765	.0056	.0001	.0000	.0000	.0000	.0000
6	.9743	.6070	.1594	.0171	.0007	.0000	.0000	.0000	.0000
7	.9923	.7608	.2813	.0434	.0026	.0000	.0000	.0000	.0000
8	.9981	.8714	.4314	.0939	.0081	.0001	.0000	.0000	.0000
9	.9996	.9390	.5887	.1762	.0214	.0007	.0000	.0000	.0000
10	.9999	.9745	.7303	.2914	.0494	.0027	.0000	.0000	.0000
11	1.0000	.9906	.8406	.4310	.1003	.0081	.0001	.0000	.0000
12	1.0000	.9970	.9155	.5784	.1808	.0210	.0006	.0000	.0000
13	1.0000	.9992	.9599	.7144	.2923	.0479	.0021	.0000	.0000
14	1.0000	.9999	.9830	.8245	.4277	.0968	.0063	.0000	.0000
15	1.0000	1.0000	.9936	.9028	.5722	.1751	.0169	.0002	.0000
16	1.0000	1.0000	.9978	.9518	.7076	.2852	.0400	.0009	.0000
17	1.0000	1.0000	.9993	.9788	.8191	.4212	.0844	.0031	.0000
18	1.0000	1.0000	.9998	.9917	.8998	.5686	.1593	.0095	.0000
19	1.0000	1.0000	.9999	.9970	.9505	.7082	.2696	.0256	.0001
20	1.0000	1.0000	1.0000	.9990	.9785	.8234	.4112	.0611	.0005
21	1.0000	1.0000	1.0000	.9998	.9918	.9057	.5685	.1287	.0021
22	1.0000	1.0000	1.0000	.9999	.9973	.9562	.7186	.2393	.0079
23	1.0000	1.0000	1.0000	1.0000	.9992	.9825	.8405	.3931	.0259
24	1.0000	1.0000	1.0000	1.0000	.9998	.9940	.9234	.5726	.0733
25	1.0000	1.0000	1.0000	1.0000	.9999	.9981	.9698	.7449	.1756
26	1.0000	1.0000	1.0000	1.0000	1.0000	.9985	.9906	.8774	.3527
27	1.0000	1.0000	1.0000	1.0000	1.0000	.9998	.9978	.9559	.5888
28	1.0000	1.0000	1.0000	1.0000	1.0000	1.0000	.9996	.9896	.8165
29	1.0000	1.0000	1.0000	1.0000	1.0000	1.0000	.9999	.9989	.9578
30	1.0000	1.0000	1.0000	1.0000	1.0000	1.0000	1.0000	1.0000	1.0000

From National Bureau of Standards, *Tables of Binomial Distribution*, Applied Mathematics Series, No. 6, 1950. With the permission of the publisher.

appendix C
random digits

61 19 69 04 46	26 45 74 77 74	51 92 43 37 29	65 39 45 95 93	42 58 26 05 27
15 47 44 52 66	95 27 07 99 53	59 36 78 38 48	82 39 61 01 18	33 21 15 94 66
94 55 72 85 73	67 89 75 43 87	54 62 24 44 31	91 19 04 25 92	92 92 74 59 73
42 48 11 62 13	97 34 40 87 21	16 86 84 87 67	03 07 11 20 59	25 70 14 66 70
23 52 37 83 17	73 20 88 98 37	68 93 59 14 16	26 25 22 96 63	05 52 28 25 62
04 49 35 24 94	75 24 63 38 24	45 86 25 10 25	61 96 27 93 35	65 33 71 24 72
00 54 99 76 54	64 05 18 81 59	96 11 96 38 96	54 69 28 23 91	23 28 72 95 29
35 96 31 53 07	26 89 80 93 54	33 35 13 54 62	77 97 45 00 24	90 10 33 93 33
59 80 80 83 91	45 42 72 68 42	83 60 94 97 00	13 02 12 48 92	78 56 52 01 06
46 05 88 52 36	01 39 00 22 86	77 28 14 40 77	93 91 08 36 47	70 61 74 29 41
32 17 90 05 97	87 37 92 52 41	05 56 70 70 07	86 74 31 71 57	85 39 41 18 38
69 23 46 14 06	20 11 74 52 04	15 95 66 00 00	18 74 39 24 23	97 11 89 63 38
19 56 54 14 30	01 75 87 53 79	40 41 92 15 85	66 67 43 68 06	84 96 28 52 07
45 15 51 49 38	19 47 60 72 46	43 66 79 45 43	59 04 79 00 33	20 82 66 95 41
94 86 43 19 94	36 16 81 08 51	34 88 88 15 53	01 54 03 54 56	05 01 45 11 76
98 08 62 48 26	45 24 02 84 04	44 99 90 88 96	39 09 47 34 07	35 44 13 18 80
33 18 51 62 32	41 94 15 09 49	89 43 54 85 81	88 69 54 19 94	37 54 87 30 43
80 95 10 04 06	96 38 27 07 74	20 15 12 33 87	25 01 62 52 98	94 62 46 11 71
79 75 24 91 40	71 96 12 82 96	69 86 10 25 91	74 85 22 05 39	00 38 75 95 79
18 63 33 25 37	98 14 50 65 71	31 01 02 46 74	05 45 56 14 27	77 93 89 19 36
74 02 94 39 02	77 55 73 22 70	97 79 01 71 19	52 52 75 80 21	80 81 45 17 48
54 17 84 56 11	80 99 33 71 43	05 33 51 29 69	56 12 71 92 55	36 04 09 03 24
11 66 44 98 83	52 07 98 48 27	59 38 17 15 39	09 97 33 34 40	88 46 12 33 56
48 32 47 79 28	31 24 96 47 10	02 29 53 68 70	32 30 75 75 46	15 02 00 99 94
69 07 49 41 38	87 63 79 19 76	35 58 40 44 01	10 51 82 16 15	01 84 87 69 38
09 18 82 00 97	32 82 53 95 27	04 22 08 63 04	83 38 98 73 74	64 27 85 80 44
90 04 58 54 97	51 98 15 06 54	94 93 88 19 97	91 87 07 61 50	68 47 66 46 59
73 18 95 02 07	47 67 72 62 69	62 29 06 44 64	27 12 46 70 18	41 36 18 27 60
75 76 87 64 90	20 97 18 17 49	90 42 91 22 72	95 37 50 58 71	93 82 34 31 78
54 01 64 40 56	66 28 13 10 03	00 68 22 73 98	20 71 45 32 95	07 70 61 78 13
08 35 86 99 10	78 54 24 27 85	13 66 15 88 73	04 61 89 75 53	31 22 30 84 20
28 30 60 32 64	81 33 31 05 91	40 51 00 78 93	32 60 46 04 75	94 11 90 18 40
53 84 08 62 33	81 59 41 36 28	51 21 59 02 90	28 46 66 87 95	77 76 22 07 91
91 75 75 37 41	61 61 36 22 69	50 26 39 02 12	55 78 17 65 14	83 48 34 70 55
89 41 59 26 94	00 39 75 83 91	12 60 71 76 46	48 94 97 23 06	94 54 13 74 03
77 51 30 38 20	86 83 42 99 01	68 41 48 27 74	51 90 81 39 80	72 89 35 55 07
19 50 23 71 74	69 97 92 02 88	55 21 02 97 73	74 28 77 52 51	65 34 46 74 15
21 81 85 93 13	93 27 88 17 57	05 68 67 31 56	07 08 28 50 46	31 85 33 84 52
51 47 46 64 99	68 10 72 36 21	94 04 99 13 45	42 83 60 91 91	08 00 74 54 49
99 55 96 83 31	62 53 52 41 70	69 77 71 28 30	74 81 97 81 42	43 86 07 28 34
33 71 34 80 07	93 58 47 28 69	51 92 66 47 21	58 30 32 98 22	93 17 49 39 72
85 27 48 68 93	11 30 32 92 70	28 83 43 41 37	73 51 59 04 00	71 14 84 36 43

84 13 38 96 40	44 03 55 21 66	73 85 27 00 91	61 22 26 05 61	62 32 71 84 23
56 73 21 62 34	17 39 59 61 31	10 12 39 16 22	85 49 65 75 60	81 60 41 88 80
65 13 85 68 06	87 64 88 52 61	34 31 36 58 61	45 87 52 10 69	85 64 44 72 77
38 00 10 21 76	81 71 91 17 11	71 60 29 29 37	74 21 96 40 49	65 58 44 96 98
37 40 29 63 97	01 30 47 75 86	56 27 11 00 86	47 32 46 26 05	40 03 03 74 38
97 12 54 03 48	87 08 33 14 17	21 81 53 92 50	75 23 76 20 47	15 50 12 95 78
21 82 64 11 34	47 14 33 40 72	64 63 88 59 02	49 13 90 64 41	03 85 65 45 52
73 13 54 27 42	95 71 90 90 35	85 79 47 42 96	08 78 98 81 56	64 69 11 92 02
07 63 87 79 29	03 06 11 80 72	96 20 74 41 56	23 82 19 95 38	04 71 36 69 94
60 52 88 34 41	07 95 41 98 14	59 17 52 06 95	05 53 35 21 39	61 21 20 64 55
83 59 63 56 55	06 95 89 29 83	05 12 80 97 19	77 43 35 37 83	92 30 15 04 98
10 85 06 27 46	99 59 91 05 07	13 49 90 63 19	53 07 57 18 39	06 41 01 93 62
39 82 09 89 52	43 62 26 31 47	64 42 18 08 14	43 80 00 93 51	31 02 47 31 67
59 58 00 64 78	75 56 97 88 00	88 83 55 44 86	23 76 80 61 56	04 11 10 84 08
38 50 80 73 41	23 79 34 87 63	90 82 29 70 22	17 71 90 42 07	95 95 44 99 53
30 69 27 06 68	94 68 81 61 27	56 19 68 00 91	82 06 76 34 00	05 46 26 92 00
65 44 39 56 59	18 28 82 74 37	49 63 22 40 41	08 33 76 56 76	96 29 99 08 36
27 26 75 02 64	13 19 27 22 94	07 47 74 46 06	17 98 54 89 11	97 34 13 03 58
91 30 70 69 91	19 07 22 42 10	36 69 95 37 28	28 82 53 57 93	28 97 66 62 52
68 43 49 46 88	84 47 31 36 22	62 12 69 84 08	12 84 38 25 90	09 81 59 31 46
48 90 81 58 77	54 74 52 45 91	35 70 00 47 54	83 82 45 26 92	54 13 05 51 60
06 91 34 51 97	42 67 27 86 01	11 88 30 95 28	63 01 19 89 01	14 97 44 03 44
10 45 51 60 19	14 21 03 37 12	91 34 23 78 21	88 32 58 08 51	43 66 77 08 83
12 88 39 73 43	65 02 76 11 84	04 28 50 13 92	17 97 41 50 77	90 71 22 67 69
21 77 83 09 76	38 80 73 69 61	31 64 94 20 96	63 28 10 20 23	08 81 64 74 49
19 52 35 95 15	65 12 25 96 59	86 28 36 82 58	69 57 21 37 98	16 43 59 15 29
67 24 55 26 70	35 58 31 65 63	79 24 68 66 86	76 46 33 42 22	26 65 59 08 02
60 58 44 73 77	07 50 03 79 92	45 13 42 65 29	26 76 08 36 37	41 32 64 43 44
53 85 34 13 77	36 06 69 48 50	58 83 87 38 59	49 36 47 33 31	96 24 04 36 42
24 63 73 87 36	74 38 48 93 42	52 62 30 79 92	12 36 91 86 01	03 74 28 38 73
83 08 01 24 51	38 99 22 28 15	07 75 95 17 77	97 37 72 75 85	51 97 23 78 67
16 44 42 43 34	36 15 19 90 73	27 49 37 09 39	85 13 03 25 52	54 84 65 47 59
60 79 01 81 57	57 17 86 57 62	11 16 17 85 76	45 81 95 29 79	65 13 00 48 60
03 99 11 04 61	93 71 61 68 94	66 08 32 46 53	84 60 95 82 32	88 61 81 91 61
38 55 59 55 54	32 88 65 97 80	08 35 56 08 60	29 73 54 77 62	71 29 92 38 53
17 54 67 37 04	92 05 24 62 15	55 12 12 92 81	59 07 60 79 36	27 95 45 89 09
32 64 35 28 61	95 81 90 68 31	00 91 19 89 36	76 35 59 37 79	80 86 30 05 14
69 57 26 87 77	39 51 03 59 05	14 06 04 06 19	29 54 96 96 16	33 56 46 07 80
24 12 26 65 91	27 69 90 64 94	14 84 54 66 72	61 95 87 71 00	90 89 97 57 54
61 19 63 02 31	92 96 26 17 73	41 83 95 53 82	17 26 77 09 43	78 03 87 02 67
30 53 22 17 04	10 27 41 22 02	39 68 52 33 09	10 06 16 88 29	55 98 66 64 85
03 78 89 75 99	75 86 72 07 17	74 41 65 31 66	35 20 83 33 74	87 53 90 88 23
48 22 86 33 79	85 78 34 76 19	53 15 26 74 33	35 66 35 29 72	16 81 86 03 11
60 36 59 46 53	35 07 53 39 49	42 61 42 92 97	01 91 82 83 16	98 95 37 32 31
83 79 94 24 02	56 62 33 44 42	34 99 44 13 74	70 07 11 47 36	09 95 81 80 65

32 96 00 74 05	36 40 98 32 32	99 38 54 16 00	11 13 30 75 86	15 91 70 62 53
19 32 25 38 45	57 62 05 26 06	66 49 76 86 46	78 13 86 65 59	19 64 09 94 13
11 22 09 47 47	07 39 93 74 08	48 50 92 39 29	27 48 24 54 76	85 24 43 51 59
31 75 15 72 60	68 98 00 53 39	15 47 04 83 55	88 65 12 25 96	03 15 21 91 21
88 49 29 93 82	14 45 40 45 04	20 09 49 89 77	74 84 39 34 13	22 10 97 85 08
30 93 44 77 44	07 48 18 38 28	73 78 80 65 33	28 59 72 04 05	94 20 52 03 80
22 88 84 88 93	27 49 99 87 48	60 53 04 51 28	74 02 28 46 17	82 03 71 02 68
78 21 21 69 93	35 90 29 13 86	44 37 21 54 86	65 74 11 40 14	87 48 13 72 20
41 84 98 45 47	46 85 05 23 26	34 67 75 83 00	74 91 06 43 45	19 32 58 15 49
46 35 23 30 49	69 24 89 34 60	45 30 50 75 21	61 31 83 18 55	14 41 37 09 51
11 08 79 62 94	14 01 33 17 92	59 74 76 72 77	76 50 33 45 13	39 66 37 75 44
52 70 10 83 37	56 30 38 73 15	16 52 06 96 76	11 65 49 98 93	02 18 16 81 61
57 27 53 68 98	81 30 44 85 85	68 65 22 73 76	92 85 25 58 66	88 44 80 35 84
20 85 77 31 56	70 28 42 43 26	79 37 59 52 20	01 15 96 32 67	10 62 24 83 91
15 63 38 49 24	90 41 59 36 14	33 52 12 66 65	55 82 34 76 41	86 22 53 17 04
92 69 44 82 97	39 90 40 21 15	59 58 94 90 67	66 82 14 15 75	49 76 70 40 37
77 61 31 90 19	88 15 20 00 80	20 55 49 14 09	96 27 74 82 57	50 81 69 76 16
38 68 83 24 86	45 13 46 35 45	59 40 47 20 59	43 94 75 16 80	43 85 25 96 93
25 16 30 18 89	70 01 41 50 21	41 29 06 73 12	71 85 71 59 57	68 97 11 14 03
65 25 10 76 29	37 23 93 32 95	05 87 00 11 19	92 78 42 63 40	18 47 76 56 22
36 81 54 36 25	18 63 73 75 09	82 44 49 90 05	04 92 17 37 01	14 70 79 39 97
64 39 71 16 92	05 32 78 21 62	20 24 78 17 59	45 19 72 53 32	83 74 52 25 67
04 51 52 56 24	95 09 66 79 46	48 46 08 55 58	15 19 11 87 82	16 93 03 33 61
83 76 16 08 73	43 25 38 41 45	60 83 32 59 83	01 29 14 13 49	20 36 80 71 26
14 38 70 63 45	80 85 40 92 79	43 52 90 63 18	38 38 47 47 61	41 19 63 74 80
51 32 19 22 46	80 08 87 70 74	88 72 25 67 36	66 16 44 94 31	66 91 93 16 78
72 47 20 00 08	80 89 01 80 02	94 81 33 19 00	54 15 58 34 36	35 35 25 41 31
05 46 65 53 06	93 12 81 84 64	74 45 79 05 61	72 84 81 18 34	79 98 26 84 16
39 52 87 24 84	82 47 42 55 93	48 54 53 52 47	18 61 91 36 74	18 61 11 92 41
81 61 61 87 11	53 34 24 42 76	75 12 21 17 24	74 62 77 37 07	58 31 91 59 97
07 58 61 61 20	82 64 12 28 20	92 90 41 31 41	32 39 21 97 63	61 19 96 79 40
90 76 70 42 35	13 57 41 72 00	69 90 26 37 42	78 46 42 25 01	18 62 79 08 72
40 18 82 81 93	29 59 38 86 27	94 97 21 15 98	62 09 53 67 87	00 44 15 89 97
34 41 48 21 57	86 88 75 50 87	19 15 20 00 23	12 30 28 07 83	32 62 46 86 91
63 43 97 53 63	44 98 91 68 22	36 02 40 08 67	76 37 84 16 05	65 96 17 34 88
67 04 90 90 70	93 39 94 55 47	94 45 87 42 84	05 04 14 98 07	20 28 83 40 60
79 49 50 41 46	52 16 29 02 86	54 15 83 42 43	46 97 83 54 82	59 36 29 59 38
91 70 43 05 52	04 73 72 10 31	75 05 19 30 29	47 66 56 43 82	99 78 29 34 78
94 01 54 68 74	32 44 44 82 77	59 82 09 61 63	64 65 42 58 43	41 14 54 28 20
74 10 88 82 22	88 57 07 40 15	25 70 49 10 35	01 75 51 47 50	48 96 83 86 03
62 88 08 78 73	95 16 05 92 21	22 30 49 03 14	72 87 71 73 34	39 28 30 41 49
11 74 81 21 02	80 58 04 18 67	17 71 05 96 21	06 55 40 78 50	73 95 07 95 52
17 94 40 56 00	60 47 80 33 43	25 85 25 89 05	57 21 63 96 18	49 85 69 93 26
66 06 74 27 92	95 04 35 26 80	46 78 05 64 87	09 97 15 94 81	37 00 62 21 86
54 24 49 10 30	45 54 77 08 18	59 84 99 61 69	61 45 92 16 47	87 41 71 71 98

30 94 55 75 89	31 73 25 72 60	47 67 00 76 54	46 37 62 53 66	94 74 64 95 80
69 17 03 74 03	86 99 59 03 07	94 30 47 18 03	26 82 50 55 11	12 45 99 13 14
08 34 58 89 75	35 84 18 57 71	08 10 55 99 87	87 11 22 14 76	14 71 37 11 81
27 76 74 35 84	85 30 18 89 77	29 49 06 97 14	73 03 54 12 07	74 69 90 93 10
13 02 51 43 38	54 06 61 52 43	47 72 46 67 33	47 43 14 39 05	31 04 85 66 99
80 21 73 62 92	98 52 52 43 35	24 43 22 48 96	43 27 75 88 74	11 46 61 60 82
10 87 56 20 04	90 39 16 11 05	57 41 10 63 68	53 85 63 07 43	08 67 08 47 41
54 12 75 73 26	26 62 91 90 87	24 47 28 87 79	30 54 02 78 86	61 73 27 54 54
60 31 14 28 24	37 30 14 26 78	45 99 04 32 42	17 37 45 20 03	70 70 77 02 14
49 73 97 14 84	92 00 39 80 86	76 66 87 32 09	59 20 21 19 73	02 90 23 32 50
78 62 65 15 94	16 45 39 46 14	39 01 49 70 66	83 01 20 98 32	25 57 17 76 28
66 69 21 39 86	99 83 70 05 82	81 23 24 49 87	09 50 49 64 12	90 19 37 95 68
44 07 12 80 91	07 36 29 77 03	76 44 74 25 37	98 52 49 78 31	65 70 40 95 14
41 46 88 51 49	49 55 41 79 94	14 92 43 96 50	95 29 40 05 56	70 48 10 69 05
94 55 93 75 59	49 67 85 31 19	70 31 20 56 82	66 98 63 40 99	74 47 42 07 40
41 61 57 03 60	64 11 45 86 60	90 85 06 46 18	80 62 05 17 90	11 43 63 80 72
50 27 39 31 13	41 79 48 68 61	24 78 18 96 83	55 41 18 56 67	77 53 59 98 92
41 39 68 05 04	90 67 00 82 89	40 90 20 50 69	95 08 30 67 83	28 10 25 78 16
25 80 72 42 60	71 52 97 89 20	72 68 20 73 85	90 72 65 71 66	98 88 40 85 83
06 17 09 79 65	88 30 29 80 41	21 44 34 18 08	68 98 48 36 20	89 74 79 88 82
60 80 85 44 44	74 41 28 11 05	01 17 62 88 38	36 42 11 64 89	18 05 95 10 61
80 94 04 48 93	10 40 83 62 22	80 58 27 19 44	92 63 84 03 33	67 05 41 60 67
19 51 69 01 20	46 75 97 16 43	13 17 75 52 92	21 03 68 28 08	77 50 19 74 27
49 38 65 44 80	23 60 42 35 54	21 78 54 11 01	91 17 81 01 74	29 42 09 04 38
06 31 28 89 40	15 99 56 93 21	47 45 86 48 09	98 18 98 18 51	29 65 18 42 15
60 94 20 03 07	11 89 79 26 74	40 40 56 80 32	96 71 75 42 44	10 70 14 13 93
92 32 99 89 32	78 28 44 63 47	71 20 99 20 61	39 44 89 31 36	25 72 20 85 64
77 93 66 35 74	31 38 45 19 24	85 56 12 96 71	58 13 71 78 20	22 75 13 65 18
38 10 17 77 56	11 65 71 38 97	95 88 95 70 67	47 64 81 38 85	70 66 99 34 06
39 64 16 94 57	91 33 92 25 02	92 61 38 97 19	11 94 75 62 03	19 32 42 05 04
84 05 44 04 55	99 39 66 36 80	67 66 76 06 31	69 18 19 68 45	38 52 51 16 00
47 46 80 35 77	57 64 96 32 66	24 70 07 15 94	14 00 42 31 53	69 24 90 57 47
43 32 13 13 70	28 97 72 38 96	76 47 96 85 62	62 34 20 75 89	08 89 90 59 85
64 28 16 18 26	18 55 56 49 37	13 17 33 33 65	78 85 11 64 99	87 06 41 30 75
66 84 77 04 95	32 35 00 29 85	86 71 63 87 46	26 31 37 74 63	55 38 77 26 81
72 46 13 32 30	21 52 95 34 24	92 58 10 22 62	78 43 86 62 76	18 39 67 35 38
21 03 29 10 50	13 05 81 62 18	12 47 05 65 00	15 29 27 61 39	59 52 65 21 13
95 36 26 70 11	06 65 11 61 36	01 01 60 08 57	55 01 85 63 74	35 82 47 17 08
40 71 29 73 80	10 40 45 54 52	34 03 06 07 26	75 21 11 02 71	36 63 36 84 24
58 27 56 17 64	97 58 65 47 16	50 25 94 63 45	87 19 54 60 92	26 78 76 09 39
89 51 41 17 88	68 22 42 34 17	73 95 97 61 45	30 34 24 02 77	11 04 97 20 49
15 47 25 06 69	48 13 93 67 32	46 87 43 70 88	73 46 50 98 19	58 86 93 52 20
12 12 08 61 24	51 24 74 43 02	60 88 35 21 09	21 43 73 67 86	49 22 67 78 37

appendix D
table of square roots

	.00	.01	.02	.03	.04	.05	.06	.07	.08	.09
1.0	1.0000	1.0050	1.0100	1.0149	1.0198	1.0247	1.0296	1.0344	1.0392	1.0440
1.1	1.0488	1.0536	1.0583	1.0630	1.0677	1.0724	1.0770	1.0817	1.0863	1.0909
1.2	1.0954	1.1000	1.1045	1.1091	1.1136	1.1180	1.1225	1.1269	1.1314	1.1358
1.3	1.1402	1.1446	1.1489	1.1533	1.1576	1.1619	1.1662	1.1705	1.1747	1.1790
1.4	1.1832	1.1874	1.1916	1.1958	1.2000	1.2042	1.2083	1.2124	1.2166	1.2207
1.5	1.2247	1.2288	1.2329	1.2369	1.2410	1.2450	1.2490	1.2530	1.2570	1.2610
1.6	1.2649	1.2689	1.2728	1.2767	1.2806	1.2845	1.2884	1.2923	1.2961	1.3000
1.7	1.3038	1.3077	1.3115	1.3153	1.3191	1.3229	1.3266	1.3304	1.3342	1.3379
1.8	1.3416	1.3454	1.3491	1.3528	1.3565	1.3601	1.3638	1.3675	1.3711	1.3748
1.9	1.3784	1.3820	1.3856	1.3892	1.3928	1.3964	1.4000	1.4036	1.4071	1.4107
2.0	1.4142	1.4177	1.4213	1.4248	1.4283	1.4318	1.4353	1.4387	1.4422	1.4457
2.1	1.4491	1.4526	1.4560	1.4595	1.4629	1.4663	1.4697	1.4731	1.4765	1.4799
2.2	1.4832	1.4866	1.4900	1.4933	1.4967	1.5000	1.5033	1.5067	1.5100	1.5133
2.3	1.5166	1.5199	1.5232	1.5264	1.5297	1.5330	1.5362	1.5395	1.5427	1.5460
2.4	1.5492	1.5524	1.5556	1.5588	1.5620	1.5652	1.5684	1.5716	1.5748	1.5780
2.5	1.5811	1.5843	1.5875	1.5906	1.5937	1.5969	1.6000	1.6031	1.6062	1.6093
2.6	1.6125	1.6155	1.6186	1.6217	1.6248	1.6279	1.6310	1.6340	1.6371	1.6401
2.7	1.6432	1.6462	1.6492	1.6523	1.6553	1.6583	1.6613	1.6643	1.6673	1.6703
2.8	1.6733	1.6763	1.6793	1.6823	1.6852	1.6882	1.6912	1.6941	1.6971	1.7000
2.9	1.7029	1.7059	1.7088	1.7117	1.7146	1.7176	1.7205	1.7234	1.7263	1.7292
3.0	1.7321	1.7349	1.7378	1.7407	1.7436	1.7464	1.7493	1.7521	1.7550	1.7578
3.1	1.7607	1.7635	1.7664	1.7692	1.7720	1.7748	1.7776	1.7804	1.7833	1.7861
3.2	1.7889	1.7916	1.7944	1.7972	1.8000	1.8028	1.8055	1.8083	1.8111	1.8138
3.3	1.8166	1.8193	1.9221	1.8248	1.8276	1.8303	1.8330	1.8358	1.8385	1.8412
3.4	1.8439	1.8466	1.8493	1.8520	1.8547	1.8574	1.8601	1.8628	1.8655	1.8682
3.5	1.8708	1.8735	1.8762	1.8788	1.8815	1.8841	1.8868	1.8894	1.8921	1.8947
3.6	1.8974	1.9000	1.9026	1.9053	1.9079	1.9105	1.9131	1.9157	1.9183	1.9209
3.7	1.9235	1.9261	1.9287	1.9313	1.9339	1.9365	1.9391	1.9416	1.9442	1.9468
3.8	1.9494	1.9519	1.9545	1.9570	1.9596	1.9621	1.9647	1.9672	1.9698	1.9723
3.9	1.9748	1.9774	1.9799	1.9824	1.9849	1.9875	1.9900	1.9925	1.9950	1.9975
4.0	2.0000	2.0025	2.0050	2.0075	2.0100	2.0125	2.0149	2.0174	2.0199	2.0224
4.1	2.0248	2.0273	2.0298	2.0322	2.0347	2.0372	2.0396	2.0421	2.0445	2.0469
4.2	2.0494	2.0518	2.0543	2.0567	2.0591	2.0616	2.0640	2.0664	2.0688	2.0712
4.3	2.0736	2.0761	2.0785	2.0809	2.0833	2.0857	2.0881	2.0905	2.0928	2.0952
4.4	2.0976	2.1000	2.1024	2.1048	2.1071	2.1095	2.1119	2.1142	2.1166	2.1190
4.5	2.1213	2.1237	2.1260	2.1284	2.1307	2.1331	2.1354	2.1378	2.1401	2.1424
4.6	2.1448	2.1471	2.1494	2.1517	2.1541	2.1564	2.1587	2.1610	2.1633	2.1656
4.7	2.1679	2.1703	2.1726	2.1749	2.1772	2.1794	2.1817	2.1840	2.1863	2.1886
4.8	2.1909	2.1932	2.1954	2.1977	2.2000	2.2023	2.2045	2.2068	2.2091	2.2113
4.9	2.2136	2.2159	2.2181	2.2204	2.2226	2.2249	2.2271	2.2293	2.2316	2.2338
5.0	2.2361	2.2383	2.2405	2.2428	2.2450	2.2472	2.2494	2.2517	2.2539	2.2561
5.1	2.2583	2.2605	2.2627	2.2650	2.2672	2.2694	2.2716	2.2738	2.2760	2.2782

	.00	.01	.02	.03	.04	.05	.06	.07	.08	.09
5.2	2.2804	2.2825	2.2847	2.2869	2.2891	2.2913	2.2935	2.2956	2.2978	2.3000
5.3	2.3022	2.3043	2.3065	2.3087	2.3108	2.3130	2.3152	2.3173	2.3195	2.3216
5.4	2.3238	2.3259	2.3281	2.3302	2.3324	2.3345	2.3367	2.3388	2.3409	2.3431
5.5	2.3452	2.3473	2.3495	2.3516	2.3537	2.3558	2.3580	2.3601	2.3622	2.3643
5.6	2.3664	2.3685	2.3707	2.3728	2.3749	2.3770	2.3791	2.3812	2.3833	2.3854
5.7	2.3875	2.3896	2.3917	2.3937	2.3958	2.3979	2.4000	2.4021	2.4042	2.4062
5.8	2.4083	2.4104	2.4125	2.4145	2.4166	2.4187	2.4207	2.4228	2.4249	2.4269
5.9	2.4290	2.4310	2.4331	2.4352	2.4372	2.4393	2.4413	2.4434	2.4454	2.4474
6.0	2.4495	2.4515	2.4536	2.4556	2.4576	2.4597	2.4617	2.4637	2.4658	2.4678
6.1	2.4698	2.4718	2.4739	2.4759	2.4779	2.4799	2.4819	2.4839	2.4860	2.4880
6.2	2.4900	2.4920	2.4940	2.4960	2.4980	2.5000	2.5020	2.5040	2.5060	2.5080
6.3	2.5100	2.5120	2.5140	2.5159	2.5179	2.5199	2.5219	2.5239	2.5259	2.5278
6.4	2.5298	2.5318	2.5338	2.5357	2.5377	2.5397	2.5417	2.5436	2.5456	2.5475
6.5	2.5495	2.5515	2.5534	2.5554	2.5573	2.5593	2.5612	2.5632	2.5652	2.5671
6.6	2.5690	2.5710	2.5729	2.5749	2.5768	2.5788	2.5807	2.5826	2.5846	2.5865
6.7	2.5884	2.5904	2.5923	2.5942	2.5962	2.5981	2.6000	2.6019	2.6038	2.6058
6.8	2.6077	2.6096	2.6115	2.6134	2.6153	2.6173	2.6192	2.6211	2.6230	2.6249
6.9	2.6268	2.6287	2.6306	2.6325	2.6344	2.6363	2.6382	2.6401	2.6420	2.6439
7.0	2.6458	2.6476	2.6495	2.6514	2.6533	2.6552	2.6571	2.6589	2.6608	2.6627
7.1	2.6646	2.6665	2.6683	2.6702	2.6721	2.6739	2.6758	2.6777	2.6796	2.6814
7.2	2.6833	2.6851	2.6870	2.6889	2.6907	2.6926	2.6944	2.6963	2.6981	2.7000
7.3	2.7019	2.7037	2.7055	2.7074	2.7092	2.7111	2.7129	2.7148	2.7166	2.7185
7.4	2.7203	2.7221	2.7240	2.7258	2.7276	2.7295	2.7313	2.7331	2.7350	2.7368
7.5	2.7386	2.7404	2.7423	2.7441	2.7459	2.7477	2.7495	2.7514	2.7532	2.7550
7.6	2.7568	2.7586	2.7604	2.7622	2.7641	2.7659	2.7677	2.7695	2.7713	2.7731
7.7	2.7749	2.7767	2.7785	2.7803	2.7821	2.7839	2.7857	2.7875	2.7893	2.7911
7.8	2.7928	2.7946	2.7964	2.7982	2.8000	2.8018	2.8036	2.8054	2.8071	2.8089
7.9	2.8107	2.8125	2.8142	2.8160	2.8178	2.8196	2.8213	2.8231	2.8249	2.8267
8.0	2.8284	2.8302	2.8320	2.8337	2.8355	2.8373	2.8390	2.8408	2.8425	2.8443
8.1	2.8460	2.8478	2.8496	2.8513	2.8531	2.8548	2.8566	2.8583	2.8601	2.8618
8.2	2.8636	2.8653	2.8671	2.8688	2.8705	2.8723	2.8740	2.8758	2.8775	2.8792
8.3	2.8810	2.8827	2.8844	2.8862	2.8879	2.8896	2.8914	2.8931	2.8948	2.8965
8.4	2.8983	2.9000	2.9017	2.9034	2.9052	2.9069	2.9086	2.9103	2.9120	2.9138
8.5	2.9155	2.9172	2.9189	2.9206	2.9223	2.9240	2.9257	2.9275	2.9292	2.9309
8.6	2.9326	2.9343	2.9360	2.9377	2.9394	2.9411	2.9428	2.9445	2.9462	2.9479
8.7	2.9496	2.9513	2.9530	2.9547	2.9563	2.9580	2.9597	2.9614	2.9631	2.9648
8.8	2.9665	2.9682	2.9698	2.9715	2.9732	2.9749	2.9766	2.9783	2.9799	2.9816
8.9	2.9833	2.9850	2.9866	2.9883	2.9900	2.9917	2.9933	2.9950	2.9967	2.9983
9.0	3.0000	3.0017	3.0033	3.0050	3.0067	3.0083	3.0100	3.0116	3.0133	3.0150
9.1	3.0166	3.0183	3.0199	3.0216	3.0232	3.0249	3.0265	3.0282	3.0299	3.0315
9.2	3.0332	3.0348	3.0364	3.0381	3.0397	3.0414	3.0430	3.0447	3.0463	3.0480
9.3	3.0496	3.0512	3.0529	3.0545	3.0561	3.0578	3.0594	3.0610	3.0627	3.0643
9.4	3.0659	3.0676	3.0692	3.0708	3.0725	3.0741	3.0757	3.0773	3.0790	3.0806
9.5	3.0822	3.0838	3.0854	3.0871	3.0887	3.0903	3.0919	3.0935	3.0952	3.0968

	.0	.1	.2	.3	.4	.5	.6	.7	.8	.9
9.6	3.0984	3.1000	3.1016	3.1032	3.1048	3.1064	3.1081	3.1097	3.1113	3.1129
9.7	3.1145	3.1161	3.1177	3.1193	3.1209	3.1225	3.1241	3.1257	3.1273	3.1289
9.8	3.1305	3.1321	3.1337	3.1353	3.1369	3.1385	3.1401	3.1417	3.1432	3.1448
9.9	3.1464	3.1480	3.1496	3.1512	3.1528	3.1544	3.1559	3.1575	3.1591	3.1607
10.	3.1623	3.1780	3.1937	3.2094	3.2249	3.2404	3.2558	3.2711	3.2863	3.3015
11.	3.3166	3.3317	3.3466	3.3615	3.3764	3.3912	3.4059	3.4205	3.4351	3.4496
12.	3.4641	3.4785	3.4928	3.5071	3.5214	3.5355	3.5496	3.5637	3.5777	3.5917
13.	3.6056	3.6194	3.6332	3.6469	3.6606	3.6742	3.6878	3.7014	3.7148	3.7283
14.	3.7417	3.7550	3.7683	3.7815	3.7947	3.8079	3.8210	3.8341	3.8471	3.8601
15.	3.8730	3.8859	3.8987	3.9115	3.9243	3.9370	3.9497	3.9623	3.9749	3.9875
16.	4.0000	4.0125	4.0249	4.0373	4.0497	4.0620	4.0743	4.0866	4.0988	4.1110
17.	4.1231	4.1352	4.1473	4.1593	4.1713	4.1833	4.1952	4.2071	4.2190	4.2308
18.	4.2426	4.2544	4.2661	4.2778	4.2895	4.3012	4.3128	4.3243	4.3359	4.3474
19.	4.3589	4.3704	4.3818	4.3932	4.4045	4.4159	4.4272	4.4385	4.4497	4.4609
20.	4.4721	4.4833	4.4944	4.5056	4.5166	4.5277	4.5387	4.5497	4.5607	4.5717
21.	4.5826	4.5935	4.6043	4.6152	4.6260	4.6368	4.6476	4.6583	4.6690	4.6797
22.	4.6904	4.7011	4.7117	4.7223	4.7329	4.7434	4.7539	4.7645	4.7749	4.7854
23.	4.7958	4.8062	4.8166	4.8270	4.8374	4.8477	4.8580	4.8683	4.8785	4.8888
24.	4.8990	4.9092	4.9193	4.9295	4.9396	4.9497	4.9598	4.9699	4.9800	4.9900
25.	5.0000	5.0100	5.0200	5.0299	5.0398	5.0498	5.0596	5.0695	5.0794	5.0892
26.	5.0990	5.1088	5.1186	5.1284	5.1381	5.1478	5.1575	5.1672	5.1769	5.1865
27.	5.1962	5.2058	5.2154	5.2249	5.2345	5.2440	5.2536	5.2631	5.2726	5.2820
28.	5.2915	5.3009	5.3104	5.3198	5.3292	5.3385	5.3479	5.3572	5.3666	5.3759
29.	5.3852	5.3944	5.4037	5.4129	5.4222	5.4314	5.4406	5.4498	5.4589	5.4681
30.	5.4772	5.4863	5.4955	5.5045	5.5136	5.5227	5.5317	5.5408	5.5498	5.5588
31.	5.5678	5.5767	5.5857	5.5946	5.6036	5.6125	5.6214	5.6303	5.6391	5.6480
32.	5.6569	5.6657	5.6745	5.6833	5.6921	5.7009	5.7096	5.7184	5.7271	5.7359
33.	5.7446	5.7533	5.7619	5.7706	5.7793	5.7879	5.7966	5.8052	5.8138	5.8224
34.	5.8310	5.8395	5.8481	5.8566	5.8652	5.8737	5.8822	5.8907	5.8992	5.9076
35.	5.9161	5.9245	5.9330	5.9414	5.9498	5.9582	5.9666	5.9749	5.9833	5.9917
36.	6.0000	6.0083	6.0166	6.0249	6.0332	6.0415	6.0498	6.0581	6.0663	6.0745
37.	6.0828	6.0910	6.0992	6.1074	6.1156	6.1237	6.1319	6.1400	6.1482	6.1563
38.	6.1644	6.1725	6.1806	6.1887	6.1968	6.2048	6.2129	6.2209	6.2290	6.2370
39.	6.2450	6.2530	6.2610	6.2690	6.2769	6.2849	6.2929	6.3008	6.3087	6.3166
40.	6.3246	6.3325	6.3403	6.3482	6.3561	6.3640	6.3718	6.3797	6.3875	6.3953
41.	6.4031	6.4109	6.4187	6.4265	6.4343	6.4420	6.4498	6.4576	6.4653	6.4730
42.	6.4807	6.4885	6.4962	6.5038	6.5115	6.5192	6.5269	6.5345	6.5422	6.5498
43.	6.5574	6.5651	6.5727	6.5803	6.5879	6.5955	6.6030	6.6106	6.6182	6.6257
44.	6.6332	6.6408	6.6483	6.6558	6.6633	6.6708	6.6783	6.6858	6.6933	6.7007
45.	6.7082	6.7157	6.7231	6.7305	6.7380	6.7454	6.7528	6.7602	6.7676	6.7750
46.	6.7823	6.7897	6.7971	6.8044	6.8118	6.8191	6.8264	6.8337	6.8411	6.8484
47.	6.8557	6.8629	6.8702	6.8775	6.8848	6.8920	6.8993	6.9065	6.9138	6.9210
48.	6.9282	6.9354	6.9426	6.9498	6.9570	6.9642	6.9714	6.9785	6.9857	6.9929
49.	7.0000	7.0071	7.0143	7.0214	7.0285	7.0356	7.0427	7.0498	7.0569	7.0640
50.	7.0711	7.0781	7.0852	7.0922	7.0993	7.1063	7.1134	7.1204	7.1274	7.1344
51.	7.1414	7.1484	7.1554	7.1624	7.1694	7.1764	7.1833	7.1903	7.1972	7.2042
52.	7.2111	7.2180	7.2250	7.2319	7.2388	7.2457	7.2526	7.2595	7.2664	7.2732

	.0	.1	.2	.3	.4	.5	.6	.7	.8	.9
53.	7.2801	7.2870	7.2938	7.3007	7.3075	7.3144	7.3212	7.3280	7.3348	7.3417
54.	7.3485	7.3553	7.3621	7.3689	7.3756	7.3824	7.3892	7.3959	7.4027	7.4095
55.	7.4162	7.4229	7.4297	7.4364	7.4431	7.4498	7.4565	7.4632	7.4699	7.4766
56.	7.4833	7.4900	7.4967	7.5033	7.5100	7.5166	7.5233	7.5299	7.5366	7.5432
57.	7.5498	7.5565	7.5631	7.5697	7.5763	7.5829	7.5895	7.5961	7.6026	7.6092
58.	7.6158	7.6223	7.6289	7.6354	7.6420	7.6485	7.6551	7.6616	7.6681	7.6746
59.	7.6811	7.6877	7.6942	7.7006	7.7071	7.7136	7.7201	7.7266	7.7330	7.7395
60.	7.7460	7.7524	7.7589	7.7653	7.7717	7.7782	7.7846	7.7910	7.7974	7.8038
61.	7.8102	7.8166	7.8230	7.8294	7.8358	7.8422	7.8486	7.8549	7.8613	7.8677
62.	7.8740	7.8804	7.8867	7.8930	7.8994	7.9057	7.9120	7.9183	7.9246	7.9310
63.	7.9373	7.9436	7.9498	7.9561	7.9624	7.9687	7.9750	7.9812	7.9875	7.9937
64.	8.0000	8.0062	8.0125	8.0187	8.0250	8.0312	8.0374	8.0436	8.0498	8.0561
65.	8.0623	8.0685	8.0747	8.0808	8.0870	8.0932	8.0994	8.1056	8.1117	8.1179
66.	8.1240	8.1302	8.1363	8.1425	8.1486	8.1548	8.1609	8.1670	8.1731	8.1792
67.	8.1854	8.1915	8.1976	8.2037	8.2098	8.2158	8.2219	8.2280	8.2341	8.2401
68.	8.2462	8.2523	8.2583	8.2644	8.2704	8.2765	8.2825	8.2885	8.2946	8.3006
69.	8.3066	8.3126	8.3187	8.3247	8.3307	8.3367	8.3427	8.3487	8.3546	8.3606
70.	8.3666	8.3726	8.3785	8.3845	8.3905	8.3964	8.4024	8.4083	8.4143	8.4202
71.	8.4261	8.4321	8.4380	8.4439	8.4499	8.4558	8.4617	8.4676	8.4735	8.4794
72.	8.4853	8.4912	8.4971	8.5029	8.5088	8.5147	8.5206	8.5264	8.5323	8.5381
73.	8.5440	8.5499	8.5557	8.5615	8.5674	8.5732	8.5790	8.5849	8.5907	8.5965
74.	8.6023	8.6081	8.6139	8.6197	8.6255	8.6313	8.6371	8.6429	8.6487	8.6545
75.	8.6603	8.6660	8.6718	8.6776	8.6833	8.6891	8.6948	8.7006	8.7063	8.7121
76.	8.7178	8.7235	8.7293	8.7350	8.7407	8.7464	8.7521	8.7579	8.7636	8.7693
77.	8.7750	8.7807	8.7864	8.7920	8.7977	8.8034	8.8091	8.8148	8.8204	8.8261
78.	8.8318	8.8374	8.8431	8.8487	8.8544	8.8600	8.8657	8.8713	8.8769	8.8826
79.	8.8882	8.8938	8.8994	8.9051	8.9107	8.9163	8.9219	8.9275	8.9331	8.9387
80.	8.9443	8.9499	8.9554	8.9610	8.9666	8.9722	8.9778	8.9833	8.9889	8.9944
81.	9.0000	9.0056	9.0111	9.0167	9.0222	9.0277	9.0333	9.0388	9.0443	9.0499
82.	9.0554	9.0609	9.0664	9.0719	9.0774	9.0830	9.0885	9.0940	9.0995	9.1049
83.	9.1104	9.1159	9.1214	9.1269	9.1324	9.1378	9.1433	9.1488	9.1542	9.1597
84.	9.1652	9.1706	9.1761	9.1815	9.1869	9.1924	9.1978	9.2033	9.2087	9.2141
85.	9.2195	9.2250	9.2304	9.2358	9.2412	9.2466	9.2520	9.2574	9.2628	9.2682
86.	9.2736	9.2790	9.2844	9.2898	9.2952	9.3005	9.3059	9.3113	9.3167	9.3220
87.	9.3274	9.3327	9.3381	9.3434	9.3488	9.3541	9.3595	9.3648	9.3702	9.3755
88.	9.3808	9.3862	9.3915	9.3968	9.4021	9.4074	9.4128	9.4181	9.4234	9.4287
89.	9.4340	9.4393	9.4446	9.4499	9.4552	9.4604	9.4657	9.4710	9.4763	9.4816
90.	9.4868	9.4921	9.4974	9.5026	9.5079	9.5131	9.5184	9.5237	9.5289	9.5341
91.	9.5394	9.5446	9.5499	9.5551	9.5603	9.5656	9.5708	9.5760	9.5812	9.5864
92.	9.5917	9.5969	9.6021	9.6073	9.6125	9.6177	9.6229	9.6281	9.6333	9.6385
93.	9.6437	9.6488	9.6540	9.6592	9.6644	9.6695	9.6747	9.6799	9.6850	9.6902
94.	9.6954	9.7005	9.7057	9.7108	9.7160	9.7211	9.7263	9.7314	9.7365	9.7417
95.	9.7468	9.7519	9.7570	9.7622	9.7673	9.7724	9.7775	9.7826	9.7877	9.7929
96.	9.7980	9.8031	9.8082	9.8133	9.8184	9.8234	9.8285	9.8336	9.8387	9.8438
97.	9.8489	9.8539	9.8590	9.8641	9.8691	9.8742	9.8793	9.8843	9.8894	9.8944
98.	9.8995	9.9045	9.9096	9.9146	9.9197	9.9247	9.9298	9.9348	9.9398	9.9448
99.	9.9499	9.9549	9.9599	9.9649	9.9700	9.9750	9.9800	9.9850	9.9900	9.9950

appendix E
cumulative
normal
distribution

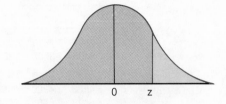

Table entries are cumulative probabilities represented in the shaded area above.

z	.00	.01	.02	.03	.04	.05	.06	.07	.08	.09
.0	.5000	.5040	.5080	.5120	.5160	.5199	.5239	.5279	.5319	.5359
.1	.5398	.5438	.5478	.5517	.5557	.5596	.5636	.5675	.5714	.5753
.2	.5793	.5832	.5871	.5910	.5948	.5987	.6026	.6064	.6103	.6141
.3	.6179	.6217	.6255	.6293	.6331	.6368	.6406	.6443	.6480	.6517
.4	.6554	.6591	.6628	.6664	.6700	.6736	.6772	.6808	.6844	.6879
.5	.6915	.6950	.6985	.7019	.7054	.7088	.7123	.7157	.7190	.7224
.6	.7257	.7291	.7324	.7357	.7389	.7422	.7454	.7486	.7517	.7549
.7	.7580	.7611	.7642	.7673	.7704	.7734	.7764	.7794	.7823	.7852
.8	.7881	.7910	.7939	.7967	.7995	.8023	.8051	.8078	.8106	.8133
.9	.8159	.8186	.8212	.8238	.8264	.8289	.8315	.8340	.8365	.8389
1.0	.8413	.8438	.8461	.8485	.8508	.8531	.8554	.8577	.8599	.8621
1.1	.8643	.8665	.8686	.8708	.8729	.8749	.8770	.8790	.8810	.8830
1.2	.8849	.8869	.8888	.8907	.8925	.8944	.8962	.8980	.8997	.9015
1.3	.9032	.9049	.9066	.9082	.9099	.9115	.9131	.9147	.9162	.9177
1.4	.9192	.9207	.9222	.9236	.9251	.9265	.9279	.9292	.9306	.9319
1.5	.9332	.9345	.9357	.9370	.9382	.9394	.9406	.9418	.9429	.9441
1.6	.9452	.9463	.9474	.9484	.9495	.9505	.9515	.9525	.9535	.9545
1.7	.9554	.9564	.9573	.9582	.9591	.9599	.9608	.9616	.9625	.9633
1.8	.9641	.9649	.9656	.9664	.9671	.9678	.9686	.9693	.9699	.9706
1.9	.9713	.9719	.9726	.9732	.9738	.9744	.9750	.9756	.9761	.9767
2.0	.9772	.9778	.9783	.9788	.9793	.9798	.9803	.9808	.9812	.9817
2.1	.9821	.9826	.9830	.9834	.9838	.9842	.9846	.9850	.9854	.9857
2.2	.9861	.9864	.9868	.9871	.9875	.9878	.9881	.9884	.9887	.9890
2.3	.9893	.9896	.9898	.9901	.9904	.9906	.9909	.9911	.9913	.9916
2.4	.9918	.9920	.9922	.9925	.9927	.9929	.9931	.9932	.9934	.9936
2.5	.9938	.9940	.9941	.9943	.9945	.9946	.9948	.9949	.9951	.9952
2.6	.9953	.9955	.9956	.9957	.9959	.9960	.9961	.9962	.9963	.9964
2.7	.9965	.9966	.9967	.9968	.9969	.9970	.9971	.9972	.9973	.9974
2.8	.9974	.9975	.9976	.9977	.9977	.9978	.9979	.9979	.9980	.9981
2.9	.9981	.9982	.9982	.9983	.9984	.9984	.9985	.9985	.9986	.9986
3.0	.9987	.9987	.9987	.9988	.9988	.9989	.9989	.9989	.9990	.9990
3.1	.9990	.9991	.9991	.9991	.9992	.9992	.9992	.9992	.9993	.9993
3.2	.9993	.9993	.9994	.9994	.9994	.9994	.9994	.9995	.9995	.9995
3.3	.9995	.9995	.9995	.9996	.9996	.9996	.9996	.9996	.9996	.9997
3.4	.9997	.9997	.9997	.9997	.9997	.9997	.9997	.9997	.9997	.9998

z	1.282	1.645	1.960	2.326	2.575	3.090	3.291	3.981	4.417
$F(z)$	.90	.95	.975	.99	.995	.999	.9995	.99995	.999995

Adopted by permission from A. M. Mood, INTRODUCTION TO THE THEORY OF STATIS-
TICS, Table II, New York: McGraw-Hill Book Company, 1950.

appendix F
percentage points of the *t* distribution

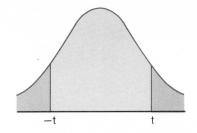

ν	$Q = .4$ $2Q = .8$	.25 .5	.1 .2	.05 .1	.025 .05	.01 .02	.005 .01	.0025 .005	.001 .002	.0005 .001
1	.325	1.000	3.078	6.314	12.706	31.821	63.657	127.32	318.31	636.62
2	.289	.816	1.886	2.920	4.303	6.965	9.925	14.089	22.327	31.598
3	.277	.765	1.638	2.353	3.182	4.541	5.841	7.453	10.214	12.924
4	.271	.741	1.533	2.132	2.776	3.747	4.604	5.598	7.173	8.610
5	.267	.727	1.476	2.015	2.571	3.365	4.032	4.773	5.893	6.869
6	.265	.718	1.440	1.943	2.447	3.143	3.707	4.317	5.208	5.959
7	.263	.711	1.415	1.895	2.365	2.998	3.499	4.029	4.785	5.408
8	.262	.706	1.397	1.860	2.306	2.896	3.355	3.833	4.501	5.041
9	.261	.703	1.383	1.833	2.262	2.821	3.250	3.690	4.297	4.781
10	.260	.700	1.372	1.812	2.228	2.764	3.169	3.581	4.144	4.587
11	.260	.697	1.363	1.796	2.201	2.718	3.106	3.497	4.025	4.437
12	.259	.695	1.356	1.782	2.179	2.681	3.055	3.428	3.930	4.318
13	.259	.694	1.350	1.771	2.160	2.650	3.012	3.372	3.852	4.221
14	.258	.692	1.345	1.761	2.145	2.624	2.977	3.326	3.787	4.140
15	.258	.691	1.341	1.753	2.131	2.602	2.947	3.286	3.733	4.073
16	.258	.690	1.337	1.746	2.120	2.583	2.921	3.252	3.686	4.015
17	.257	.689	1.333	1.740	2.110	2.567	2.898	3.222	3.646	3.965
18	.257	.688	1.330	1.734	2.101	2.552	2.878	3.197	3.610	3.922
19	.257	.688	1.328	1.729	2.093	2.539	2.861	3.174	3.579	3.883
20	.257	.687	1.325	1.725	2.086	2.528	2.845	3.153	3.552	3.850
21	.257	.686	1.323	1.721	2.080	2.518	2.831	3.135	3.527	3.819
22	.256	.686	1.321	1.717	2.074	2.508	2.819	3.119	3.505	3.792
23	.256	.685	1.319	1.714	2.069	2.500	2.807	3.104	3.485	3.767
24	.256	.685	1.318	1.711	2.064	2.492	2.797	3.091	3.467	3.745
25	.256	.684	1.316	1.708	2.060	2.485	2.787	3.078	3.450	3.725
26	.256	.684	1.315	1.706	2.056	2.479	2.779	3.067	3.435	4.707
27	.256	.684	1.314	1.703	2.052	2.473	2.771	3.057	3.421	3.690
28	.256	.683	1.313	1.701	2.048	2.467	2.763	3.047	3.408	3.674
29	.256	.683	1.311	1.699	2.045	2.462	2.756	3.038	3.396	3.659
30	.256	.683	1.310	1.697	2.042	2.457	2.750	3.030	3.385	3.646
40	.255	.681	1.303	1.684	2.021	2.423	2.704	2.971	3.307	3.551
60	.254	.679	1.296	1.671	2.000	2.390	2.660	2.915	3.232	3.460
120	.254	.677	1.289	1.658	1.980	2.358	2.617	2.860	3.160	3.373
∞	.253	.674	1.282	1.645	1.960	2.326	2.576	2.807	3.090	3.291

Q is the upper-tail area of the distribution for ν degrees of freedom, appropriate for use in a single-tail test. For a two-tail test, $2Q$ must be used.

This table is reproduced with the kind permission of Prof. E. S. Pearson and the trustees of Biometrika from E. S. Pearson and H. O. Hartley (eds.), *The Biometrika Tables for Statisticians*, vol. 1, third edition, *Biometrika*, 1966

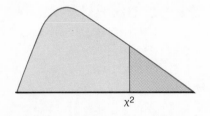

ν \ Q	0.995	0.990	0.975	0.950	0.900	0.750	0.500
1	392704.10^{-10}	157088.10^{-9}	982069.10^{-9}	393214.10^{-8}	0.0157908	0.1015308	0.454936
2	0.0100251	0.0201007	0.0506356	0.102587	0.210721	0.575364	1.38629
3	0.0717218	0.114832	0.215795	0.351846	0.584374	1.212534	2.36597
4	0.206989	0.297109	0.484419	0.710723	1.063623	1.92256	3.35669
5	0.411742	0.554298	0.831212	1.145476	1.61031	2.67460	4.35146
6	0.675727	0.872090	1.23734	1.63538	2.20413	3.45460	5.34812
7	0.989256	1.239043	1.68987	2.16735	2.83311	4.25485	6.34581
8	1.34441	1.64650	2.17973	2.73264	3.48954	5.07064	7.34412
9	1.73493	2.08790	2.70039	3.32511	4.16816	5.89883	8.34283
10	2.15586	2.55821	3.24697	3.94030	4.86518	6.73720	9.34182
11	2.60322	3.05348	3.81575	4.57481	5.57778	7.58414	10.3410
12	3.07382	3.57057	4.40379	5.22603	6.30380	8.43842	11.3403
13	3.56503	4.10692	5.00875	5.89186	7.04150	9.29907	12.3398
14	4.07467	4.66043	5.62873	6.57063	7.78953	10.1653	13.3393
15	4.60092	5.22935	6.26214	7.26094	8.54676	11.0365	14.3389
16	5.14221	5.81221	6.90766	7.96165	9.31224	11.9122	15.3385
17	5.69722	6.40776	7.56419	8.67176	10.0852	12.7919	16.3382
18	6.26480	7.01491	8.23075	9.39046	10.8649	13.6753	17.3379
19	6.84397	7.63273	8.90652	10.1170	11.6509	14.5620	18.3377
20	7.43384	8.26040	9.59078	10.8508	12.4426	15.4518	19.3374
21	8.03365	8.89720	10.28293	11.5913	13.2396	16.3444	20.3372
22	8.64272	9.54249	10.9823	12.3380	14.0415	17.2396	21.3370
23	9.26043	10.19567	11.6886	13.0905	14.8480	18.1373	22.3369
24	9.88623	10.8564	12.4012	13.8484	15.6587	19.0373	23.3367
25	10.5197	11.5240	13.1197	14.6114	16.4734	19.9393	24.3366
26	11.1602	12.1981	13.8439	15.3792	17.2919	20.8434	25.3365
27	11.8076	12.8785	14.5734	16.1514	18.1139	21.7494	26.3363
28	12.4613	13.5647	15.3079	16.9279	18.9392	22.6572	27.3362
29	13.1211	14.2565	16.0471	17.7084	19.7677	23.5666	28.3361
30	13.7867	14.9535	16.7908	18.4927	20.5992	24.4776	29.3360
40	20.7065	22.1643	24.4330	26.5093	29.0505	33.6603	39.3353
50	27.9907	29.7067	32.3574	34.7643	37.6886	42.9421	49.3349
60	35.5345	37.4849	40.4817	43.1880	46.4589	52.2938	59.3347
70	43.2752	45.4417	48.7576	51.7393	55.3289	61.6983	69.3345
80	51.1719	53.5401	57.1532	60.3915	64.2778	71.1445	79.3343
90	59.1963	61.7541	65.6466	69.1260	73.2911	80.6247	89.3342
100	67.3276	70.0649	74.2219	77.9295	82.3581	90.1332	99.3341
X	−2.5758	−2.3263	−1.9600	−1.6449	−1.2816	−0.6745	0.0000

ν \ Q	0.250	0.100	0.050	0.025	0.010	0.005	0.001
1	1.32330	2.70554	3.84146	5.02389	6.63490	7.87944	10.828
2	2.77259	4.60517	5.99146	7.37776	9.21034	10.5966	13.816
3	4.10834	6.25139	7.81473	9.34840	11.3449	12.8382	16.266
4	5.38527	7.77944	9.48773	11.1433	13.2767	14.8603	18.467
5	6.62568	9.23636	11.0705	12.8325	15.0863	16.7496	20.515
6	7.84080	10.6446	12.5916	14.4494	16.8119	18.5476	22.458
7	9.03715	12.0170	14.0671	16.0128	18.4753	20.2777	24.322
8	10.2189	13.3616	15.5073	17.5345	20.0902	21.9550	26.125
9	11.3888	14.6837	16.9190	19.0228	21.6660	23.5894	27.877
10	12.5489	15.9872	18.3070	20.4832	23.2093	25.1882	29.588
11	13.7007	17.2750	19.6751	21.9200	24.7250	26.7568	31.264
12	14.8454	18.5493	21.0261	23.3367	26.2170	28.2995	32.909
13	15.9839	19.8119	22.3620	24.7356	27.6882	29.8195	34.528
14	17.1169	21.0641	23.6848	26.1189	29.1412	31.3194	36.123
15	18.2451	22.3071	24.9958	27.4884	30.5779	32.8013	37.697
16	19.3689	23.5418	26.2962	28.8454	31.9999	34.2672	39.252
17	20.4887	24.7690	27.5871	30.1910	33.4087	35.7185	40.790
18	21.6049	25.9894	28.8693	31.5264	34.8053	37.1565	42.312
19	22.7178	27.2036	30.1435	32.8523	36.1909	38.5823	43.820
20	23.8277	28.4120	31.4104	34.1696	37.5662	39.9968	45.315
21	24.9348	29.6151	32.6706	35.4789	38.9322	41.4011	46.797
22	26.0393	30.8133	33.9244	36.7807	40.2894	42.7957	48.268
23	27.1413	32.0069	35.1725	38.0756	41.6384	44.1813	49.728
24	28.2412	33.1962	36.4150	39.3641	42.9798	45.5585	51.179
25	29.3389	34.3816	37.6525	40.6465	44.3141	46.9279	52.618
26	30.4346	35.5632	38.8851	41.9232	45.6417	48.2899	54.052
27	31.5284	36.7412	40.1133	43.1945	46.9629	49.6449	55.476
28	32.6205	37.9159	41.3371	44.4608	48.2782	50.9934	56.892
29	33.7109	39.0875	42.5570	45.7223	49.5879	52.3356	58.301
30	34.7997	40.2560	43.7730	46.9792	50.8922	53.6720	59.703
40	45.6160	51.8051	55.7585	59.3417	63.6907	66.7660	73.402
50	56.3336	63.1671	67.5048	71.4202	76.1539	79.4900	86.661
60	66.9815	74.3970	79.0819	83.2977	88.3794	91.9517	99.607
70	77.5767	85.5270	90.5312	95.0232	100.425	104.215	112.317
80	88.1303	96.5782	101.879	106.629	112.329	116.321	124.839
90	98.6499	107.565	113.145	118.136	124.116	128.299	137.208
100	109.141	118.498	124.342	129.561	135.807	140.169	149.449
X	+0.6745	+1.2816	+1.6449	+1.9600	+2.3263	+2.5758	+3.0902

appendix H
percentage points
of the F distribution

Upper 10 percent points

ν_2 \ ν_1	1	2	3	4	5	6	7	8	9	10	12	15	20	24	30	40	60	120	∞
1	39.86	49.50	53.59	55.83	57.24	58.20	58.91	59.44	59.86	60.19	60.71	61.22	61.74	62.00	62.26	62.53	62.79	63.06	63.33
2	8.53	9.00	9.16	9.24	9.29	9.33	9.35	9.37	9.38	9.39	9.41	9.42	9.44	9.45	9.46	9.47	9.47	9.48	9.49
3	5.54	5.46	5.39	5.34	5.31	5.28	5.27	5.25	5.24	5.23	5.22	5.20	5.18	5.18	5.17	5.16	5.15	5.14	5.13
4	4.54	4.32	4.19	4.11	4.05	4.01	3.98	3.95	3.94	3.92	3.90	3.87	3.84	3.83	3.82	3.80	3.79	3.78	3.76
5	4.06	3.78	3.62	3.52	3.45	3.40	3.37	3.34	3.32	3.30	3.27	3.24	3.21	3.19	3.17	3.16	3.14	3.12	3.10
6	3.78	3.46	3.29	3.18	3.11	3.05	3.01	2.98	2.96	2.94	2.90	2.87	2.84	2.82	2.80	2.78	2.76	2.74	2.72
7	3.59	3.26	3.07	2.96	2.88	2.83	2.78	2.75	2.72	2.70	2.67	2.63	2.59	2.58	2.56	2.54	2.51	2.49	2.47
8	3.46	3.11	2.92	2.81	2.73	2.67	2.62	2.59	2.56	2.54	2.50	2.46	2.42	2.40	2.38	2.36	2.34	2.32	2.29
9	3.36	3.01	2.81	2.69	2.61	2.55	2.51	2.47	2.44	2.42	2.38	2.34	2.30	2.28	2.25	2.23	2.21	2.18	2.16
10	3.29	2.92	2.73	2.61	2.52	2.46	2.41	2.38	2.35	2.32	2.28	2.24	2.20	2.18	2.16	2.13	2.11	2.08	2.06
11	3.23	2.86	2.66	2.54	2.45	2.39	2.34	2.30	2.27	2.25	2.21	2.17	2.12	2.10	2.08	2.05	2.03	2.00	1.97
12	3.18	2.81	2.61	2.48	2.39	2.33	2.28	2.24	2.21	2.19	2.15	2.10	2.06	2.04	2.01	1.99	1.96	1.93	1.90
13	3.14	2.76	2.56	2.43	2.35	2.28	2.23	2.20	2.16	2.14	2.10	2.05	2.01	1.98	1.96	1.93	1.90	1.88	1.85
14	3.10	2.73	2.52	2.39	2.31	2.24	2.19	2.15	2.12	2.10	2.05	2.01	1.96	1.94	1.91	1.89	1.86	1.83	1.80
15	3.07	2.70	2.49	2.36	2.27	2.21	2.16	2.12	2.09	2.06	2.02	1.97	1.92	1.90	1.87	1.85	1.82	1.79	1.76
16	3.05	2.67	2.46	2.33	2.24	2.18	2.13	2.09	2.06	2.03	1.99	1.94	1.89	1.87	1.84	1.81	1.78	1.75	1.72
17	3.03	2.64	2.44	2.31	2.22	2.15	2.10	2.06	2.03	2.00	1.96	1.91	1.86	1.84	1.81	1.78	1.75	1.72	1.69
18	3.01	2.62	2.42	2.29	2.20	2.13	2.08	2.04	2.00	1.98	1.93	1.89	1.84	1.81	1.78	1.75	1.72	1.69	1.66
19	2.99	2.61	2.40	2.27	2.18	2.11	2.06	2.02	1.98	1.96	1.91	1.86	1.81	1.79	1.76	1.73	1.70	1.67	1.63
20	2.97	2.59	2.38	2.25	2.16	2.09	2.04	2.00	1.96	1.94	1.89	1.84	1.79	1.77	1.74	1.71	1.68	1.64	1.61
21	2.96	2.57	2.36	2.23	2.14	2.08	2.02	1.98	1.95	1.92	1.87	1.83	1.78	1.75	1.72	1.69	1.66	1.62	1.59
22	2.95	2.56	2.35	2.22	2.13	2.06	2.01	1.97	1.93	1.90	1.86	1.81	1.76	1.73	1.70	1.67	1.64	1.60	1.57
23	2.94	2.55	2.34	2.21	2.11	2.05	1.99	1.95	1.92	1.89	1.84	1.80	1.74	1.72	1.69	1.66	1.62	1.59	1.55
24	2.93	2.54	2.33	2.19	2.10	2.04	1.98	1.94	1.91	1.88	1.83	1.78	1.73	1.70	1.67	1.64	1.61	1.57	1.53
25	2.92	2.53	2.32	2.18	2.09	2.02	1.97	1.93	1.89	1.87	1.82	1.77	1.72	1.69	1.66	1.63	1.59	1.56	1.52
26	2.91	2.52	2.31	2.17	2.08	2.01	1.96	1.92	1.88	1.86	1.81	1.76	1.71	1.68	1.65	1.61	1.58	1.54	1.50
27	2.90	2.51	2.30	2.17	2.07	2.00	1.95	1.91	1.87	1.85	1.80	1.75	1.70	1.67	1.64	1.60	1.57	1.53	1.49
28	2.89	2.50	2.29	2.16	2.06	2.00	1.94	1.90	1.87	1.84	1.79	1.74	1.69	1.66	1.63	1.59	1.56	1.52	1.48
29	2.89	2.50	2.28	2.15	2.06	1.99	1.93	1.89	1.86	1.83	1.78	1.73	1.68	1.65	1.62	1.58	1.55	1.51	1.47
30	2.88	2.49	2.28	2.14	2.05	1.98	1.93	1.88	1.85	1.82	1.77	1.72	1.67	1.64	1.61	1.57	1.54	1.50	1.46
40	2.84	2.44	2.23	2.09	2.00	1.93	1.87	1.83	1.79	1.76	1.71	1.66	1.61	1.57	1.54	1.51	1.47	1.42	1.38
60	2.79	2.39	2.18	2.04	1.95	1.87	1.82	1.77	1.74	1.71	1.66	1.60	1.54	1.51	1.48	1.44	1.40	1.35	1.29
120	2.75	2.35	2.13	1.99	1.90	1.82	1.77	1.72	1.68	1.65	1.60	1.55	1.48	1.45	1.41	1.37	1.32	1.26	1.19
∞	2.71	2.30	2.08	1.94	1.85	1.77	1.72	1.67	1.63	1.60	1.55	1.49	1.42	1.38	1.34	1.30	1.24	1.17	1.00

Upper 5 percent points

$\nu_2 \backslash \nu_1$	1	2	3	4	5	6	7	8	9	10	12	15	20	24	30	40	60	120	∞
1	161.4	199.5	215.7	224.6	230.2	234.0	236.8	238.9	240.5	241.9	243.9	245.9	248.0	249.1	250.1	251.1	252.2	253.3	254.3
2	18.51	19.00	19.16	19.25	19.30	19.33	19.35	19.37	19.38	19.40	19.41	19.43	19.45	19.45	19.46	19.47	19.48	19.49	19.50
3	10.13	9.55	9.28	9.12	9.01	8.94	8.89	8.85	8.81	8.79	8.74	8.70	8.66	8.64	8.62	8.59	8.57	8.55	8.53
4	7.71	6.94	6.59	6.39	6.26	6.16	6.09	6.04	6.00	5.96	5.91	5.86	5.80	5.77	5.75	5.72	5.69	5.66	5.63
5	6.61	5.79	5.41	5.19	5.05	4.95	4.88	4.82	4.77	4.74	4.68	4.62	4.56	4.53	4.50	4.46	4.43	4.40	4.36
6	5.99	5.14	4.76	4.53	4.39	4.28	4.21	4.15	4.10	4.06	4.00	3.94	3.87	3.84	3.81	3.77	3.74	3.70	3.67
7	5.59	4.74	4.35	4.12	3.97	3.87	3.79	3.73	3.68	3.64	3.57	3.51	3.44	3.41	3.38	3.34	3.30	3.27	3.23
8	5.32	4.46	4.07	3.84	3.69	3.58	3.50	3.44	3.39	3.35	3.28	3.22	3.15	3.12	3.08	3.04	3.01	2.97	2.93
9	5.12	4.26	3.86	3.63	3.48	3.37	3.29	3.23	3.18	3.14	3.07	3.01	2.94	2.90	2.86	2.83	2.79	2.75	2.71
10	4.96	4.10	3.71	3.48	3.33	3.22	3.14	3.07	3.02	2.98	2.91	2.85	2.77	2.74	2.70	2.66	2.62	2.58	2.54
11	4.84	3.98	3.59	3.36	3.20	3.09	3.01	2.95	2.90	2.85	2.79	2.72	2.65	2.61	2.57	2.53	2.49	2.45	2.40
12	4.75	3.89	3.49	3.26	3.11	3.00	2.91	2.85	2.80	2.75	2.69	2.62	2.54	2.51	2.47	2.43	2.38	2.34	2.30
13	4.67	3.81	3.41	3.18	3.03	2.92	2.83	2.77	2.71	2.67	2.60	2.53	2.46	2.42	2.38	2.34	2.30	2.25	2.21
14	4.60	3.74	3.34	3.11	2.96	2.85	2.76	2.70	2.65	2.60	2.53	2.46	2.39	2.35	2.31	2.27	2.22	2.18	2.13
15	4.54	3.68	3.29	3.06	2.90	2.79	2.71	2.64	2.59	2.54	2.48	2.40	2.33	2.29	2.25	2.20	2.16	2.11	2.07
16	4.49	3.63	3.24	3.01	2.85	2.74	2.66	2.59	2.54	2.49	2.42	2.35	2.28	2.24	2.19	2.15	2.11	2.06	2.01
17	4.45	3.59	3.20	2.96	2.81	2.70	2.61	2.55	2.49	2.45	2.38	2.31	2.23	2.19	2.15	2.10	2.06	2.01	1.96
18	4.41	3.55	3.16	2.93	2.77	2.66	2.58	2.51	2.46	2.41	2.34	2.27	2.19	2.15	2.11	2.06	2.02	1.97	1.92
19	4.38	3.52	3.13	2.90	2.74	2.63	2.54	2.48	2.42	2.38	2.31	2.23	2.16	2.11	2.07	2.03	1.98	1.93	1.88
20	4.35	3.49	3.10	2.87	2.71	2.60	2.51	2.45	2.39	2.35	2.28	2.20	2.12	2.08	2.04	1.99	1.95	1.90	1.84
21	4.32	3.47	3.07	2.84	2.68	2.57	2.49	2.42	2.37	2.32	2.25	2.18	2.10	2.05	2.01	1.96	1.92	1.87	1.81
22	4.30	3.44	3.05	2.82	2.66	2.55	2.46	2.40	2.34	2.30	2.23	2.15	2.07	2.03	1.98	1.94	1.89	1.84	1.78
23	4.28	3.42	3.03	2.80	2.64	2.53	2.44	2.37	2.32	2.27	2.20	2.13	2.05	2.01	1.96	1.91	1.86	1.81	1.76
24	4.26	3.40	3.01	2.78	2.62	2.51	2.42	2.36	2.30	2.25	2.18	2.11	2.03	1.98	1.94	1.89	1.84	1.79	1.73
25	4.24	3.39	2.99	2.76	2.60	2.49	2.40	2.34	2.28	2.24	2.16	2.09	2.01	1.96	1.92	1.87	1.82	1.77	1.71
26	4.23	3.37	2.98	2.74	2.59	2.47	2.39	2.32	2.27	2.22	2.15	2.07	1.99	1.95	1.90	1.85	1.80	1.75	1.69
27	4.21	3.35	2.96	2.73	2.57	2.46	2.37	2.31	2.25	2.20	2.13	2.06	1.97	1.93	1.88	1.84	1.79	1.73	1.67
28	4.20	3.34	2.95	2.71	2.56	2.45	2.36	2.29	2.24	2.19	2.12	2.04	1.96	1.91	1.87	1.82	1.77	1.71	1.65
29	4.18	3.33	2.93	2.70	2.55	2.43	2.35	2.28	2.22	2.18	2.10	2.03	1.94	1.90	1.85	1.81	1.75	1.70	1.64
30	4.17	3.32	2.92	2.69	2.53	2.42	2.33	2.27	2.21	2.16	2.09	2.01	1.93	1.89	1.84	1.79	1.74	1.68	1.62
40	4.08	3.23	2.84	2.61	2.45	2.34	2.25	2.18	2.12	2.08	2.00	1.92	1.84	1.79	1.74	1.69	1.64	1.58	1.51
60	4.00	3.15	2.76	2.53	2.37	2.25	2.17	2.10	2.04	1.99	1.92	1.84	1.75	1.70	1.65	1.59	1.53	1.47	1.39
120	3.92	3.07	2.68	2.45	2.29	2.17	2.09	2.02	1.96	1.91	1.83	1.75	1.66	1.61	1.55	1.50	1.43	1.35	1.25
∞	3.84	3.00	2.60	2.37	2.21	2.10	2.01	1.94	1.88	1.83	1.75	1.67	1.57	1.52	1.46	1.39	1.32	1.22	1.00

Upper 2.5 percent points

ν_1 / ν_2	1	2	3	4	5	6	7	8	9	10	12	15	20	24	30	40	60	120	∞
1	647.8	799.5	864.2	899.6	921.8	937.1	948.2	956.7	963.3	968.6	976.7	984.9	993.1	997.2	1001	1006	1010	1014	1018
2	38.51	39.00	39.17	39.25	39.30	39.33	39.36	39.37	39.39	39.40	39.41	39.43	39.45	39.46	39.46	39.47	30.48	39.49	39.50
3	17.44	16.04	15.44	15.10	14.88	14.73	14.62	14.54	14.47	14.42	14.34	14.25	14.17	14.12	14.08	14.04	13.99	13.95	13.90
4	12.22	10.65	9.98	9.60	9.36	9.20	9.07	8.98	8.90	8.84	8.75	8.66	8.56	8.51	8.46	8.41	8.36	8.31	8.26
5	10.01	8.43	7.76	7.39	7.15	6.98	6.85	6.76	6.68	6.62	6.52	6.43	6.33	6.28	6.23	6.18	6.12	6.07	6.02
6	8.81	7.26	6.60	6.23	5.99	5.82	5.70	5.60	5.52	5.46	5.37	5.27	5.17	5.12	5.07	5.01	4.96	4.90	4.85
7	8.07	6.54	5.89	5.52	5.29	5.12	4.99	4.90	4.82	4.76	4.67	4.57	4.47	4.42	4.36	4.31	4.25	4.20	4.14
8	7.57	6.06	5.42	5.05	4.82	4.65	4.53	4.43	4.36	4.30	4.20	4.10	4.00	3.95	3.89	3.84	3.78	3.73	3.67
9	7.21	5.71	5.08	4.72	4.48	4.32	4.20	4.10	4.03	3.96	3.87	3.77	3.67	3.61	3.56	3.51	3.45	3.39	3.33
10	6.94	5.46	4.83	4.47	4.24	4.07	3.95	3.85	3.78	3.72	3.62	3.52	3.42	3.37	3.31	3.26	3.20	3.14	3.08
11	6.72	5.26	4.63	4.28	4.04	3.88	3.76	3.66	3.59	3.53	3.43	3.33	3.23	3.17	3.12	3.06	3.00	2.94	2.88
12	6.55	5.10	4.47	4.12	3.89	3.73	3.61	3.51	3.44	3.37	3.28	3.18	3.07	3.02	2.96	2.91	2.85	2.79	2.72
13	6.41	4.97	4.35	4.00	3.77	3.60	3.48	3.39	3.31	3.25	3.15	3.05	2.95	2.89	2.84	2.78	2.72	2.66	2.60
14	6.30	4.86	4.24	3.89	3.66	3.50	3.38	3.29	3.21	3.15	3.05	2.95	2.84	2.79	2.73	2.67	2.61	2.55	2.49
15	6.20	4.77	4.15	3.80	3.58	3.41	3.29	3.20	3.12	3.06	2.96	2.86	2.76	2.70	2.64	2.59	2.52	2.46	2.40
16	6.12	4.69	4.08	3.73	3.50	3.34	3.22	3.12	3.05	2.99	2.89	2.79	2.68	2.63	2.57	2.51	2.45	2.38	2.32
17	6.04	4.62	4.01	3.66	3.44	3.28	3.16	3.06	2.98	2.92	2.82	2.72	2.62	2.56	2.50	2.44	2.38	2.32	2.25
18	5.98	4.56	3.95	3.61	3.38	3.22	3.10	3.01	2.93	2.87	2.77	2.67	2.56	2.50	2.44	2.38	2.32	2.26	2.19
19	5.92	4.51	3.90	3.56	3.33	3.17	3.05	2.96	2.88	2.82	2.72	2.62	2.51	2.45	2.39	2.33	2.27	2.20	2.13
20	5.87	4.46	3.86	3.51	3.29	3.13	3.01	2.91	2.84	2.77	2.68	2.57	2.46	2.41	2.35	2.29	2.22	2.16	2.09
21	5.83	4.42	3.82	3.48	3.25	3.09	2.97	2.87	2.80	2.73	2.64	2.53	2.42	2.37	2.31	2.25	2.18	2.11	2.04
22	5.79	4.38	3.78	3.44	3.22	3.05	2.93	2.84	2.76	2.70	2.60	2.50	2.39	2.33	2.27	2.21	2.14	2.08	2.00
23	5.75	4.35	3.75	3.41	3.18	3.02	2.90	2.81	2.73	2.67	2.57	2.47	2.36	2.30	2.24	2.18	2.11	2.04	1.97
24	5.72	4.32	3.72	3.38	3.15	2.99	2.87	2.78	2.70	2.64	2.54	2.44	2.33	2.27	2.21	2.15	2.08	2.01	1.94
25	5.69	4.29	3.69	3.35	3.13	2.97	2.85	2.75	2.68	2.61	2.51	2.41	2.30	2.24	2.18	2.12	2.05	1.98	1.91
26	5.66	4.27	3.67	3.33	3.10	2.94	2.82	2.73	2.65	2.59	2.49	2.39	2.28	2.22	2.16	2.09	2.03	1.95	1.88
27	5.63	4.24	3.65	3.31	3.08	2.92	2.80	2.71	2.63	2.57	2.47	2.36	2.25	2.19	2.13	2.07	2.00	1.93	1.85
28	5.61	4.22	3.63	3.29	3.06	2.90	2.78	2.69	2.61	2.55	2.45	2.34	2.23	2.17	2.11	2.05	1.98	1.91	1.83
29	5.59	4.20	3.61	3.27	3.04	2.88	2.76	2.67	2.59	2.53	2.43	2.32	2.21	2.15	2.09	2.03	1.96	1.89	1.81
30	5.57	4.18	3.59	3.25	3.03	2.87	2.75	2.65	2.57	2.51	2.41	2.31	2.20	2.14	2.07	2.01	1.94	1.87	1.79
40	5.42	4.05	3.46	3.13	2.90	2.74	2.62	2.53	2.45	2.39	2.29	2.18	2.07	2.01	1.94	1.88	1.80	1.72	1.64
60	5.29	3.93	3.34	3.01	2.79	2.63	2.51	2.41	2.33	2.27	2.17	2.06	1.94	1.88	1.82	1.74	1.67	1.58	1.48
120	5.15	3.80	3.23	2.89	2.67	2.52	2.39	2.30	2.22	2.16	2.05	1.94	1.82	1.76	1.69	1.61	1.53	1.43	1.31
∞	5.02	3.69	3.12	2.79	2.57	2.41	2.29	2.19	2.11	2.05	1.94	1.83	1.71	1.64	1.57	1.48	1.39	1.27	1.00

Upper 1 percent points

ν_2 \ ν_1	1	2	3	4	5	6	7	8	9	10	12	15	20	24	30	40	60	120	∞
1	4052	4999.5	5403	5625	5764	5859	5928	5981	6022	6056	6106	6157	6209	6235	6261	6287	6313	6339	6366
2	98.50	99.00	99.17	99.25	99.30	99.33	99.36	99.37	99.39	99.40	99.42	99.43	99.45	99.46	99.47	99.47	99.48	99.49	99.50
3	34.12	30.82	29.46	28.71	28.24	27.91	27.67	27.49	27.35	27.23	27.05	26.87	26.69	26.60	26.50	26.41	26.32	26.22	26.13
4	21.20	18.00	16.69	15.98	15.52	15.21	14.98	14.80	14.66	14.55	14.37	14.20	14.02	13.93	13.84	13.75	13.65	13.56	13.46
5	16.26	13.27	12.06	11.39	10.97	10.67	10.46	10.29	10.16	10.05	9.89	9.72	9.55	9.47	9.38	9.29	9.20	9.11	9.02
6	13.75	10.92	9.78	9.15	8.75	8.47	8.26	8.10	7.98	7.87	7.72	7.56	7.40	7.31	7.23	7.14	7.06	6.97	6.88
7	12.25	9.55	8.45	7.85	7.46	7.19	6.99	6.84	6.72	6.62	6.47	6.31	6.16	6.07	5.99	5.91	5.82	5.74	5.65
8	11.26	8.65	7.59	7.01	6.63	6.37	6.18	6.03	5.91	5.81	5.67	5.52	5.36	5.28	5.20	5.12	5.03	4.95	4.86
9	10.56	8.02	6.99	6.42	6.06	5.80	5.61	5.47	5.35	5.26	5.11	4.96	4.81	4.73	4.65	4.57	4.48	4.40	4.31
10	10.04	7.56	6.55	5.99	5.64	5.39	5.20	5.06	4.94	4.85	4.71	4.56	4.41	4.33	4.25	4.17	4.08	4.00	3.91
11	9.65	7.21	6.22	5.67	5.32	5.07	4.89	4.74	4.63	4.54	4.40	4.25	4.10	4.02	3.94	3.86	3.78	3.69	3.60
12	9.33	6.93	5.95	5.41	5.06	4.82	4.64	4.50	4.39	4.30	4.16	4.01	3.86	3.78	3.70	3.62	3.54	3.45	3.36
13	9.07	6.70	5.74	5.21	4.86	4.62	4.44	4.30	4.19	4.10	3.96	3.82	3.66	3.59	3.51	3.43	3.34	3.25	3.17
14	8.86	6.51	5.56	5.04	4.69	4.46	4.28	4.14	4.03	3.94	3.80	3.66	3.51	3.43	3.35	3.27	3.18	3.09	3.00
15	8.68	6.36	5.42	4.89	4.56	4.32	4.14	4.00	3.89	3.80	3.67	3.52	3.37	3.29	3.21	3.13	3.05	2.96	2.87
16	8.53	6.23	5.29	4.77	4.44	4.20	4.03	3.89	3.78	3.69	3.55	3.41	3.26	3.18	3.10	3.02	2.93	2.84	2.75
17	8.40	6.11	5.18	4.67	4.34	4.10	3.93	3.79	3.68	3.59	3.46	3.31	3.16	3.08	3.00	2.92	2.83	2.75	2.65
18	8.29	6.01	5.09	4.58	4.25	4.01	3.84	3.71	3.60	3.51	3.37	3.23	3.08	3.00	2.92	2.84	2.75	2.66	2.57
19	8.18	5.93	5.01	4.50	4.17	3.94	3.77	3.63	3.52	3.43	3.30	3.15	3.00	2.92	2.84	2.76	2.67	2.58	2.49
20	8.10	5.85	4.94	4.43	4.10	3.87	3.70	3.56	3.46	3.37	3.23	3.09	2.94	2.86	2.78	2.69	2.61	2.52	2.42
21	8.02	5.78	4.87	4.37	4.04	3.81	3.64	3.51	3.40	3.31	3.17	3.03	2.88	2.80	2.72	2.64	2.55	2.46	2.36
22	7.95	5.72	4.82	4.31	3.99	3.76	3.59	3.45	3.35	3.26	3.12	2.98	2.83	2.75	2.67	2.58	2.50	2.40	2.31
23	7.88	5.66	4.76	4.26	3.94	3.71	3.54	3.41	3.30	3.21	3.07	2.93	2.78	2.70	2.62	2.54	2.45	2.35	2.26
24	7.82	5.61	4.72	4.22	3.90	3.67	3.50	3.36	3.26	3.17	3.03	2.89	2.74	2.66	2.58	2.49	2.40	2.31	2.21
25	7.77	5.57	4.68	4.18	3.85	3.63	3.46	3.32	3.22	3.13	2.99	2.85	2.70	2.62	2.54	2.45	2.36	2.27	2.17
26	7.72	5.53	4.64	4.14	3.82	3.59	3.42	3.29	3.18	3.09	2.96	2.81	2.66	2.58	2.50	2.42	2.33	2.23	2.13
27	7.68	5.49	4.60	4.11	3.78	3.56	3.39	3.26	3.15	3.06	2.93	2.78	2.63	2.55	2.47	2.38	2.29	2.20	2.10
28	7.64	5.45	4.57	4.07	3.75	3.53	3.36	3.23	3.12	3.03	2.90	2.75	2.60	2.52	2.44	2.35	2.26	2.17	2.06
29	7.60	5.42	4.54	4.04	3.73	3.50	3.33	3.20	3.09	3.00	2.87	2.73	2.57	2.49	2.41	2.33	2.23	2.14	2.03
30	7.56	5.39	4.51	4.02	3.70	3.47	3.30	3.17	3.07	2.98	2.84	2.70	2.55	2.47	2.39	2.30	2.21	2.11	2.01
40	7.31	5.18	4.31	3.83	3.51	3.29	3.12	2.99	2.89	2.80	2.66	2.52	2.37	2.29	2.20	2.11	2.02	1.92	1.80
60	7.08	4.98	4.13	3.65	3.34	3.12	2.95	2.82	2.72	2.63	2.50	2.35	2.20	2.12	2.03	1.94	1.84	1.73	1.60
120	6.85	4.79	3.95	3.48	3.17	2.96	2.79	2.66	2.56	2.47	2.34	2.19	2.03	1.95	1.86	1.76	1.66	1.53	1.38
∞	6.63	4.61	3.78	3.32	3.02	2.80	2.64	2.51	2.41	2.32	2.18	2.04	1.88	1.79	1.70	1.59	1.47	1.32	1.00

Source: This table is reproduced with the kind permission of Prof. E. S. Pearson and the trustees of Biometrika from E. S. Pearson and H. O. Hartley (eds), *The Biometrika Tables for Statisticians*, vol. 1, third edition. *Biometrika*, 1966.

2-2 exercises

1. a) 15   b) -8 c) $-3\ 1/3$
2. a) $X_1 + X_2 + X_3 + X_4 + X_5$ b) $X_1 + X_2 + X_3 + X_4 + 4a$
   c) $(X_1 + X_2 + X_3) + (Y_1 + Y_2 + Y_3)$ d) $X_1Y_1 + X_2Y_2 + X_3Y_3$
3. a) 12   b) 30   c) 18   d) 36
4. a) 30   b) -6 c) 330
5. a) 200   b) 0   c) 700 and -700
6. a) 190   b) 750   c) 1390   d) 1580   e) 148,500
7. a) 45   b) 109   c) 190
11. a) 27   b) 32   c) 24
12. a) 18   b) 60   c) 564

2-3 exercises

2. a) Region 1: $B' \cap A$ b) Region 2: $A \cap B$ c) Region 3:
   $A' \cap B$ d) Region 4: $(A \cup B)'$
3. a) $\{d, e\}$ b) $\{e, g\}$ c) $\{c, e\}$ d) $\{e\}$
   e) $\{a, b, c, d, e, f, g, h\}$ f) $\{a, b, c, d, e, g, i, j\}$
   g) $\{d, e, f, g, h, c, i, j\}$ h) $\{a, b, c, d, e, f, g, h, i, j\}$
   i) $\{a, b, c\}$ j) $\{a, b, k\}$
4. a) $\{c\}$ b) $\{a, b, c, d, e, 1, 2, 3\}$ c) $\{c, 1\}$ d) $\{c, 3\}$
   e) $\{h\}$
6. Seventeen students will not go to graduate school, nor get married, nor work full time.
7. 35
8. a) All 11 members   b) The 11 members   c) The three women
   d) The set of two independent members
9. a) No   b) A and B
10. The subsets of U are $\emptyset$, $\{a\}$, $\{b\}$, $\{a, b\}$
    The subsets of S are $\emptyset$, $\{1\}$, $\{2\}$, $\{3\}$, $\{1, 2\}$, $\{1, 3\}$, $\{2, 3\}$, $\{1, 2, 3\}$
11. There are 16 subsets.
12. There are 2^n subsets in all. Only one subset contains no element at all—the null set. Only one subset contains exactly n elements. There are n subsets which contain exactly one element each.

2-4 exercises

1. a) No   b) No   c) Yes   d) No
3. The domain of $f(X)$ is the set of all numbers. The range of $f(X)$ is also the set of all numbers.
4. a) The range is 4, 12, 28, 52, and 84.
   b) The domain of $f(X)$ is 0, -1, 1, -2, 2, -3, and 3.
5. a) 4   b) 12   c) 19   d) $a^2 - 2a + 4$ e) $a^2 + 2a + 4$
   f) $c^4 - 2c^2 + 4$
6. a) 3   b) -3 c) 3   d) 5.4   e) $\dfrac{c^2 + 2}{c}$
7. a) 1   b) $(1 + a)^a$ c) 9

8. a) 1   b) $4 - 4a + a^2$ c) $(2 - a)^a$
9. a) 10   b) 10   c) 10
10. $g[f(Y)] = Y^2 + 4Y + 5$ $g[f(1)] = 10$ $g[f(3)] = 26$
11. a) 1   b) 0   c) 2   d) 9
12. a) No   b) $acX + ad + b$ c) $acX + bc + d$
    d) Yes, if $ad = bc$ and $b = d$; otherwise, no.

3-2 exercises

1. a) Four   b) $\{TT, TH, HT, HH\}$ c) Simple event
2. a) $\{1, 2, 3, 4, 5, 6\}$ b) Compound event
3. 16 possible sequences
4. a) 1   b) 3   c) 8   d) 1 and 8   e) 5, 6, 7, and 8
7. a) $\{KK, KQ, KJ, QK, QQ, QJ, JK, JQ, JJ\}$
   b) Compound event

3-3 exercises

1. a) 720   b) 6   c) 210   d) 24   e) 20   f) 1   g) 1
2. $A\,B\,C$ $A\,C\,B$ $B\,A\,C$ $B\,C\,A$ $C\,A\,B$ $C\,B\,A$
3. a) 42 and 42   b) $\dfrac{7!}{5!\,2!}$
4. 12 ways
6. 46,656
7. 20
8. 676,000
9. 1024
10. 120
11. $P^6_3 = 120$ $P^6_6 = 720$
12. a) 51,300   b) 142,506   c) 15,756
13. 6840
14. $\binom{11}{3} = 165$ $P^{11}_5 = 55,440$
15. 102,375
16. 2,598,960

3-4 exercises

1. 0.495
2. a) 0.175   b) 0.825
3. $\dfrac{1}{190}$
4. $\dfrac{1}{17,576,000}$
5. 190   23/38
6. $\dfrac{1}{142,506}$

7. 3/10
8. 2/7
9. $\dfrac{33}{16,660}$
10. a) 3/8   b) 3/4   c) 1/4
11. a) 1/6   b) 1/12   c) 1/6   d) 1/6

**4-2
exercises**

1. 1/2
2. 1
3. Yes, A and B are mutually exclusive. $P(A \cup B) = 1/3$
4. $P(A \cup B) = 3/4$
5. 13/36
6. No, E and F are not mutually exclusive.   4/5
7. P(spade, heart, or diamond) = 3/4 P(diamond or ace) = 4/13
8. a) 7/13   b) 1/2   c) 1/13
9. 0.88 and 0.12
10. 0.90 and 0.10
11. 0.60
12. a) 0.67   b) 0.58

**4-3
exercises**

1. a) Independent   b) Independent   c) Independent
   d) Dependent   e) Independent   f) Dependent
3. a) 1/4   b) 3/50   c) 27/1000   d) 3/100
4. a) $\dfrac{12}{371,293}$ b) $\dfrac{1}{371,293}$ c) $\dfrac{1}{256}$
5. a) 1/16   b) 1/16   c) 1/4
6. 1/4
7. 0.595
8. $P(H) = 0.6$ and $P(T) = 0.4$ a) 0.064   b) 0.784   c) 0.648
   d) 0.352
9. a) 0.81   b) 0.01   c) 0.18
10. a) Yes   b) $P(M|F) = 0.6 \; P(M|F') = 0.6 \;\; P(M) = 0.6$
    c) $P(F|M) = 0.7 \; P(F|M') = 0.7 \;\; P(F) = 0.7$
    d) $P(M'|F') = 0.4 \; P(M'|F) = 0.4 \;\; P(M') = 0.4$
    e) $P(F'|M) = 0.3 \; P(F'|M') = 0.3 \;\; P(F') = 0.3$

**4-4
exercises**

1. a) Dependent   b) Dependent   c) Dependent
   d) Independent   e) Independent   f) Dependent
   g) Dependent   h) Dependent
2. a) 0.3   b) 0.45
3. a) 2/9   b) 1/15   c) 1/120   d) 1/24

4. a) $\dfrac{1}{270{,}725}$ b) $\dfrac{1}{1{,}082{,}900}$ c) $\dfrac{33}{16{,}660}$

5. 11/30

6. 1/114

7. 0.63

8. a) 0.06   b) 0.28

9. $P(G) = 35/96$

10. a) No   b) $P(M|F) = 4/7$ $P(M'|F) = 3/7$
    $P(M|F') = 2/3$ $P(M'|F') = 1/3$

11. $P(S \cap T) = 0.06$ $P(S \cap T') = 0.14$ $P(S' \cap T) = 0.48$
    $P(S' \cap T') = 0.32$ a) 1/9   b) 16/23

12. a) 4/7   b) 2/3

4-5 exercises

2. a) X designates the number of spots on the face of the die after it is rolled.

3. a) 11/16   b) 11/16   c) 15/16   d) 7/8   e) 15/16   f) 3/8

4. $P(TTT) = 64/1000$ $P(TTH \text{ or } THT \text{ or } HTT) = 288/1000$
   $P(HHT \text{ or } HTH \text{ or } THH) = 432/1000$ $P(HHH) = 216/1000$

5. a) 1/4   b) It is the number of stereo receivers of model H (or T) sold.

   c) Random events: HHH HHT HTH THH TTH THT HTT TTT
      Values of X: 3   2   2   2   1   1   1   0

   d) X: 0 1 2 3
        $P(X)$: $\frac{1}{8}$ $\frac{3}{8}$ $\frac{3}{8}$ $\frac{1}{8}$

6. Y: 0 1 4 9
   $P(Y) = P(X)$: $\frac{1}{8}$ $\frac{3}{8}$ $\frac{3}{8}$ $\frac{1}{8}$

7. X: 0 1 2 3 4
   $P(X)$: $\frac{1}{16}$ $\frac{4}{16}$ $\frac{6}{16}$ $\frac{4}{16}$ $\frac{1}{16}$

8. a) X designates the number of spots on the face of the die after it is rolled.

   b) X: 1 2 3 4 5 6 Total
        $P(X)$: $\frac{1}{21}$ $\frac{2}{21}$ $\frac{3}{21}$ $\frac{4}{21}$ $\frac{5}{21}$ $\frac{6}{21}$ $\frac{21}{21} = 1$

9. | W | $P(W)$ |
   | --- | --- |
   | 0 | $1 - p$ |
   | 1 | p |
   | Total | 1.0 |

4-6 exercises

1. a) Bernoulli   b) Bernoulli   c) Binomial   d) Binomial

2. a) 0.375   b) 0.25   c) 0.2304

3. a) 0.125   b) 0.0625   c) 0.125

4. a) 0.5   b) 0.5

5. a) 0.1762   b) 0.0207   c) 0.9423

6. a) 0.264   b) 0.8791   c) 0.0328

7. a) 0.9785   b) 0.1809   c) 0.8110
8. 0.2684
9. 0.1840
10. 0.1754
11. a) 0.1573   b) 0.0169   c) 0.9831   d) 0.2633   e) 0.4091
12. a) 0.6082   b) 0.8671   c) 0.5970
13. a) 0.2007   b) 0.8336   c) 0.3670   d) 0.5630
14. a) 0.2501   b) 0.6020   c) 0.6481
15. a) 0.3370   b) 0.3174   c) 0.6528
16. a) 0.1681   b) 0.0024   c) 0.9693   d) 0.1630
17. a) 0.0004   b) 0.9269   c) 1
18. a) 0.2852   b) 0.6768   c) 0

5-2 exercises

5. 1/720
6. 1/120
7. 1/495
8. 1/4845
9. 2/5

5-3 exercises

7. The class interval is 15.
8. a) Values of observations: 2 3 4 5 6 7 8 9 Total
    Frequency distribution: 2 2 4 5 3 2 1 1   20
9. a) Values of observations: 5 6 7 8 9 10 11 12 Total
    Frequency distribution: 2 2 4 5 3 2 1 1   20
11. a) Midpoints: 2.0 2.3 2.6 2.9 3.2 3.5 3.8 Total
    Frequency: 2 3 7 11 7 7 3   40
12. a) Midpoints: 55 60 65 70 75 80 85 90 95 Total
    Frequency: 2 3 4 5 8 10 8 6 4   50

5-4 exercises

1. The mode
2. I prefer to use equation 5-2 because equation 5-3 requires a frequency distribution that is not yet existent.
3. a) 3.0   b) 2.9825
4. a) 78   b) 78
6. a) 5   b) 5
7. 15
8. 7
9. 2.97
10. 79
11. The class 2.75 − 3.05 with the midpoint 2.9
12. The class 77.5 − 82.5 with the midpoint 80
13. White

14. Single
15. a) 20　　b) 20　　c) 20
16. 12 passengers.　　Yes, because all or most of the 12 people may be heavier than the average.
17. a) 110　　b) 100
18. a) 112.12　　b) 114.67

5-5 exercises

4. 360
6. a) 70/6　　b) 70/6　　The variance of a set of scores is unchanged if a constant is added to, or subtracted from, each of the scores.
7. $\hat{s}^2 = 0.21625$　　$\hat{s} = 0.4648$
8. $\hat{s}^2 = 111$　　$\hat{s} = 10.5$
9. a) $s^2 = 4.22$　　$s = 2.1$
　 b) $s^2 = 4.22$　　$s = 2.1$
10. a) $\hat{s}^2 = 3.3$　　$\hat{s} = 1.8$
　 b) $\hat{s}^2 = 3.3$　　$\hat{s} = 1.8$
11. a) $\hat{s}^2 = 3.3$　　$\hat{s} = 1.8$　　b) $\hat{s}^2 = 3.3$　　$\hat{s} = 1.8$
12. $s^2 = 3.8$
13. $s^2 = 0.221795$
14. $s^2 = 113.3$
15. $s^2 = 17.2$　　$s = 4.15$
16. $s^2 = 7.5$　　$s = 2.7$

6-2 exercises

1. 1.5
2. 1
3. 3.5
4. 7
5. −$3.2
6. $1.5
7. 2.7 million dollars or 0.7 million dollars profit
8. $E(X) = 2/5$
9. The expected value of loss = $25.6
10. 2.20
11. The mean IQ is 106.7
12. $E(Y) = 28.5$

6-3 exercises

1. 0.75
2. 5.83
3. $\sigma^2 = 1.61$
4. $\sigma^2 = 2.36$
5. $\sigma^2 = 2.2$
6. $\sigma^2 = 721.05$

7. $\sigma^2 = 1.66$
8. a) A population   b) $\sigma^2 = 5$
9. $\sigma^2 = 1.25$
10. For $n = 5$, $p = 0.1$ $E(X) = 0.5$ $\sigma^2 = 0.45$
     For $n = 5$, $p = 0.9$ $E(X) = 4.5$ $\sigma^2 = 0.45$
     The variance is in fact $np(1 - p)$.

6-4 exercises

1. a) Sampling distribution of $\overline{X}$

$\overline{X}$:	2	3	4	5	6	Total
$P(\overline{X})$:	$\frac{1}{9}$	$\frac{2}{9}$	$\frac{3}{9}$	$\frac{2}{9}$	$\frac{1}{9}$	$\frac{9}{9} = 1$

    b) $\sigma_{\overline{X}}^2 = 1\ 1/3$ c) $\sigma_{\overline{X}}^2 = 2\ 2/3$
2. a) $\sigma_X^2 = 2/3$
    b) Sampling distribution of $\overline{X}$

$\overline{X}$:	1	1.5	2	2.5	3	Total
$P(\overline{X})$:	$\frac{1}{9}$	$\frac{2}{9}$	$\frac{3}{9}$	$\frac{2}{9}$	$\frac{1}{9}$	$\frac{9}{9} = 1$

    c) $\sigma_{\overline{X}}^2 = 1/3$
3. a) $\sigma^2 = 376.11$ b) $\sigma_{\overline{X}}^2 = 3.9$ c) $E(\overline{X}) = E(X) = 106.7$
4. a) Sampling distribution of $\overline{X}$

$\overline{X}$:	0	1	2	3	4	5	6	Total
$P(\overline{X})$:	$\frac{1}{16}$	$\frac{2}{16}$	$\frac{3}{16}$	$\frac{4}{16}$	$\frac{3}{16}$	$\frac{2}{16}$	$\frac{1}{16}$	$\frac{16}{16} = 1$

    b) $E(\overline{X}) = 3$ c) $\sigma_{\overline{X}}^2 = 2.5$
5. $\sigma_{\overline{Y}}^2 = 6.25$ $\sigma_{\overline{Y}} = 2.5$
6. $E(\overline{X}) = E(X) = 2000$ pounds $\sigma_{\overline{X}} = 5$
7. $E(\overline{X}) = E(X) = 40$ words $\sigma_{\overline{X}} = 2$
8. $E(\overline{X}) = E(X) = 30{,}000$ miles $\sigma_{\overline{X}} = 50$

6-5 exercises

1   8/9
2. a) 1/100   b) 1/4
3. 1/9
4. a) 0   b) 3/4
5. a) 1/4   b) 8/9
6. 0.96
7. $n = 900$
8. $n = 100$

7-2 exercises

1. a) 0.50   b) 0.9115   c) 0.0594   d) 0.1587   e) 0.8944
    f) 0.3849   g) 0.1359   h) 0.8095
2. a) 0.9772   b) 0.9265   c) 0.9608   d) 0.9505
    e) 0.1293   f) 0.8791   g) 0.2754
3. a) 0.9900   b) 0.9950   c) 0.01   d) 0.98   e) 0.99
4. a) 0.01   b) 0.005   c) 0.95   d) 0.90
5. a) 2.57   b) 2.58   c) 2.326   d) 2.575   e) 1.565
    f) -1.57 g) 0.58   h) 2.0

6. a) 1.645   b) 2.326   c) 2.81
7. a) 0.0968   b) 0.1357   c) 0.7881   d) 0.4514
   e) 0.2620
8. a) 0.7881   b) 0.9192   c) 0.2620   d) 0.0718
9. $\sigma = 50$
10. 50

7-3
exercises

1. a) 0.1151   b) 0.0548   c) 0.2743   d) 0.5793
   e) 0.5654   f) 0.1359   g) 0.3659
2. a) 0.0062   b) 0.0228   c) 0.6827   d) 0.3023
3. a) 0.8413   b) 0.0668   c) 25 students
4. 6915 chickens
5. a) 0.0013   b) 0.0668   c) 0.8562
6. a) 0.8185   b) 0.0228
7. a) 3085 workers   b) 228 workers
8. a) 0.0668   b) 6915 plants
9. a) 0.6915   b) 2.28 or two persons
10. 0.2012

7-4
exercises

1. 0.0030
2. a) 0.1587   b) 0.0668   c) 0.9319
3. a) 0.0062   b) 0.0228   c) 0.1747   d) 0.1557
4. a) 0.9999   b) 0.0668   c) 0.3085   d) 0.0228
5. $E(X) = 4.5$ $\sigma_X^2 = 8.25$ $\sigma_{\bar{X}} = 0.287$
   a) 0.8830   b) 0.1562   c) 0.7680   d) 0.8768
6. a) 0.9871   b) 0.8686   c) 0.9867   d) 0.9992
7. a) $\bar{X}$: 3   4.5   6   7.5   9   Total
   $P(\bar{X})$: $\frac{1}{9}$ $\frac{2}{9}$ $\frac{3}{9}$ $\frac{2}{9}$ $\frac{1}{9}$ $\frac{9}{9} = 1$
   b) $\bar{X}$: 3   4   5   6   7   8   9   Total
   $P(\bar{X})$: $\frac{1}{27}$ $\frac{3}{27}$ $\frac{6}{27}$ $\frac{7}{27}$ $\frac{6}{27}$ $\frac{3}{27}$ $\frac{1}{27}$ $\frac{27}{27} = 1$
   c) The distribution becomes closer to the normal pattern as the
   sample size is increased.
8. a) 0.0001   b) 0.0011   c) 0.7008   d) 0.0329

7-5
exercises

1. a) 0.50   b) 0.4938   c) 0.0994
2. a) 0.0478   b) 0.2024
3. a) 0.8413   b) 0.7881
4. a) 0.0228   b) 0.3085
5. a) 0.2119   b) 0.2743   c) 0.5403
6. 0.0062
7. a) 0.3662   b) 0.3626
8. a) 0.3543   b) 0.3833

8-2 exercises

8. a) $H_0: \mu = 500$ $H_1: \mu \neq 500$
 b) $H_0: \mu = 500$ $H_1: \mu > 500$
 c) $H_0: \mu = 500$ $H_1: \mu < 500$

8-3 exercises

2. a) $\alpha = 0.0000$ $\beta = 0.9506$ b) $\alpha = 0.0434$ $\beta = 0.7187$
   c) $\alpha = 0.1809$ $\beta = 0.0844$ d) $\alpha = 0.0845$ $\beta = 0.1808$
3. Since $Z = 1.333$, we do not reject H_0. $\beta = 0.1534$
4. Since $Z = -2$, we do not reject H_0.
5. $\beta = 0.2502$ $\beta = 0.0429$
6. Since $Z = -3.2$, we reject H_0.
7. $\beta = 0$ a) $\beta = 0.0275$ b) $\beta = 0$
8. Since $Z = 2$, we do not reject H_0.
9. Since $Z = 1.667$, we reject H_0.
10. Since $Z = -2.667$, we reject H_0.
11. Since $Z = -1.6$, we do not reject H_0.
12. Since $Z = -2$, we do not reject H_0.
13. Since $Z = -2.5$, we reject H_0.
14. Since $Z = 2.5$, we reject H_0.
15. Since $Z = 3.3$, we reject H_0.

8-4 exercises

1. Since $Z = 1$, we do not reject H_0.
2. Since $Z = -1.5$, we do not reject H_0.
3. Since $Z = -2.5$, we reject H_0.
4. Since $Z = -1.8$, we do not reject H_0.
5. Since $Z = 1.875$, we reject H_0.
6. Since $Z = 2$, we reject H_0.
7. Since $Z = -1.33$, we reject H_0.
8. Since $Z = 2$, we reject H_0.
9. Since $Z = 2.83$, we reject H_0.

8-5 exercises

1. Since $Z = 1.667$, we reject H_0.
2. Since $Z = -1.94$, we do not reject H_0.
3. Since $X = 18$, we do not reject H_0.
4. Since $Z = 1.25$, we do not reject H_0.
5. Since $Z = 1.667$, we do not reject H_0.
6. Since $Z = 1.667$, we do not reject H_0.
7. Since $X = 11$, we do not reject H_0.
8. Since $Z = 1.5$, we do not reject H_0.
9. Since $Z = 2.5$, we reject H_0.
10. a) 0.35 to 0.65   b) 0.4850 to 0.5150   c) 0.4985 to 0.5015
    d) As the values of n become larger, the proportion of heads approaches the probability of heads, namely, 1/2.

8-6
exercises

3. $99.12 < \mu < 110.88$
4. $23.85625 < \mu < 25.14375$
5. -19.304 and -0.696
6. $-2{,}187.5 < \delta < 387.5$
7. $0.06 < \delta < 5.94$
8. $-7.28 < \delta < -0.72$
9. $38.37 < \mu < 61.63$
10. $0.3775 < \rho < 0.6225$

9-2
exercises

7. a) 2.228　　b) -1.812　　c) 3.169　　d) -2.764
8. a) -5　　b) -1　　c) $s = 16$　　d) $\overline{X} = 5$
9. a) 1.318　　b) -1.318　　c) ± 1.711

9-3
exercises

2. Since $T = 2$, we do not reject H_0.
3. Since $T = 3.33$, we reject H_0.
4. Since $T = -2.0$, we reject H_0.　　The 90 percent confidence interval for μ is $2.81445 < \mu < 2.98555$
5. a) Since $T = 2$, we do not reject H_0.　　b) The 95 percent confidence limits for μ are 5.99345 and 6.20655.
6. a) Since $T = -0.833$, we do not reject H_0.　　b) The 99 percent confidence limits for μ are $1{,}512.872$ and $1{,}647.128$.
7. a) Reject H_0 if $T \geq 2.602$.　　b) Since $T = 1.56$, we do not reject H_0.
8. a) Since $T = -2.5$, we reject H_0.　　b) The 98 percent confidence interval for μ is $140.032 < \mu < 159.968$.

9-4
exercises

3. Since $T = 1.59$, we do not reject H_0.
4. a) Since $T = 3$, we reject H_0.　　b) The 99 percent confidence interval for the true difference is $0.004 < \delta < 0.196$.
5. a) Since $T = -0.98$, we do not reject H_0.　　b) The 95 percent confidence limits are $-0.673 < \delta < 0.273$.
6. a) Since $T = -0.68$, we do not reject H_0.　　b) The 95 percent confidence limits for the true difference are -8.21 and 4.21.
7. a) Since $T = 1.1$, we do not reject H_0.　　b) The 95 percent confidence interval for the true difference is $-0.96 < \delta < 2.96$.
8. Since $T = 3.25$, we reject H_0.
9. Since $T = 2.076$, we reject H_0 at $\alpha = 0.1$.

10-1
exercises

3. a) 34.1696　　b) 28.4120　　c) 12.4426　　d) 10.8508
   e) 8.26040
4. a) 0.01　　b) 0.05　　c) 0.50　　d) 0.10　　e) 0.005

10-2
exercises

1. Since $X^2 = 32.222$, we do not reject H_0.
2. Since $X^2 = 32.625$, we do not reject H_0.
3. Since $X^2 = 20.14$, we do not reject H_0.   The 95 percent confidence limits for σ^2 are $15.86 < \sigma^2 < 45.18$.
4. a) $\overline{X} = 10$ $s^2 = 5$ b) Since $X^2 = 10$, we do not reject H_0.
   c) The 90 percent confidence interval for σ^2 is $2.58 < \sigma^2 < 14.64$.
5. $X^2 = 4.58$, which is greater than the critical value 1.61031. Accordingly, we do not reject H_0.
6. Since $X^2 = 22$, we do not reject H_0.

10-3
exercises

1. Since $X^2 = 6.0$, we do not reject the hypothesis.
2. Since $X^2 = 2.20$, we do not reject the hypothesis.
3. Since $X^2 = 7.842$, we do not reject the null hypothesis.
4. Since $X^2 = 16.50$, we reject the null hypothesis.
5. Since $X^2 = 3.2$, we do not reject Mendel's Theory.
6. Since $X^2 = 2.7$, we do not reject the null hypothesis.

10-4
exercises

1. Since $X^2 = 3.125$, we do not reject the null hypothesis.
2. Since $X^2 = 0.625$, we do not reject the null hypothesis.
3. Since $X^2 = 80.3$, we reject the null hypothesis.
4. Since $X^2 = 1.584$, we do not reject the null hypothesis.
5. Since $X^2 = 1.388$, we conclude that the vaccine is not effective in preventing a cold.
6. Since $X^2 = 8.7502$, we do not reject the null hypothesis.

11-1
exercises

6. a) 2.54   b) 2.38   c) 2.03   d) 2.08   e) 2.33   f) 2.74
7. a) 1.79   b) 2.11   c) 2.45   d) 2.92
8. a) 2.85   b) 4.56   c) No, because the critical value 2.85 is larger than $70/50 = 1.40$.

11-2
exercises

1. Since $F = 1.5$, we do not reject H_0.
2. Since $F = 1.2$, we do not reject H_0.
3. Since $F = 1.2$, we do not reject H_0.
4. Since $F = 1.14$, we do not reject H_0.
5. Since $F = 1.67$, we reject H_0.
6. Since $F = 1.96$, we reject H_0.

11-3
exercises

1. Since $F = 2.31$, we do not reject the null hypothesis.
2. Since $F = 3.03$, we do not reject the null hypothesis.
3. Since $F = 22.2$, we reject the null hypothesis.
4. a) $F = 3.84$, so we do not reject the null hypothesis.

b) $F = 3.84$, so we reject the null hypothesis at $\alpha = 0.10$.

5. Since $F = 1.99$, we do not reject H_0.

6. Since $F = 3.57$, we do not reject H_0.

12-2 exercises

1. a) $Y = 0.67 + 0.67X$ b) $Y = 7.37$
2. a) $Y = 1/3 + 14/15\,X$ b) $Y = 9.67$ bushels
3. $Y = 5 + 32.5X$ $135
4. $Y = 25.4 - 0.95X$ $Y = 12.1$
5. $Y = 0.31 + 0.84X$
6. $Y = 11.3 + 1.03X$

12-3 exercises

1. a) $r = 0.94$ b) Since $T = 7.29$, we reject H_0.
2. a) $r = 14/15$ b) Since $T = 7.35$, we reject H_0.
3. a) $r = 0.98$ b) Since $T = 9.85$, we reject H_0.
4. a) $r = -0.906$ b) Since $T = -3.74$, we do not reject H_0.
5. $r = 0.87$
6. $r = 0.95$

13-2 exercises

1. Since $Z = 0.89$, we do not reject the null hypothesis.
2. Since the computed Z value is 0.5, we shall therefore not reject the null hypothesis.
3. Since $X = 3$, we do not reject H_0.
4. a) Since $X = 10$, we do not reject H_0.   b) Since $Z = 1.3$, we do not reject H_0.
5. Since $Z = 2.22$ or $Z = -2.22$, we reject the null hypothesis.
6. Since $Z = \pm 0.88$, we do not reject the null hypothesis.

13-3 exercises

1. $H = 5.7613$, we do not reject the hypothesis that no difference exists.
2. $H = 8.8318$, we reject the null hypothesis.
3. $H = 3.76419$, we do not reject the null hypothesis.
4. $H = 2.94$, we do not reject the null hypothesis.
5. $H = 4.965$, we do not reject the null hypothesis.
6. $H = 10.2274$, we reject the null hypothesis.

13-4 exercises

1. a) $R = 0.909$ b) Since $T = 6.54$, we reject the null hypothesis.
2. a) $R = 0.973$ b) Since $T = 16.85$, we reject the null hypothesis.
3. a) $R = 0.7714$ b) Since $T = 2.425$, we do not reject H_0.
4. a) $R = 0.927$ b) Since $T = 7.414$, we reject the null hypothesis.

14-2
exercises

1. a) Since $5,000,000 is the maximum of the maximums, the management should choose an investment to increase the capacity by 1,000,000 units.
   b) 1. Increase by 100,000 units:
      The expected payoff = $2,100,000.
      2. Increase by 500,000 units:
      The expected payoff = $2,300,000.
      3. Increase by 1,000,000 units:
      The expected payoff = $2,200,000.
      The decision should be to increase capacity by 500,000 units.
2. a) By the minimax rule the decision maker should decide to buy insurance.
   b) 1. The expected loss if he buys insurance = $50,002.
      2. The expected loss if he does not buy insurance = $5,000.
      Since the expected loss under (2) above is much smaller, by the Bayesian criterion the decision maker should not buy the insurance policy.
3. a) The minimax loss action for the jury is "acquit."
   b) 1. If condemning, the expected loss is 5.
      2. If acquitting, the expected loss is 2.5.
      Since the expected loss is smaller if the jury decides to "acquit," this should be the jury's action.
4. a) By the maximax criterion, the company shall produce the most expensive model.
   b) 1. The expected payoff from the most expensive model is 5.6.
      2. The expected payoff from the medium-priced model is 6.1.
      3. The expected payoff from the least expensive model is 6.25.
      Since the expected payoff from the least expensive model is the highest, the company should manufacture the least expensive.

14-3
exercises

1. a) Let p be the probability of taking action A_1 and $(1 - p)$ be the probability of taking action A_2.
   $p = 1/7 \qquad 1 - p = 6/7$
   b) The value of the game is $-$285,714.28.
2. $p = 1/4 \qquad 1 - p = 3/4$
   The expected payoff of the game is $22.5.
3. a) 1. For competitor A:
      $p = 1/2 \qquad 1 - p = 1/2$
      2. For competitor B:
      $p = 1/4 \qquad 1 - p = 3/4$
   b) 1. For competitor A: 4.5
      2. For competitor B: 4.5
4. a) Let p be the probability of wearing rainy-day outfit and $(1 - p)$ the probability of wearing non rainy-day outfit.
   b) The expected loss of the golf player is 20/6.

index

Statistics: An Intuitive Approach was set in Times Roman body type and Helvetica Light display type by Black Dot. Bookwalter did the printing and binding. The book and cover were designed by Naomi Takigawa. House of Graphics supplied the technical illustrations. Paul Kelly was the sponsoring editor, and Sara Boyd the project editor.

456/54321